Intercellular Signalling in the Mammary Gland

Intercellular Signalling in the Mammary Gland

Edited by

Colin J. Wilde, Malcolm Peaker, and Christopher H. Knight

Hannah Research Institute
Ayr, Scotland, United Kingdom

Plenum Press • New York and London

Library of Congress Cataloging-in-Publication Data

On file

QP
188
.m3
I56
1995

Proceedings of the 1994 Hannah Symposium on Intercellular Signalling in the Mammary Gland,
held April 13–15, 1994, in Ayr, Scotland, United Kingdom

ISBN 0-306-45075-5

© 1995 Plenum Press, New York
A Division of Plenum Publishing Corporation
233 Spring Street, New York, N. Y. 10013

10 9 8 7 6 5 4 3 2 1

Printed in the United States of America

PREFACE

All being done, we went to Mrs Shipmans, who is a great butter-woman; and I did see there the most of milke and cream, and the cleanest, that I ever saw in my life (29 May 1661).

Among others, Sir Wm. Petty did tell me that in good earnest, he hath in his will left such parts of his estate to him that could invent such and such things - as among others, that could discover truly the way of milk coming into the breasts of a woman...(22 March 1665).

My wife tells me that she hears that my poor aunt James hath had her breast cut off here in town - her breast having long been out of order (5 May 1665).

From the Diary of Samuel Pepys,
published as *The Shorter Pepys* (edited by R. Latham),
Penguin Books (1987)

The long-standing ultimate importance of research on the mammary gland is illustrated by the importance attached to cows' milk for human consumption, to human lactation and to breast cancer by Samuel Pepys and his contemporaries in the middle of the 17th century. Research has tended to develop in isolation in these three areas of continuing contemporary importance largely because in most countries, the underlying science of agricultural productivity is funded separately from the underlying science of human health and welfare. The intention of this Symposium was to bring together workers from these different strategic strands of research to consider a level of biological organisation where physiological and pathological processes share common features and at which considerable recent advances have been made - intercellular communication within the mammary gland.

Although paracrine and autocrine mechanisms are now trendy pursuits, they have only become so in the past decade. Previously, the attention on signalling systems within the body was mainly directed at the long-range and relatively slow endocrine system or the just as long-range, but fast, nervous system. The emphasis was on integration and control operating between organs and, more recently, on the mechanisms of hormone action on target cells. But this approach ignored another level of organisation and communication that, in the evolution of complexity, was overlayed by long-distance chemical communication through a circulatory system - the signalling between cells in close proximity that must have been involved in the emergence and evolution of multicellular organisms and that is also the basis for early embryonic development. Short-distance chemical signalling has not only a long phylogenetic history but a vital rôle both in ontogeny and in integrative control within organisms at all stages of their life.

This Symposium on Intercellular Signalling in the Mammary Gland was the first international conference to be held in the new Kevoca Conference Suite at the Hannah Research Institute. It was attended by eighty-one participants from seventeen countries. We thank the following for sponsorship: Becton Dickinson UK Ltd, Boehringer Mannheim UK, CAB International, Churchill Livingston, Enterprise Ayrshire, Hoeffer Scientific Instruments UK, ICN Biomedicals Ltd, Life Technologies Ltd, Plenum Publishing Company Ltd, Universal Biologicals Ltd.

The range of topics covered by the talks and posters at the Symposium ranged from the autocrine control of milk secretion by a newly-discovered protein in milk to the genes involved in cellular interactions within the mammary gland, from the intramammary control of milk secretion in women and macropod marsupials to the control of mammary development by locally-produced growth factors, from the interactions of endocrine factors with intramammary control to the interactions between mammary epithelial cells and the extracellular matrix on which they sit, from the cellular mechanisms of milk secretion and their local control to the molecular basis of apoptosis, and from the regulation of gene expression and transformation of mammary cells to the occurrence of gap junctions and synchronization of secretory events within a mammary alveolus. The levels of biological organisation covered the whole animal interacting with its offspring, through organs, tissues and cells to the control of gene expression.

A consistent feature of studying the mammary gland has been the wider biological importance of discoveries made. This Symposium was no exception and raised a number of questions on the evolutionary 'design' and essential features of intercellular communication systems, for example, on the effective distance of communication, on amplification and re-transmission of signals, on the chemicals used for signalling and on whether some chemicals are emitted to test the extracellular environment of a cell rather than to control other cells (*i.e.* endocrine, autocrine or paracrine 'radar'). These are the sorts of questions that can be answered using the mammary gland and its constituent tissues.

Finally, turning from general principles of intercellular communication back to a particular matter concerning milk secretion: was Sir William Petty's Prize ever claimed?

Colin J. Wilde
Malcolm Peaker
Christopher H. Knight

Hannah Research Institute
Ayr, Scotland

CONTENTS

MAMMARY DEVELOPMENT: GROWTH AND GROWTH FACTORS

MILK SECRETION

APPENDIX I

THE ENVIRONMENT OF THE MAMMARY SECRETORY CELL

Christopher H. Knight

Hannah Research Institute
Ayr KA6 5HL, U.K.

INTRODUCTION

Any discussion of intercellular signalling requires knowledge of the cell types present. In the case of the mammary gland, it quickly becomes apparent that this knowledge must extend to non-cellular components of the environment. Whereas most exocrine glands 'secrete it and forget it', the mammary gland retains its product and is, therefore, susceptible to a continuing interaction with that product. So, the stored milk is another important element in the environment which surrounds the secretory cell. This review will consider cells, matrix and milk and the interactions they all have with the secretory epithelial cell.

MAMMARY STRUCTURE

Mammary structure has been reviewed frequently (eg Mepham, 1987; Howlett and Bissell, 1990). Contained within the mammary gland are very many cell types and a great variety of extracellular molecules, giving rise to a complexity which belies some rather basic observations. The cell type of most immediate interest is the secretory epithelial cell, and the arrangement of these cells into spherical alveoli is well known. An essential characteristic of the secretory cell is its polarity, which brings the various exterior surfaces of the cell into direct contact with different elements of the overall environment; the apical surface with the cell's own secretion, the basal surface with basement membrane and the lateral surfaces with neighbouring epithelial cells. Basal and apical membrane surface areas are increased by infoldings and microvilli respectively, lateral cell:cell contact is through tight junctions, gap junctions and desmosomes. The extent of these various associations is not static; synthesis, secretion and expulsion of milk are dynamic processes which result in considerable changes in cell shape, and thus in the surface areas available for information exchange and in the integrity of cell:cell contacts.

Intercellular Signalling in the Mammary Gland
Edited by C.J. Wilde *et al.*, Plenum Press, New York, 1995

Moving outward, myoepithelial cells surrounding individual alveoli are in tight juxtaposition with the secretory cell basement membrane. Mammary tissue is highly vascularized, so capillary endothelial cells are abundant and in close proximity to secretory cells. There is an extensive lymph system, consisting of small lymphatic vessels surrounding lobules of alveoli and draining to one or more lymph nodes often deep within the glandular tissue. Lymphocytes and monocytes are common infiltrators, again achieving close proximity to secretory cells, and fibroblasts are present in considerable numbers. The immature gland contains a lot of adipose tissue, and the stroma of the differentiated gland retains many adipocytes (but not the fat).

This complex cellular environment is only one part of the story. The various cells lie in a collagenous connective tissue framework or 'glue', the extracellular matrix (ECM). It is now well recognized that ECM is not merely a sleeping partner, but is very much involved in the regulation of normal mammary function (Streuli, 1993). ECM is itself a highly complex mixture of many molecules, including various collagens, fibronectin, laminin, glycosaminoglycans and others. Much of the ECM is of mesenchymal origin, hence the abundant fibroblasts. The basement membrane of the secretory cell is part of the ECM, and this is produced at least in part by the secretory cell itself.

The mammary gland is a rather unusual exocrine gland in that it stores its product for variable periods of time. Thus, another structural feature is the storage system. At one extreme this consists purely of the spaces within alveoli and the bore of the ducts which carry the milk directly to the skin (monotremes) or nipple (rabbits, women). At the other extreme (dairy species), the small ducts lead into larger galactophores which end in a number of large, elastic sinuses; the gland cistern. The teat contains a second storage cistern which can be quite large, particularly in goats. Stored milk contains many bioactive factors so, just like the ECM, it is apparent that storage properties must also be taken into account when formulating theories about control of mammary function.

INFORMATION NETWORKS

Communication between cells occurs in many different ways (for review, see Baulieu, 1990). The 'simplest' level of information exchange thus far proposed occurs entirely within a cell. This is intracrine control (Bern, 1990), whereby a signal is produced and acts upon an internal receptor without leaving the cell of origin. *In situ* hybridization studies have revealed the presence of mRNA for prolactin in mammary secretory cells (Steinmetz, Grant, and Malven, 1993), and it is known that the prolactin receptor becomes internalized into these same cells (Nolin, 1979). Together, these observations indicate the possible existence of an intracrine mechanism. However, the biological significance of such a system has not been established. On the one hand, it appears to provide no opportunity for reaction to extracellular events, indeed, by providing a local source of a hormone, it might actually prevent such reaction by masking the external signal. On the other hand, concerted regulation of the many biochemical processes within the secretory cell must involve intracellular information exchange. The question is, does this always emanate from external cues, or does intracellular monitoring sometimes initiate the signal?

Juxtacrine control has been proposed as a term to describe communication by neighbouring cells which are in physical contact with each other. Molecular diffusion through gap junctions has been recognized for very many years and probably accounts for most if not all 'juxtacrine' phenomena; it is difficult to see how a secreted signal could specifically target only the immediate neighbour whilst ignoring both its parent cell and other cells in close proximity. In mammary tissue, juxtacrine interaction may occur between neighbouring secretory cells (Ehmann *et al.*, 1994) and also, potentially, between secretory cells and myoepithelial cells, since these also establish gap junction connections. The essential feature of juxtacrine control is the requirement for physical contact and by implication the information network remains localised; transfer of chemical information between a number of successive cells will occur, but only slowly (in signalling terms).

Autocrine control involves a secreted signal feeding-back by local diffusion to affect the parent cell and other cells of the same type. Although this is primarily recognized as a localised network, there is no theoretical reason why an autocrine signal (in the sense of secretory cell and target cell being of the same type) could not operate systemically in the case of paired organs such as the mammary gland. The feedback inhibitor of lactation (FIL) is an autocrine mechanism operating within the mammary gland (Wilde *et al.*, 1995b).

Another local mechanism is paracrine control. In this case, a signal produced by one cell type diffuses locally to affect different cell types. Given the cellular complexity of the mammary gland, the scope for paracrine control is quite enormous, as will be seen elsewhere in this volume. As one example, consider the interplay between epithelium and mesenchyme during mammary morphogenesis. The epithelium induces local differentiation of mesenchyme, which then starts to synthesise steroid receptors. It is through these receptors that systemic hormones then stimulate the mesenchymal cells to produce chemical signals which regulate epithelial outgrowth locally (Kratochwil, 1986).

Endocrine control is too familiar to require further explanation. One essential difference between paracrinology and endocrinology is in the extent of the network; millimetres or less versus several metres or, in modern computer jargon, an ethernet versus the 'information superhighway'. By the same reckoning, pheromonal information exchange (which can occur over distances of several kilometres) must constitute interplanetary communication! The mammary gland is a classic endocrine target tissue but is also now recognized to be a producer of endocrine signals in its own right (Peaker, 1991). In the sense of producing signals (hormones in milk) which may affect another member of the species (Peaker and Neville, 1991), the mammary gland could also be deemed to have a pheromonal role (pheromonal in the widest sense of the word). It is unlikely (although possible) that chemical information passes in the opposite direction in the same, milk-borne, way. However, the suckling young has a much better means of communicating with its mother and specifically her mammary gland; stimulation of neuroendocrine reflexes. Once again, this is a familiar concept, but it does serve to emphasise the very considerable extent of intercellular signalling in which the mammary gland is involved.

Strictly, signalling between ECM and secretory cells is not intercellular in nature. However, it is quite apparent that chemical constituents of the ECM are essential for normal mammary morphogenesis (Faulkin and DeOme, 1960) and differentiation (Streuli, 1995). The relationship is bidirectional; secretory cells are responsible for producing some elements of the basement membrane, and by so doing influence their own immediate environment. One role of ECM is physical support and maintenance of cell shape; *in vitro* studies in particular have established this as an essential prerequisite for differentiated function. So, ECM signals and ECM shapes. Another possible role that has been largely ignored is suppression; by isolating organ from organ ECM provides a barrier to irrelevant or undesired information. Given its state of relative seclusion outside the body cavity this role may not be particularly relevant in the case of the mammary gland.

There is one more mammary signalling function about which we presently know very little. In the goat (but not, curiously, in the guinea pig) removal of one mammary gland results in compensatory growth of the other gland (Knight, 1987; Knight and Peaker, 1981). Clearly, a size-specific signal of some sort is produced. This may be interrogated by a remote site which then responds with a 'grow' or 'don't grow' signal, as appropriate. Alternatively, the mammary gland may monitor its own output ('biological radar'; see Peaker, 1992) and initiate locally controlled growth when the returned signal falls below an expected value.

INTERACTION BETWEEN NETWORKS

In the simplest analysis of hormone action, the hormone interacts directly with the target cell via a specific surface receptor and initiates a cascade of intracellular events leading to the response. This direct action should be reproducible *in vitro* in simple culture of the isolated, defined cell type. In actuality, several of the hormones which are believed to affect mammary function *in vivo* turn out to have no direct action *in vitro*. In the developing gland, the classical example is oestrogen; clearly mammogenic *in vivo* but still not unequivocally shown to have a direct mitogenic action (Forsyth, 1989). Incidentally, the secretory cell can produce its own oestrogen by aromatization of androgens (Peaker and Taylor, 1990), although it is not known whether this has any local mitogenic role. The more biologically relevant point is that hormones such as oestrogen interact with cellular components of the mammary gland to stimulate local production of growth factors which then elicit a mitogenic response by paracrine action. There are many different candidate growth factors, and a similar number of reviews of the subject! (Forsyth, 1989; Grosse, 1995; Oka *et al.*, 1991; Plaut, 1993; Sandowski *et al.*, 1993).

The intruiging endocrine blind-spot of the lactating ruminant gland concerns growth hormone's mode of action. There can be no doubting the galactopoietic effect of GH *in vivo*. *In vitro*, GH fails to stimulate secretory cell proliferation or differentiation (Collier *et al.*, 1993; Shamay *et al.*, 1988). However, GH did produce a local mammogenic response when infused via the streak canal into individual glands of pregnant heifers (Collier *et al.*, 1993). This does not establish direct action, and indeed it is not certain that mammary epithelium is capable of responding directly to GH since it appears to lack the necessary receptor (Akers, 1985). The receptor is present on mammary stromal elements which also possess mRNA for IGF1 (Hauser *et al.*, 1990), and growth factors including IGF1 administered locally also elicit a growth response (Collier *et al.*, 1993). Therefore, one likely scenario is that the mammogenic effect of GH is mediated by IGF1, a recognized mammary mitogen (Winder and Forsyth, 1986). We have confirmed GH's mammogenic effect in goats using *in vivo* bromodeoxyuridine labelling to quantify cell proliferation (Knight *et al.*, 1994). Cell number was increased more than 40% by treatment for 6 weeks *prepartum*, but the same treatment administered starting at parturition had no significant effect. This was not because lactating tissue is incapable of proliferation; a different stimulus (hemimastectomy to induce compensatory growth) was effective *postpartum* but had little action prior to parturition (CH Knight, JR Brown and K Sejrsen, unpublished). It may quite simply be that the major change in the stromal:parenchymal ratio which occurs during the transition to lactation results in the loss of an effective paracrine interaction, however, more fundamental changes in sensitivity of stromal tissue to GH and/or secretory tissue to IGF1 certainly cannot be ruled out.

This does not answer the question of how GH increases milk output. The evidence for IGF1 mediation is equivocal; local infusion into the pudic artery produces a small increase under particular milking strategies (Prosser *et al.*, 1990), but this appears to be an acceleration of a frequent milking response rather than an actual galactopoietic action of IGF1 *per se* (Prosser and Davis, 1992). Stimulation of mammary blood flow may be the key; perhaps GH increases stromal production of vasodilatory factors (prostaglandins, for instance; Nielsen *et al.*, 1994). This is speculative, but if it proves to be so then the milk yield response would be due to a paracrine action in which the secretory cells are not themselves involved. Although GH's systemic actions are considerable, it is generally believed that they are not in themselves sufficent to increase milk yield; systemic 'push' is not enough, the mammary gland must 'pull' (Flint, 1995; Knight *et al.*, 1994). We have failed to achieve a local galactopoietic action of GH infused into one mammary gland of goats via the streak canal (Sejrsen and Knight, 1993). Although paracellular transfer was facilitated by pharmacological disruption of tight junction integrity, it is far from certain that the infused hormone reached the basal side of the secretory cell in significant quantities. Recently, a local galactopoietic effect of GH has been demonstrated in rats (Flint, 1995; Flint and Gardner, 1994), which cannot be mimicked by IGF1 (or IGF2, IGF analogues or complexes of IGFs and binding protein; Flint *et al.*, 1994). The timecourse of the response indicated a local paracrine action, but clearly this did not involve IGF1. The problem remains and, just for good measure, it is now apparent that the mammary gland is capable of producing its own GH, at least in dogs (Selman *et al.*, 1994).

In these examples, the primary action is endocrine. There is also one circumstance in which the reverse kind of interaction occurs; the primary change occurs locally in the gland to alter its subsequent responsiveness to an endocrine signal. Milk secretion is under negative feedback control by an autocrine protein termed the feedback inhibitor of lactation (FIL: Wilde *et al.*, 1995b). More or less frequent removal of milk increases or decreases secretion rate, respectively, through alterations in the exposure of secretory tissue to FIL; this is considered in detail elsewhere in these proceedings (Wilde *et al.*, 1995a). FIL's primary action is at the level of intracellular protein trafficking prior to exocytosis (Rennison *et al.*, 1993). Although the majority of protein synthesized by the mammary gland is casein and other secreted milk proteins, cellular proteins including hormone receptors are also produced. One such is the prolactin receptor. This receptor has a short half life and is internalised together with its ligand as the first step of prolactin signal transduction. As a consequence, there is considerable vectorial trafficking of newly synthesized receptor towards the basal cell membrane (Costlow, 1986). There is now evidence to show that this process is downregulated by FIL, presumably in just the same way as milk protein secretion is reduced. *In vivo*, accumulation of milk in individual unsuckled glands of rabbits for periods of 24h or more leads to a localised reduction of prolactin binding (Bennett *et al.*, 1992), an effect which can be mimicked (within a shorter period of 12h) by intraductal injection of FIL (CN Bennett, CH Knight and CJ Wilde, unpublished). *In vitro*, FIL added to cultures of isolated mouse mammary epithelial cells causes a downregulation of cell surface prolactin binding with no decrease in total binding to permeabilised cells (Bennett *et al.*, 1990). In other words, FIL increases receptor internalisation, effectively rendering the cell less responsive to circulating prolactin (Bennett, 1993). In this way, local modulation of a systemic endocrine signal provides an extra level of control over mammary function. It is not believed that this mechanism is responsible for the acute effect of FIL on milk secretion, however, it is quite conceivable that reduced prolactin sensitivity is responsible for or involved in a subsequent FIL action, that of reduced mammary differentiation (Wilde *et al.*, 1991).

THE MILK-SIDE ENVIRONMENT

Milk contains very many bioactive factors; hormones, growth factors and metabolites which are known to be capable of effecting specific biological actions on specific target cells. That is not to say that their presence in milk necessarily fulfills any biological role

(Peaker and Neville, 1991). Probably, the 'expected' biological effect would be one occurring in the neonate which ingests the milk (Koldovsky, 1989). Some of the hormones in milk do appear to be specifically produced by mammary tissue, whereas others are actively transported across the epithelium from blood to milk; perhaps in this case the gland is performing an excretory rather than signalling function. Here, I shall consider a different possibility, a role within the mammary gland, as an effector of secretory cell development or function.

Up till now, our discussion has been concerned very largely with the basal side of the secretory cell. Any interaction with bioactive factors in milk will occur on the apical membrane, so in recent experiments we have looked for hormone receptors on that membrane. Apical membranes can be isolated from milk by differential centrifugation (Shennan, 1992). The fraction obtained shows transport properties which are very different from basal membrane but typical of apical membrane, so one can be confident of a highly enriched apical membrane fraction (Shennan, 1992). Binding of prolactin to membranes prepared in this way was measured (Knight *et al.*, 1993). Specific binding was observed in almost all preparations (18/22), but always at a modest level. This was not substantially improved by solubilising the membranes, and as yet the extent of binding has precluded any quantitative assessment of affinity. Assuming that the observation is a genuine one, it indicates that milk-borne prolactin could potentially affect the secretory cell. On the other hand, it may merely serve to explain how the prolactin gets into milk in the first place! (Fleet, Forsyth, and Taylor, 1992).

In looking for a clear and unequivocal example of a biological action being exerted from milk onto mammary gland, the obvious choice is FIL. The historical development of the concept is discussed elsewhere in these proceedings (Peaker, 1995), and a later chapter will describe the FIL protein itself and its intracellular mode of action (Wilde *et al.*, 1995a). A major gap in our present understanding concerns the mechanics of FIL's action. It is produced within the secretory cell but undoubtedly exerts its action from without; the mechanism is autocrine, not intracrine. This can be demonstrated quite easily by introducing milk fractions containing FIL into the mammary gland through the streak canal *in vivo* or into the culture dish containing rabbit mammary explants *in vitro*; milk yield falls and lactose and casein synthesis is inhibited (Wilde *et al.*, 1995a). The site within the mammary gland where FIL is effective has also been determined. In goats, more frequent removal of milk by normal milking or suckling increases milk yield, but frequent drainage of milk through an intraductal catheter does not (Henderson and Peaker, 1987). The difference relates to the extent of milk removal; in the latter case, only milk present in the cistern will drain through the catheter, whereas during normal milking the milk ejection reflex ensures that some (probably most) of the milk still present in alveolar tissue is also removed. The conclusion is straightforward; FIL acts on the secretory tissue, and is therefore effective in alveolar milk but not in cisternal milk. There is clear evidence of concentration dependence in FIL's action, and the mathematical mechanics of how the required drop in concentration is achieved after milking are considered elsewhere (Peaker, 1995; Wilde *et al.*, 1995a). For these models to operate, the absolute amount of FIL present within the lumen of secretory alveoli must be reduced by milking, and the amount by which FIL concentration falls will depend on the extent to which this is achieved. It follows that milk distribution, compartmentation and movement within the mammary gland during the interval between milkings is of considerable importance as a determinant of FIL's effectiveness, and hence will influence milk yield.

MAMMARY STORAGE CHARACTERISTICS

From a milk storage standpoint, the udder of dairy species can be considered rather simplistically as a two compartment structure; secreted milk is stored either within the lumen of secretory alveolar tissue (alveolar milk) or within the large ducts and the gland and teat cisterns (cisternal milk). Distribution and movement of milk between these compartments can be measured by fractional milking, obtaining first the cisternal milk and then the alveolar milk, this latter including a portion termed residual milk (that milk left behind after a normal milking which can be removed with the aid of exogenous oxytocin). To remove just cisternal milk it is necessary to either passively avoid or else actively block the milk ejection reflex. We have developed a pharmacological approach to the latter, using a biologically inert analogue of oxytocin (CAP: Ferring, Malmo, Sweden) to competitively block the action of endogenous oxytocin at the receptor level (Knight *et al.*, 1994). Following intravenous injection of CAP, the goat or cow is milked as normal but no milk ejection occurs so only cisternal milk is obtained. CAP has a short biological half-life so a second milking approximately 10 minutes later will yield alveolar milk. An alternative approach is to drain cisternal milk through an intraductal catheter without physically 'milking'. With care, the milk ejection reflex is avoided; this can be confirmed by simultaneous recording of intramammary pressure (Knight and Dewhurst, 1994a). A direct comparison of CAP and drainage methods in goats gave very similar values for cistern milk volume (Knight *et al.*, 1994).

Milk distribution has been measured at various intervals after milking in goats (Peaker and Blatchford, 1988) and in cows (Knight and Dewhurst, 1994a). One hour after a normal milking, the cistern of late lactation dairy cows contained around 500ml of milk, whereas in peak lactation cows the value was less than one-third of this, a significant difference. Secretion rate was, of course, higher in the peak lactation group, so the difference must be due to an increase in cisternal compliance as lactation progresses. The prediction from this observation is that post-milking FIL inactivation is likely to be greater in late lactation cows. It follows that efficient milking is of supreme importance in cows at peak lactation. Cisternal milk volume did not increase significantly in either group between 2 and 6h post-milking; further secretion during this time all accumulated within alveolar tissue. Thereafter, cisternal and alveolar milk volumes both increased until the last measurement at 20h post-milking, so movement of milk into the cistern was not simply by 'overflow' from a full alveolar compartment. At its peak, alveolar milk constituted 96% and 80% of total milk in peak and late lactation cows, respectively, but these peak values were achieved at 1h and 4h post-milking. By 8h (a typical milking interval in commercial practice) the alveolar percentage was very similar in the two groups, but varied quite considerably between individual cows. This variation was not related to residual milk volume or to milk distribution at short post-milking intervals, but was closely correlated with milk distribution at longer intervals such as 20h. In other words, by 8h post-milking, cows could be ranked according to their storage characteristics in a way which was relevant to extended milking intervals (Knight and Dewhurst, 1994a). In the goat, milk moves rather more easily into the cistern. Cisternal and alveolar filling progressed linearly for the first 6h post-milking, and all subsequent accumulation was in the cisternal compartment (Peaker and Blatchford, 1988). The cistern of the goat is relatively more capacious and compliant than that of the cow.

We have examined relationships between milk storage characteristics and milk yield responses to different milking frequencies (Dewhurst and Knight, 1994; Knight and Dewhurst, 1994b). Twice daily milking is the norm for commercial dairying in most developed countries; milking more frequently (perhaps using fully automated robotic milking systems) will increase yield, whereas if extensification is preferred the labour saving in milking once daily might outweigh the decreased yield. We anticipated that different cows would respond in different ways to such changes. 'Large cisterned' cows (ie those which store relatively more milk in the cistern) should be more tolerant of once daily milking, but less responsive to frequent milking, by virtue of being less affected by FIL during twice daily milking. Both relationships have now been confirmed and statistically proven. Cisternal milk proportion and milk yield responses to one week of half-udder thrice-daily milking were determined in 14 cows; response was assessed relative to yield of the normally milked half udder (Dewhurst and Knight, 1994). There was a significant negative relationship ($r=0.70$, $P<0.01$), confirming that greater cisternal storage equates with poorer reponse to more frequent milking. This has also been demonstrated in goats (Knight *et al.*, 1989). In a separate experiment, 10 cows were milked once daily for one week. The immediate decrease in yield (ie during the first 48h of once daily milking) was again highly correlated with cisternal milk proportion ($r=0.81$, $P<0.01$), the lowest decrease being associated with the highest cistern proportion (Knight and Dewhurst, 1994b). In neither experiment was there any relationship between either pretreatment milk yield or residual milk volume and the yield response, so the effect is truly related to storage characteristics.

Development of the mammary epithelium has been extensively studied for many years. Whilst our knowledge may not be totally complete it is, to put it bluntly, light-years ahead of our understanding of the growth of the cisternal storage tissues. This is a fruitful area for future work. Initial findings suggest that the cistern does 'grow' (stretch?) during the course of lactation and with increasing parity (Dewhurst and Knight, 1993), and it appears that this process may be manipulated quite easily. Repeated short periods of once daily milking increase subsequent cistern proportion in both cows and goats (CH Knight and JR Brown, unpublished observations). If this treatment is applied to only half of the udder and then, at a later date, both halves are milked once daily, the yield decrease is lower in the previously treated half, just as we would predict.

CONCLUSIONS

The mammary gland is a complex organ controlled by a multiplicity of interacting systems. Local control can operate in isolation from systemic influences, and can also modulate the effect of at least one systemic hormone. Endocrine signals can affect secretory cells directly, but perhaps just as often they act indirectly via paracrine interaction. This highlights the importance of the cellular environment of the secretory cell, but possibly just as important is the non-cellular environment; the correct extracellular matrix is essential for normal morphogenesis and function. Finally, the milk-side environment (apical membrane, milk storage characteristics) is important in the operation of the local autocrine control mechanism.

ACKNOWLEDGEMENTS

Many people contributed over several years to work referred to in this review. I am grateful to them all. Part of the work was funded by the Scottish Office Agricultural and Fisheries Department.

REFERENCES

Akers, R.M., 1985, Lactogenic hormones: binding sites, mammary growth, secretory cell differentiation and milk biosynthesis in ruminants, *J. Dairy Sci.* 68:501.

Baulieu, E.-E., 1990, Hormones, a complex communications network, *in:* "Hormones. From molecules to disease", Baulieu E.-E. and Kelly, P.A., eds., p. 3, Hermann, New York.

Bennett, C.N., 1993 Autocrine mechanisms modulating enocrine regulation of mammary gland function. *Ph.D, University of Glasgow.*

Bennett, C.N., Knight, C.H. and Wilde, C.J., 1990, Regulation of mammary prolactin binding by secreted milk proteins, *J. Endocr.* 127 Suppl:141.

Bennett, C.N., Wilde, C.J. and Knight, C.H., 1992, Changes in prolactin binding with milk accumulation in the lactating rabbit, *J. Endocr.* 132 suppl:292.

Bern, H., 1990, The 'new' endocrinology: its scope and its impact. *Amer. Zool.* 30:877.

Collier, R.J., McGrath, M.F., Byatt, J.C. and Zurfluh, L.L., 1993, Regulation of mammary growth by peptide hormones: involvement of receptors, growth factors and binding proteins. *Livest. Prod. Sci.* 34:21.

Costlow, M.E., 1986, Prolactin interaction with its receptor and the relationship to the subsequent regulation of metabolic processes, *in:* "Actions of prolactin on molecular processes", Rillema J.A., ed. p. 5, CRC Press, Florida.

Dewhurst, R.J. and Knight, C.H., 1993, An investigation of the changes in sites of milk storage in the bovine udder over two lactation cycles, *Anim. Prod.* 57:379.

Dewhurst, R.J. and Knight, C.H., 1994, Relationship between milk storage characteristics and the short-term response of dairy cows to thrice-daily milking, *Anim. Prod.* 58:181.

Ehmann, U.K., Calderwood, S.K. and Stevenson, M.A., 1994, Juxtacrine growth stimulation of mouse mammary cells in culture, *in:* "Intercellular signalling in the mammary gland", Wilde, C.J., Peaker, M. and Knight, C.H. eds., Plenum Publishing Company, New York.

Faulkin, L.J. and DeOme, K.B., 1960, Regulation of growth and spacing of gland elements in the mammary fat pad of the C3H mouse, *J. Natl. Cancer Inst.* 24:953.

Fleet, I.R., Forsyth, I.A. and Taylor, J.C., 1992, Transfer of prolactin into milk in the goat. *J. Endocr.* 132 suppl:254.

Flint, D.J., 1995, Regulation of milk secretion and composition by growth hormone and prolactin, *in:* "Intercellular signalling in the mammary gland", Wilde, C.J., Peaker, M. and Knight, C.H. eds., Plenum Publishing Company, New York.

Flint, D.J. and Gardner, M.J., 1994, Evidence that growth hormone stimulates milk synthesis by direct action on the mammary gland and that prolactin exerts effects on milk secretion by maintenance of mammary DNA content and tight junction status, *Endocr.* in press.

Flint, D.J., Tonner, E., Beattie, J. and Gardner, M.J., 1994, Several insulin-like growth factor-I analogues and complexes of insulin-like growth factors-I and -II with insulin-like growth factor-binding protein-3 fail to mimic the effect of growth hormone upon lactation in the rat, *J. Endocr.* 140:211.

Forsyth, I.A., 1989, Growth factors in mammary gland function, *J. Reprod. Fert.* 85:759.

Grosse, R., 1995, Control of mammary gland development by growth factors, *in* "Intercellular signalling in the mammary gland", Wilde, C.J., Peaker, M. and Knight, C.H. eds., Plenum Publishing Company, New York.

Hauser, S.D., McGrath, M.F., Collier, R.J. and Krivi, G.G., 1990, Cloning and in vivo expression of bovine growth hormone receptor mRNA, *Mol. Cell. Endocr.* 72:187.

Henderson, A.J. and Peaker, M., 1987, Effect of removing milk from the mammary ducts and alveoli, or of diluting stored milk, on the rate of milk secretion in the goat, *Quart. J. Exp. Physiol.* 72:13.

Howlett, A.R. and Bissell, M.J., 1990, Regulation of mammary epithelial cell function: a role for stromal and basement membrane matrices, *Protoplasma.* 159:85.

Knight, C.H., 1987, Compensatory changes in mammary development and function after hemimastectomy in lactating goats, *J. Reprod. Fert.* 79:343.

Knight, C.H., Brosnan, T., Wilde, C.J. and Peaker, M., 1989, Evidence for a relationship between gross mammary anatomy and the increase in milk yield obtained during thrice daily milking in goats. *J. Reprod. Fertil. Abstr. Ser.* 3:32.

Knight, C.H., Brown, J.R. and Sejrsen, K., 1994, A comparison of growth hormone-induced mammogenesis in pregnant and lactating goats, *Endocr. Metab.* 1 suppl B:52.

Knight, C.H. and Dewhurst, R.D., 1994a, Milk accumulation and distribution in the bovine udder during the interval between milkings, *J. Dairy Res.* 61:167.

9

Knight, C.H. and Dewhurst, R.D., 1994b, Once daily milking of dairy cows; relationship between yield loss and mammary cistern capacity, *J. Dairy Res.* in press.

Knight, C.H., France, J. and Beever, D.E., 1994, Nutrient metabolism and utilisation in the mammary gland, *Livest. Prod. Sci.* 39:129.

Knight, C.H. and Peaker, M., 1981, Lack of compensatory mammary growth following hemimastectomy in the guinea pig, *Comp. Bichem. Physiol.* 70A:427.

Knight, C.H., Stelwagen, K., Farr, V.C. and Davis, S.R., 1994, Use of an oxytocin analogue to determine cisternal and alveolar milk pool sizes in goats, *J. Dairy Sci.* in press.

Knight, C.H., Stewart, E.J. and Bennett, C.N., 1993, Prolactin binding by mammary secretory cell apical 'milk-side' membrane, *J. Reprod. Fert. Abstr. Ser.* 11:103.

Koldovsky, O., 1989, Search for role of milk-borne biologically active peptides for the suckling, *J. Nutr.* 119:1543.

Kratochwil, K., 1986, Hormone action and epithelial-stromal interaction: mutual dependence, *Horm. Cell Regn.* 139:9.

Mepham, T.B., 1987, *"Physiology of Lactation"*. Open University Press, Milton Keynes.

Nielsen, M.O., Fleet, I.R., Jakobsen, K. and Heap, R.B., 1994, The local effect of prostacyclin, prostaglandin E2 and prostaglandin F2alpha on mammary blood flow of lactating goats, *J. Endocr.* in press.

Nolin, J.M., 1979, The prolactin incorporation cycle of the milk secretory cell, *J. Histochem. Cytochem.* 278:1203.

Oka, T., Yoshimura, M., Lavandero, S., Wada, K. and Ohba, Y., 1991, Control of growth and differentiation of the mammary gland by growth factors, *J. Dairy Sci.* 74:2788.

Peaker, M., 1991, Production of hormones by the mammary gland: short review, *Endocr. Reg.* 25:10.

Peaker, M., 1992, Commentary: Chemical signalling systems: the rules of the game, *J. Endocr.* 135:1.

Peaker, M., 1995, Autocrine control of milk secretion: development of the concept, *in:* "Intercellular signalling in the mammary gland", Wilde, C.J., Peaker, M. and Knight, C.H. eds., Plenum Publishing Company, New York.

Peaker, M. and Blatchford, D.R., 1988, Distribution of milk in the goat mammary gland and its relation to the rate and control of milk secretion, *J. Dairy Res.* 55:41.

Peaker, M. and Neville, M.C., 1991, Hormones in milk: chemical signals to the offspring. *J. Endocr.* 131:1.

Peaker, M. and Taylor, E., 1990, Oestrogen production by the goat mammary gland: transient aromatase activity during late pregnancy, *J. Endocr.* 125:R1.

Plaut, K., 1993, Role of epidermal growth factor and transforming growth factors in mammary development and lactation, *J. Dairy Sci.* 76:1526.

Prosser, C. and Davis, S.R., 1992, Milking frequency alters the milk yield and mammary blood flow response to intra-mammary infusion of insulin-like growth factor-I in the goat, *J. Endocr.* 135:311.

Prosser, C.G., Fleet, I.R., Corps, A.N., Froesch, E.R. and Heap, R.B., 1990, Increase in milk secretion and mammary blood flow by intra-arterial infusion of insulin-like growth factor-1 into the mammary gland of the goat, *J. Endocr.* 126:437.

Rennison, M.E., Kerr, M., Addey, C.V.P., Handel, S.E., Turner, M.D., Wilde, C.J. and Burgoyne, R.D., 1993, Inhibition of constitutive protein secretion from lactating mouse mammary epithelial cells by FIL feedback inhibitor of lactation, a secreted milk protein, *J. Cell Sci.* 106:641.

Sandowski, Y., Peri, I. and Gertler, A., 1993, Partial purification and characterization of putative paracrine/autocrine bovine mammary epithelium growth factors, *Livest. Prod. Sci.* 35:35.

Sejrsen, K. and Knight, C.H., 1993, Unilateral intramammary infusion of GH does not support a local galactopoietic action of growth hormone in goats, *Proc. Nutr. Soc.* 52:278A.

Selman, P.J., Mol, J.A., Rutteman, G.R., Van Garderen, E. and Rijnberk, A., 1994, Progestin-induced growth hormone excess in the dog originates in the mammary gland, *Endocr.* 134:287.

Shamay, A., Cohen, N., Niwa, M. and Gertler, A., 1988, Effect of insulin-like growth factor I on dexoyribonucleic acid synthesis and galactopoiesis in bovine undifferentiated and lactating mammary tissue in vitro, *Endocr.*, 123, 804-809.

Shennan, D.B., 1992, K^+ and Cl^- transport by mammary secretory cell apical membrane vesicles isolated from milk, *J. Dairy Res.* 59:339.

Steinmetz, R.W., Grant, A.L. and Malven, P.V., 1993, Transcription of prolactin gene in milk secretory cells of the rat mammary gland, *J. Endocr.* 136:271.

Streuli, C., 1993, Extracellular matrix and gene expression in mammary epithelium. *Sem. Cell Biol.* 4:307.

Streuli, C., 1995, Mechanisms of extracellular matrix-induced control of milk protein secretion, *in:* "Intercellular signalling in the mammary gland", Wilde, C.J., Peaker, M. and Knight, C.H. eds., Plenum Publishing Company, New York.

Wilde, C.J., Addey, C.V.P., Boddy-Finch, L.M. and Peaker, M., 1995a, Autocrine control of milk secretion: from concept to application, *in:* "Intercellular signalling in the mammary gland", Wilde, C.J., Peaker, M. and Knight, C.H. eds., Plenum Publishing Company, New York.

Wilde, C.J., Addey, C.V.P., Boddy, L.M. and Peaker, M., 1995b, Autocrine regulation of milk secretion by a protein in milk, *Biochem. J.* in press.

Wilde, C.J., Blatchford, D.R. and Peaker, M., 1991, Regulation of mouse mammary cell differentiation by extracellular milk proteins, *Exp. Physiol.* 76:379.

Winder, S.J. and Forsyth, I.A., 1986, Insulin-like growth factor 1 IGF-1 is a potent mitogen for ovine mammary epithelial cells, *J. Endocr.* 108 suppl:141.

ROLE OF BOVINE PLACENTAL LACTOGEN IN INTERCELLULAR SIGNALLING DURING MAMMARY GROWTH AND LACTATION

Robert J. Collier, John C. Byatt,
Michael F. McGrath and Philip J. Eppard

Monsanto Company, The Agricultural Group
Animal Sciences Division
St. Louis, Mo, USA

INTRODUCTION

In dairy cattle, the majority of mammary growth takes place during pregnancy and as such, it might be expected that signals from the placenta would be involved in controlling mammogenesis. Placental lactogen (PL) is produced only during pregnancy and may be one of the endocrine factors that controls mammary growth. It is also generally recognized that mammary growth is controlled not only by endocrine signals, but also by paracrine and/or autocrine factors. The *in vitro* and *in vivo* mammogenic effects of some of these growth factors are discussed, along with possible interactions between the locally acting factors and endocrine hormones such as bovine PL (bPL). In addition, it is common for dairy cattle to be pregnant during much of lactation. Thus, the possible influences of bPL on lactation are also discussed.

FACTORS REGULATING MAMMARY GROWTH

Milk yield is a function of the number of secretory cells in the mammary gland and the activity of these cells during lactation. Although both of these factors contribute to an animal's overall lactation potential, cell number establishes the finite limitation. Control of bovine mammary epithelial cell proliferation has been studied mostly with *in vitro* culture systems and from these studies a number of growth factors have been implicated in control of mammary growth.

The somatomedins are clearly important mediators of mammary epithelial growth in rodents and humans (Osborne *et al.*, 1990) and the same is true in ruminants. Insulin-like growth factors (IGF) I and II both stimulate proliferation of bovine mammary epithelial cells grown in collagen gels (McGrath *et al.*, 1991; Peri *et al.*, 1992). Proliferation of ovine mammary epithelial cell clumps grown on attached collagen gels was also stimulated by IGFs I and II (Winder *et al.*, 1989). In all of these cases, IGF-I was at least 10-fold more potent than IGF-II, suggesting that IGF-II was mediating its effect through the IGF type I receptor. Both type I and type II receptors are present in bovine mammary tissue during pregnancy (Dehoff *et al.*, 1988). However, it is primarily type I receptors that increase in number during pregnancy and somatomedin effects on mammary growth are mediated through this receptor. A mammogenic role for IGF-I may also be inferred from the increased population of receptors (type I) for this growth factor during pregnancy (Collier *et al.*, 1989).

Current evidence suggests that the mammary stroma, but not epithelial cells synthesize IGF-I during pregnancy in cattle (Hauser *et al.*, 1990). However, the source of a truncated form of IGF-I (des-3-IGF-I; Francis *et al.*, 1986), is presently unknown. This molecule is of interest because it is approximately 4-fold more potent than full length IGF-I with respect to stimulation of mammary epithelial cell proliferation (McGrath *et al.*, 1991). In addition, the presence of des-IGF-I in colostrum (Francis *et al.*, 1986), but not in milk (Shimamoto *et al.*, 1992) suggests that this growth factor is either produced in the mammary gland during pregnancy or that it is specifically concentrated in this organ. In either case, the presence of des-IGF-I may be one of the factors that regulates mammary growth. The increased potency of des-IGF-I appears to be due to the fact that the truncated molecule is not bound by several of the somatomedin binding proteins (IGFBP) including IGFBP-2 (Ross *et al.*, 1989). Bovine mammary epithelial cells secrete both IGFBP-2 and -3 (McGrath *et al.*, 1991) and secretion of these binding proteins is stimulated by IGF-I and to a lesser extent by des-IGF-I. Thus, control of IGFBP secretion and differential affinity for these binding proteins suggests a mechanism for regulating local IGF activity.

Another class of growth factors involved in control of mammary growth are epidermal growth factor (EGF) and/or transforming growth factor-α (TGF-α) (Plaut, 1993). Both human EGF and bovine TGF-α stimulate proliferation of mammary epithelial cells from mid-pregnant heifers (Tou *et al.*, 1990). In addition, sialoadenectomy of athymic mice decreased the ability of transplanted bovine mammary tissue to increase DNA synthesis in response to estradiol and progesterone (Sheffield and Yuh, 1988). This inhibition was reversed by treatment with EGF, suggesting a role for EGF and/or TGF-α in regulating mammogenesis and hormonal responsiveness. Consistent with responsiveness of mammary epithelial cells to EGF and TGF-α is the presence of high affinity receptors for these growth factors in mammary tissue from both pregnant and lactating cows (Spitzer and Grosse, 1987). The number of receptors was at least two-fold higher in tissue from pregnant animals. These receptors appear to be expressed predominantly in alveolar epithelial cells (Glimm *et al.*, 1992).

Bovine TGF-α was first identified in pituitary gland (Samsoondar *et al.*, 1986). Subsequently, Zurfluh *et al.* (1990) obtained the sequence for this growth factor and demonstrated that mRNA for TGF-α is present in several bovine tissues including mammary gland. Cattle also produce an EGF-like molecule that is different to TGF-α (Byatt *et al.*, 1990), but the structure of this growth factor has not been determined. Certainly EGF-like activity, measured by radioreceptor assay, in blood and tissues such as salivary gland is several orders of magnitude lower in cattle than in mice. Therefore EGF and/or TGF-α appear to be locally acting factors in bovine mammary gland, although it is unknown which cell type(s) secrete this factor(s).

In addition to mitogenic growth factors there are also regulatory factors that can inhibit growth of the mammary gland. Members of the TGF-ß family inhibit ductal growth in mice and may regulate functional differentiation of epithelial cells (Daniel and Robinson, 1992). Transforming growth factors-ß$_1$ and ß$_2$ have been identified in bovine milk and message for these growth factors is localized in both mammary epithelial and intralobular stromal cells of bovine mammary gland (Maier *et al.*, 1991). However, it is presently unknown how these regulatory growth factors interact with the mitogenic growth factors to control mammary growth and development in cattle.

A mammary-derived growth inhibitor (MDGI) has been purified from lactating bovine mammary gland (Kurtz *et al.*, 1989) that is not structurally related to TGF-ß. The MDGI is produced by mammary epithelial cells and inhibits proliferation of several normal and transformed mammary epithelial cell lines in a dose-dependent and reversible manner (Grosse and Langen, 1990). Furthermore, concentrations of both the MDGI protein and mRNA are higher in fully differentiated lactating glands than in developing glands of pregnant animals (Kurtz *et al.*, 1989). The bovine MDGI has extensive sequence homology with fatty-acid binding proteins and can function as a lipid carrier (Böhmer *et al.*, 1988). How the lipid-binding properties of MDGI relate to its ability to inhibit cell proliferation is unknown. One possibility is that substrate for prostaglandin synthesis (i.e. arachadonic acid) may be regulated by binding to MDGI. There is both direct and circumstantial evidence implicating prostaglandins or their precursors as necessary in rodent mammary development (Imagawa *et al.*, 1988). Bovine mammary epithelial cells also appear to synthesize prostaglandins (McGrath *et al.*, 1990). Furthermore, the cyclooxygenase inhibitor, indomethacin, reduced mitogenic activity normally stimulated by IGF-I plus EGF (McGrath *et al.*, 1990). Mammary-derived growth inhibitor may therefore serve to modulate the autocrine or paracrine role of stimulatory prostaglandins.

EFFECT OF INDIVIDUAL GROWTH FACTORS ON MAMMARY GROWTH AND MILK YIELD

Since the regulation of mammary growth clearly involves several peptides with either stimulatory or inhibitory actions the obvious question is whether or not any one of these is limiting. Furthermore, transitory growth responses following treatment with an individual growth factor may not result in a change in cumulative mammary growth *postpartum*.

This issue was addressed in separate *in vivo* studies utilizing IGF-I or EGF, two growth factors known to directly stimulate proliferation of bovine mammary epithelial cells *in vitro* and *in vivo* (Collier *et al.*, 1992). In the first study, 60 pregnant nonlactating Holstein cows were randomly assigned to one of three treatment groups. Treatments consisted of sterile medium chain triglyceride oil (MCT; control), 250 µg IGF-I in MCT oil, or 750 µg IGF-I in MCT oil. These treatments were administered by intramammary infusion through the streak canal, beginning 45 days prior to anticipated calving date. Each cow received infusions at three day intervals from days 45 to 27 prior to expected calving (six infusions total). All animals received a prophylactic antibiotic treatment (Cefa-Lac, Bristol Meyers) on days 25, 24 and 23 prior to expected calving. Animals were blocked on start date and lactation number and assigned randomly within block to treatment. Milk production and composition were *postpartum*. Mean milk composition and production for the first nine weeks *postpartum* are shown in Table 1. Neither milk yield nor composition was altered in the lactation following intramammary infusion of IGF-I during late pregnancy in cows.

A similar experimental design was used in a second study to examine the effect of EGF, administered during pregnancy, on subsequent milk yield. Pregnant, nonlactating Holstein cows received six intramammary infusions of excipient (MCT, n=19) or recombinant human epidermal growth factor (hEGF) at doses of 250 µg (n=20) or 1000 μg/infusion (n=20) at three day intervals beginning 45 days prior to expected calving. Milk yield and composition were monitored for the first 63 days of lactation (Table 2). Similar to the results following intramammary infusion of IGF-I, there was no effect of hEGF infusion during late pregnancy on subsequent milk yield and composition. Following 63 days of lactation, half of the animals in each treatment group were given daily injection of recombinant bovine somatotropin (rbST; 25 mg) for 14 days. This protocol examined the effect of intramammary infusion of hEGF during late pregnancy on subsequent

Table 1. Effect of intramammary infusions of insulin-like growth factor-I (IGF-I) administered *prepartum* on least-squares means for milk production and composition for weeks 1 to 9 *postpartum*.

		IGF-I Treatments	
Item[1]	Control	250 μg	750 μg
Milk (kg/day)	39.0 ± 1.6	39.6 ± 1.6	39.3 ± 1.6
3.5% Fat-corrected milk (kg/day)	41.1 ± 1.6	42.1 ± 1.6	40.7 ± 1.5
Fat (%)	3.76 ± 0.1	3.71 ± 0.1	3.64 ± 0.1
Protein (%)	3.04 ± .05	3.07 ± .05	3.03 ± .05
Lactose (%)	4.95 ± .04	5.02 ± .04	5.02 ± .04
Somatic cell count[2] (x 1000)	69	35	59

[1]Number of animals = 20 cows per treatment group. Least-squares means ± standard error presented.
[2]Statistically analyzed using log transformation. Value presented is the antilog of least-squares means.

Table 2. Effect of intramammary infusions of human epidermal growth factor (hEGF) administered *prepartum* on least-squares means for milk production and composition for weeks 1 to 9 *postpartum*.

		hEGF Treatments	
Item[1]	Control	250 μg	1000 μg
Milk (kg/day)	39.3 ± 1.7	37.0 ± 1.7	37.5 ± 1.7
3.5% Fat-corrected milk (kg/day)	39.6 ± 1.8	37.9 ± 1.7	37.5 ± 1.8
Fat (%)	3.63 ± .07	3.71 ± .07	3.5 ± .07
Protein (%)	2.85 ± .06	2.9 ± .05	3.0 ± .06
Lactose (%)	5.19 ± .05	5.17 ± .05	5.11 ± .05
Somatic Cell Count[2] (x 1000)	93	95	93

[1]Number of animals = (control: 19; 250 μg hEGF: 20; 750 μg hEGF: 20). Least-squares means ± standard error presented.
[2]Statistically analysed using log transformation. Value presented is the antilog of least-squares means.

galactopoietic response during established lactation. Milk yield in the bST-treated group was increased by 9.9 kg/day over the 14 day challenge period. However, response to bST was not affected by previous intramammary infusion of hEGF. Failure to demonstrate a significant increase in milk yield in these studies indicated that the mammary growth response to EGF and IGF-I previously demonstrated *in vivo* (Collier *et al.*, 1992) was probably transitory and resulted in no significant difference in mammary cell number at parturition.

It is likely that interaction of mammary growth factors at the local level is regulated by systemic factors. Failure of individual growth factors to stimulate more than transitory growth may reflect lack of coordination. Thus, mammary growth is a homeorhetic process requiring intervention further up the cascade to result in cumulative changes in mammary growth. Molecules which may fit the category of homeorhetic regulator of mammary growth would be limited to those that are unique to pregnancy or which have been already implicated as homeorhetic regulators and are present during pregnancy. One such molecule is placental lactogen. Placental lactogen has both lactogenic and somatogenic properties. Thus, the ability of prolactin and somatotropin, known mammary growth promotants, to act as coordinators of mammary growth also exists since there are periods of mammary growth outside the paradigm of pregnancy and both molecules are present during pregnancy.

LOCAL AND SYSTEMIC EFFECTS OF PROLACTIN, SOMATOTROPIN AND PLACENTAL LACTOGEN ON MAMMARY GROWTH

The classic pituitary ablation and hormonal replacement experiments of Cowie (Cowie *et al.*, 1966) suggested a mammogenic role of prolactin. Almost a decade later, Smith and Schanbacher (1973) demonstrated that lactation could be successfully induced in barren cows by a seven day treatment with estradiol and progesterone. This prompted subsequent studies to refine the technique and elucidate the limiting factors for successful artificial induction of lactation (Erb, 1976). It was concluded from a number of these studies with sheep, goats and cattle that low circulating concentrations of prolactin limited mammary growth stimulated by administration of ovarian steroids (Hooley *et al.*, 1978; Hart and Morant, 1980; Schams *et al.*, 1984). However, in these studies mammary development was assessed either morphologically or from lactation data; mammary cell number was not estimated. Nonetheless, there is a positive correlation between serum prolactin levels in dairy heifers and growth of mammary parenchyma (Peticlerc *et al.*, 1985).

When mammary DNA was measured in dairy heifers treated with bromocriptine during the periparturient period (Akers *et al.*, 1981) it was shown that prolactin was necessary for mammary differentiation, but did not affect mammary growth. Our own data also supports the concept that high levels of prolactin are not required for mammary growth in cattle. We used a seven day treatment with estradiol and progesterone to stimulate mammary growth in peripubertal dairy heifers (Byatt *et al.*, 1994a). Groups of heifers were then treated with saline (negative control), recombinant prolactin or recombinant bPL for the next seven days. All animals were concurrently treated with bromocriptine for this 14 day period. Total mammary DNA was measured on day 15. Although the DNA content of mammary glands was not measured prior to steroid-treatment, it was evident from palpation that parenchymal tissue of control heifers was increased during the course of the experiment even though endogenous prolactin levels were maintained below 2 ng/ml. More importantly, there was no difference in total mammary DNA between the control group and the group treated with 80 mg/day prolactin, indicating that a high level of circulating prolactin did not stimulate additional mammary growth. There was also little evidence of mammary differentiation in the control group. By contrast, glands from prolactin-treated heifers contained copious mammary secretion. In the same experiment, treatment with bPL (80 mg/day) increased total mammary DNA by 50%. This result suggests firstly, that bPL is mammogenic and that it may be one of the factors that stimulates mammary growth during pregnancy. Secondly, although bPL is able to bind to prolactin receptors with high affinity (Scott *et al.*, 1992) this mammogenic effect is probably not mediated through lactogenic receptors. Treatment with bPL did not increase the circulating concentration of IGF-I. Thus, bPL-stimulated mammary growth was apparently not mediated via endocrine IGF-I although changes in binding proteins were not fully evaluated.

In a subsequent study, a similar experimental design was used to determine if bPL could be used to increase milk yield during an artificially induced lactation (Byatt *et al.*, 1994b). Pubertal dairy heifers were given injections of ovarian steroids for seven days to initiate mammary growth. Animals received either saline (control) or bPL (40 mg/day) for 18 days. During this 18 day period, production of endogenous prolactin was again suppressed by treatment with bromocriptine. Mammary differentiation and lactation were then initiated (days 19-23) by treatment with dexamethasone (10 mg/day) and recombinant prolactin (80 mg/day). Despite the fact that endogenous prolactin levels were maintained below 2 ng/ml during the period when the bulk of mammary growth was occurring (days 1-18), over half of the control heifers produced >10 kg milk per day. This study confirms that high concentrations of prolactin are not required to support mammary growth. Furthermore, milk yield of the bPL-treated group was 20% greater than that of the control group (Figure 1). Although not conclusive, this result supports the concept that bPL stimulates mammogenesis.

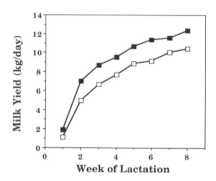

Figure 1. Milk yield of non-pregnant heifers during the first nine weeks of steroid-induced lactation. Non-pregnant, pubertal heifers were treated for seven days with oestradiol (0.05 mg/kg/day) and progesterone (0.25 mg/kg/day) to induce mammary growth. Lactation was initiated 18 days after starting steroid treatment by injection of recombinant bovine prolactin (80 mg/day) for five days and dexamethasone (15 mg/day) for three days. During the mammary growth phase of the study (days 0 to 17) heifers were treated by subcutaneous injection, with water (control, n=11, □) or recombinant bovine placental lactogen (rbPL, n=12, ■). Points indicate the average milk yield of the groups for the previous seven days. Numerical differences between the two groups were not significant (*P* >0.2).

Athough these studies indicate that bPL is mammogenic and that this effect is not mediated by lactogenic receptors, it is still uncertain how bPL mediates this growth. Since bPL binds to somatotropin receptors with similar affinity to bST it is possible that mammary growth is mediated via somatogenic receptors. However, evidence that mammary growth can be stimulated by bST is presently inconclusive. In heifers, mammary parenchymal tissue is increased by treatment with bST during the pre-pubertal period (Sejrsen *et al.*,

1986; Sandles *et al.*, 1987) although the additional growth does not necessarily increase subsequent milk production (Sandles *et al.*, 1987). Administration of bST to primigravid heifers during the last trimester of pregnancy increased subsequent milk production at a dose of 20 mg/day but not in the higher 40 mg/day treatment group (Stelwagen *et al.*, 1992). Finally, treatment of ewes with bST during the last third of pregnancy increased milk production during the subsequent lactation (Stelwagen *et al.*, 1993) although there was no apparent increase in parenchymal tissue prior to initiation of lactation. In addition, the rate of [³H]-thymidine incorporation into parenchymal tissue was not stimulated by treatment with bST. Similarly, administration of bST to late pregnant goats did not stimulate DNA synthesis by parenchymal tissue (Lee and Forsyth, 1988).

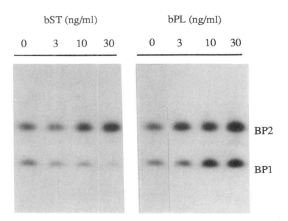

Figure 2. Western ligand-blots of conditioned media from primary bovine mammary fibroblasts following incubation with recombinant bovine placental lactogen (rbPL) or bovine somatotropin (rbST). Fibroblasts isolated from the mammary gland of a mid-pregnant heifer were passaged in serum-free media containing 1 µg/ml of oestradiol, progesterone, insulin, bovine somatotropin, bovine prolactin and bovine placental lactogen. Cells were washed overnight in serum-free media before addition of serum-free media containing either rbPL or rbST. The IGF-binding proteins secreted by the cells over five days were qualitatively measured by Western ligand-blotting. Based on molecular weight, the identity of the two IGF-binding proteins were tentatively identified as binding protein 2 (BP2) and binding protein 3 (BP3).

Any mammogenic actions of bPL and bST *in vivo* are not mediated via a direct effect, because neither bST nor bPL stimulate proliferation of mammary epithelial cells *in vitro* (Skarda *et al.*, 1982; Collier *et al.*, 1992). Whereas lactogenic (prolactin) binding sites have been identified on ruminant mammary tissue (Akers and Keys, 1984; Gertler *et al.*, 1984) there is no measurable specific binding of bST (Keys and Djiane, 1988). Despite the fact that somatogenic binding sites cannot be detected in bovine mammary tissue there is low level expression of mRNA for the somatotropin receptor in both epithelial and stromal components from the mid-pregnant gland (Hauser *et al.*, 1990). By contrast, message for IGF-I was localized predominantly in the stromal elements (consisting mostly of fibroblasts). Therefore, epithelial proliferation may well be modulated by paracrine factors such as IGF-I produced by the stroma. We have cultured fibroblasts isolated from mammary gland of a mid-pregnant heifer to determine if bST and bPL affect secretion of somatomedins and/or somatomedin binding proteins. Presently, the effect of bST and bPL on secretion of IGF-I are inconclusive, however, both hormones altered the ratio of secreted IGF binding proteins. Preliminary results indicate that both bST and bPL increase secretion of IGFBP-2 (Collier *et al.*, 1992). However, secretion of a low molecular weight binding protein, tentatively identified as IGFBP-1, is decreased by bST whereas bPL increased the concentration of this binding protein in the medium (Figure 2). Thus, the local concentration of IGF binding proteins could potentially modulate the local activity of somatomedins.

Although bPL binds to both lactogenic and somatogenic receptors there is also a specific receptor for this hormone (Galosy et al., 1991; Kessler et al., 1991) that has been putatively identified in uterine endometrium and corpus luteum (Lucy et al., 1994). However, it is presently unknown whether specific PL receptors are present in the mammary gland. It appears that the bPL receptor is structurally very similar to the somatotropin receptor (Byatt et al., 1992a; Lucy et al., 1994). A cDNA probe for the extracellular domain of the bovine somatotropin receptor also recognizes a slightly larger message (4.7 vs. 4.4 kb) in tissues such as endometrium and corpus luteum that show specific binding for bPL, but do not bind bST. It will be necessary to determine if the mRNA for the somatotropin receptor detected in mammary epithelial cells could in fact be message for the specific bPL receptor.

POSSIBLE EFFECTS OF PLACENTAL LACTOGEN ON ESTABLISHED LACTATION

Current reproductive management of dairy cattle is designed around a theoretical 12 month calving interval (305 days lactation plus 60 day dry period). In practice, average calving interval for dairy cattle in the U.S. is 13 to 14 months. Nonetheless, concurrent pregnancy and lactation is the norm for cattle that can be successfully rebred. It is recognized that persistency of lactation is lower in pregnant than non-pregnant cattle and that milk production declines more rapidly after the sixth month of gestation (Rook and Campling, 1965). We have examined data from studies where bPL was administered to lactating cattle (Byatt et al., 1992b; J.C. Byatt, unpublished observations) to determine if this placental hormone may affect lactation.

Milk yield of non-pregnant lactating cows and heifers was increased by seven or nine day treatment with recombinant bPL (Figure 3). However, bPL was several-fold less potent than bST. Therefore, physiological concentrations of bPL probably have a negligable stimulatory effect on milk yield. Based on a half-life of 7.5 min (Byatt et al., 1992c) and serum concentration of 1 ng/ml (Byatt et al., 1987) we have calculated the daily production of bPL to be approximately 5 mg/day. Other actions of bPL have been observed at physiological doses (5-10 mg/day) that could have negative effects on milk production. Firstly, treatment with bPL for nine days decreased circulating concentrations of endogenous bST by 40% (Byatt et al., 1992b). This effect was repeated in a subsequent study of similar design (J.C. Byatt, unpublished observations). The mechanism by which bPL decreases circulating bST levels is unknown. It may be mediated via direct negative feedback on pituitary somatotrophs leading to reduced secretion of bST. Given that endogenous bST levels are positively correlated with milk yield (Bonczek et al., 1988) we postulate that endogenous bPL may decrease bST secretion in the latter half of gestation and thus contribute to declining milk yields. Secondly, bPL is a partial ST agonist; thus unlike bST it apparently does not stimulate mobilization of lipid or alter insulin responsiveness of tissues (Byatt et al., 1992b). Studies with the ST receptor suggest that bPL may in fact be a better competitor for this receptor than bST (Staten et al., 1993). Bovine PL could theoretically antagonize some of the repartitioning effects of bST. Teleologically, these two actions make sense in that more metabolites could be partitioned for foetal growth, even though concurrent pregnancy and lactation is an artificially imposed condition. Another

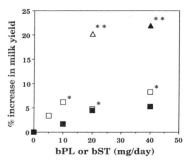

Figure 3. Percentage increase in milk yield of heifers and cows treated with either recombinant bovine placental lactogen (rbPL) or bovine somatotropin (rbST) during lactation. Mid-lactation cows (□ △) were treated with either rbPL (□) or rbST (△) for nine days. Mid-lactating heifers (■, ▲) were treated with rbPL (■) or rbST (▲) for seven days. Average milk yield for the cows and heifers was 34.3 and 28.5 kg/day, respectively. Asterisks indicate when milk yield was greater than control (*$P < 0.05$; ** $P < 0.001$).

Table 3. Effect of recombinant bovine placental lactogen (rbPL) or bovine somatotropin (rbST) on percent increase in dry matter intake of non-pregnant lactating cows and heifers.

	Percentage increase in dry matter intake	
Treatment	Experiment 1	Experiment 2
rbPL (5 mg/day)	1.3	-
rbPL (10 mg/day)	4.1*	5.2*
rbPL (20 mg/day)	1.6	3.3
rbPL (40 mg/day)	4.5*	2.5
rbST (20 mg/day)	1.0	-
rbST (40 mg/day)	-	0.5

*Indicates that value was different ($P < 0.05$) from control. In Experiment 1 (Byatt *et al.*, 1992a) mid-lactation cows were treated for nine days with either rbPL or rbST. In Experiment 2 (J.C. Byatt, unpublished observations) mid-lactation heifers were treated for seven days with either rbPL or rbST. Otherwise, the experimental design of Experiment 2 was essentially identical to that described by Byatt *et al.* (1992a).

factor which suggests that one of the functions of bPL is to support foetal growth, is the effect of this hormone on dry matter intake. Feed intake was increased 4-5 % by bPL (Byatt *et al.*, 1992b; J.C. Byatt, unpublished observations; Table 3). This appeared to be a non-somatogenic effect because bST has no short-term effect on this parameter. The net effect of all of these actions of bPL may be to improve energy balance during pregnancy and thus help support foetal growth.

REFERENCES

Akers, R.M., Bauman, D.E., Capuco, A.V., Goodman, G.T. and Tucker, H.A., 1981, Prolactin regulation of milk secretion and biochemical differentiation of mammary epithelial cells in periparturient cows, *Endocrinology* 109:23.

Akers, R.M. and Keys, J.E., 1984, Characterization of lactogenic hormone binding to membranes from ovine and bovine mammary gland and liver, *J. Dairy Sci.* 67:2224.

Böhmer, F.D., Mieth, M., Reichmann, G., Taube, C., Grosse, R. and Hollenberg, M.D., 1988, A polypeptide growth inhibitor isolated from lactating bovine mammary gland (MDGI) is a lipid-carrying protein, *J. Cell. Biochem.* 38:199.

Bonczek, R.R., Young, C.W., Wheaton, J.E. and Miller, K.P., 1988, Responses of somatotropin, insulin, prolactin, and thyroxine to selection for milk yield in Holsteins, *J. Dairy Sci.* 71:2470.

Byatt, J.C., Larson, B.R., Baganoff, M.P., McGrath, M.F. and Collier, R.J., 1990, Purification and partial characterization of a bovine epidermal growth factor-like polypeptide, *Biochem. Int.* 20:1179.

Byatt, J.C., Wallace, C.R., Bremel, R.D., Collier, R.J. and Bolt, D.J., 1987, The concentration of bovine placental lactogen and the incidence of different forms in foetal cotyledons and in foetal serum, *Domest. Anim. Endocrinol.* 4:231.

Byatt, J.C., Staten, N.R., Krivi, G.G. and Collier, R.J., 1992a, Characterization of binding sites for bovine placental lactogen, Program of Serono Symposium on Trophoblast Cells, Las Vegas, NV p. 27 (Abstr.).

Byatt, J.C., Eppard, P.J., Munyakazi, L., Sorbet, R.H., Veenhuizen, J.J., Curran, D.F. and Collier, R.J., 1992b, Stimulation of milk yield and feed intake by bovine placental lactogen in the dairy cow, *J. Dairy Sci.* 75:1216.

Byatt, J.C., Eppard, P.J., Veenhuizen, J.J., Sorbet, R.H., Buonomo, F.C., Curran, D.F. and Collier, R.J. 1992c, Serum half-life and *in vivo* actions of recombinant bovine placental lactogen in the dairy cow, *J. Endocrinol.* 132:185.

Byatt, J.C., Eppard, P.J., Veenhuizen, J.J., Curran, T.L., Curran, D.F., McGrath, M.F. and Collier, R.J., 1994a, Stimulation of mammogenesis and lactogenesis by recombinant bovine placental lactogen in steroid-primed dairy heifers, *J. Endocrinol.* 140:33.

Byatt, J.C., Hauser, D.S., Krivi, G.G., Siegel, N.R., Smith, C.E. and Stafford, J.M., 1994b, *European Patent* 0 306 470 B1.

Collier, R.J., Ganguli, S., Menke, P.T., Buonomo, F.C., McGrath, F.C., Kotts, C.E. and Krivi, G.G., 1989, Changes in insulin and somatomedin receptors and uptake of insulin, IGF-I and IGF-II during mammary growth, lactogenesis and lactation, *in*: "Biotechnology in Growth Regulation", Heap, R.B., Prosser, C.G. and Lamming, G.E., eds., Butterworths, London.

Collier, R.J., McGrath, M.F., Byatt, J.C. and Zurfluh, L.L., 1992, Regulation of mammary growth by petide hormones: involvement of receptors, growth factors and binding proteins, *Livestock Prod. Sci.* 35:21.

Cowie, A.T., Tindal, J.S. and Yokoyama, A., 1966, The induction of mammary growth in the hypophysectomized goat, *J. Endocrinol.* 34:185.

Daniel, C.W. and Robinson, S.D., 1992, Regulation of mammary growth and function by TGF-β, *Mol. Reprod. Dev.* 32:145.

Dehoff, M.H., Elgin, R.G., Collier, R.J. and Clemmons, D.R., 1988, Both type I and II insulin-like growth factor receptor binding increase during lactogenesis in bovine mammary tissue, *Endocrinology* 122:2412.

Erb, R.E., 1976, Hormonal control of mammogenesis and onset of lactation in cows - a review, *J. Dairy Sci.* 60:155.

Francis, G.L., Read, L.C., Ballard, F.J., Bagley, C.J., Upton, F.M., Gravestock, P.M. and Wallace, J.C., 1986, Purification and partial sequence analysis of insulin-like growth factor-I from bovine colostrum, *Biochem. J.* 233:207.

Galosy, S.S., Gertler, A., Elberg, G. and Laird, D.M., 1991, Distinct placental lactogen and prolactin (lactogen) receptors in bovine endometrium, *Mol. Cell. Endocrinol.* 78:229.

Gertler, A., Ashkenazi, A. and Madar, Z., 1984, Binding sites of human growth hormone and ovine and bovine prolactins in the mammary gland and the liver of lactating dairy cows, *Mol. Cell. Endocrinol.* 34:51.

Glimm, D.R., Baracos, V.E. and Kennelly, J.J., 1992, Northern and *in situ* hybridization analysis of the effects of somatotropin on bovine mammary gene expression, *J. Dairy Sci.* 75:2687.

Grosse, R. and Langen, P., 1990, Mammary-derived growth inhibitor, *in*: "Handbook of experimental pharmacology, II. Peptide growth factors and their receptors II", Sporn, M.B. and Roberts, A.B., eds., Springer-Verlag, New York, NY 95:249.

Hart, I.C. and Morant, S.V., 1980, Roles of prolactin, growth hormone, insulin and thyroxine in steroid-induced lactation in goats, *J. Endocrinol.* 84:343.

Hauser, S.D., McGrath, M.F., Collier, R.J. and Krivi, G.G., 1990, Cloning and *in vivo* expression of bovine growth hormone receptor mRNA, *Mol. Cell. Endocrinol.* 72:187.

Hooley, R.D., Campbell, J.J. and Findlay, J.K., 1978, The importance of prolactin for lactation in the ewe, *J. Endocrinol.* 79:301.

Imagawa, W., Bandyopadhyay, G.K., Wallace, D. and Nandi, S., 1988, Growth stimulation by PGE_2 and EGF activates cyclic AMP-dependent and -independent pathways in primary cultures of mouse mammary epithelial cells, *J. Cell. Physiol.* 135:509.

Kessler, M.A., Duello, T.M. and Schuler, L.A., 1991, Expression of prolactin-related hormones in the early bovine conceptus and potential for paracrine effect on the endometrium, *Endocrinology* 129:1885.

Keys, J.E. and Djiane, J., 1988, Prolactin and growth hormone binding in mammary and liver tissue of lactating cows, *J. Receptor Res.* 8:731.

Kurtz, A., Vogel, F., Funa, K., Heldin, C.H. and Grosse, R., 1989, Developmental regulation of mammary-derived growth inhibitor expression in bovine mammary tissue, *J. Cell. Biol.* 110:1779.

Lee, P.D. and Forsyth, I.A., 1988, The effects of raising growth hormone concentration *pre-partum* on milk yield in goats, *J. Endocrinol.* 117(Suppl.):37.

Lucy, M.C., Byatt, J.C., Curran, T.L., Curran, D.F. and Collier, R.J., 1994, Placental lactogen and somatotropin: hormone binding to the corpus luteum and effects on the growth and functions of the ovary in heifers, *Biol. Reprod.* 50:1136.

Maier, R., Schmid, P., Cox, D., Bilbe, G. and McMaster, G.K., 1991, Localization of transforming growth factor-β1, β2, and β3 gene expression in bovine mammary gland, *Mol. Cell. Endocrinol.* 82:191.

McGrath, M.F., Collier, R.J., Clemmons, D.R., Busby, W.H., Sweeny, C.A. and Krivi, G.G., 1991, The direct *in vitro* effect of insulin-like growth factors (IGFs) on normal bovine mammary cell proliferation and production of IGF binding proteins, *Endocrinology.* 129:671.

McGrath, M.F., Kaempfe, L.A., Sweeny, C.A. and Collier, R.J., 1990, The production and function of prostaglandin 2 (PGE_2) by bovine mammary cells grown in collagen gel culture, *J. Dairy Sci.*, 73(Suppl. 1):214 (Abstr.).

Osborne, C.K., Clemmons, D.R. and Arteaga, C.L., 1990, Regulation of breast cancer growth by insulin-like growth factors, *J. Steroid Biochem. Mol. Biol.* 37:805.

Peri, I., Shamay, A., McGrath, M.F., Collier, R.J. and Gertler, A., 1992, Comparative mitogenic and galactopoietic effects of IGF-I, IGF-II and des-3-IGF-I in bovine mammary gland *in vitro*, *Cell Biol. Int. Rep.* 16:359.

Petitclerc, D., Kineman, R.D., Zinn, S.A. and Tucker, H.A., 1985, Mammary growth response of Holstein heifers to photoperiod, *J. Dairy Sci.* 68:86.

Plaut, K., 1993, Role of epidermal growth factors in mammary development and lactation, *J. Dairy Sci.* 76:1526.

Rook, J.A.F. and Campling, R.C., 1965, Effect of stage and number of lactation on the yield and composition of cow's milk, *J. Dairy Sci.* 32:45.

Ross, M., Francis, G.L., Szabo, L., Wallace, J.C. and Ballard, F.J., 1989, Insulin-like growth factor (IGF)-binding proteins inhibit the biological activities of IGF-I and IGF-II but not des-(1-3)-IGF-I, *Biochem. J.* 258:267.

Samsoondar, J., Kobrin, M.S. and Kudlow, J.E., 1986, α-Transforming growth factor secreted by untransformed bovine anterior pituitary cells in culture. I. Purification from conditioned media, *J. Biol. Chem.* 261:14408.

Sandles, L.D., Peel C.J. and Temple-Smith, P.D., 1987, Mammogenesis and first lactation milk yields of identical-twin heifers following pre-pubertal administration of bovine growth hormone, *Anim. Prod.* 45:349.

Schams, D., Rüsse, I., Schallenberger, E., Prokopp, S. and Chan, J.S.D., 1984, The role of steroid hormones, prolactin and placental lactogen on mammary gland development in ewes and heifers, *J. Endocrinol.* 102:121.

Scott, P., Kessler, M.A. and Schuler, L.A., 1992, Molecular cloning of the bovine prolactin receptor and distribution of prolactin and growth hormone receptor transcripts in foetal and utero-placental tissues, *Mol. Cell. Endocrinol.* 89:47.

Sejrsen, K., Foldager, J., Sorensen, M.T., Akers, R.M. and Bauman, D.E., 1986, Effect of exogenous bovine somatotropin on pubertal mammary development in heifers, *J. Dairy Sci.* 69:1528.

Sheffield, L.G. and Yuh, I.S., 1988, Influence of epidermal growth factor on growth of bovine mammary tissue in athymic nude mice, *Domest. Anim. Endocrinol.* 5:141.

Shimamoto, G.T., Byatt, J.C., Jennings, J.G., Comens-Keller, P.G. and Collier, R.J., 1992, Destripeptide insulin-like growth factor-I in milk from bovine somatotropin treated cows, *Pediatr. Res.* 32:296.

Skarda, J., Urbanová, E., Bečka, S., Houdebine, L.-M., Delouis, C., Píchová, D., Předa, J., and Bílek, J., 1982, Effect of bovine growth hormone on development of goat mammary tissue in organ culture, *Endocrinol. Exp.* 16:19.

Smith, K.L. and Schanbacher, F.L., 1973, Hormone induced lactation in the bovine. I. Lactational performance following injections of 17β-estradiol and progesterone, *J. Dairy Sci.* 56:738.

Spitzer, E. and Grosse, R., 1987, EGF receptors on plasma membranes purified from bovine mammary gland of lactating and pregnant animals, *Biochem. Int.* 14:581.

Staten, N.R., Byatt, J.C. and Krivi, G.G., 1993, Ligand-specific dimerization of the extracellular domain of the bovine growth hormone receptor, *J. Biol. Chem.* 268:18467.

Stelwagen, K., Grieve, D.G., McBride, B.W. and Rehman, J.D., 1992, Growth and subsequent lactation in primigravid Holstein heifers after *prepartum* bovine somatotropin treatment, *J. Dairy Sci.* 75:463.

Stelwagen, K., Grieve, D.G., Walton, J.S., Ball, J.L. and McBride, B.W., 1993, Effect of *prepartum* bovine somatotropin in primigravid ewes on mammogenesis, milk production and hormone concentrations, *J. Dairy Sci.* 76:992.

Tou, J.S., McGrath, M.F., Zupec, M.E., Byatt, J.C., Violand, B.N., Kaempfe, L.A. and Vineyard, B.D., 1990, Chemical synthesis of bovine transforming growth factor-α: synthesis, characterization and biological activity, *Biochem. Biophys. Res. Commun.* 167:484.

Winder, S.J., Turvey, A. and Forsyth, I.A., 1989, Stimulation of DNA synthesis in cultures of ovine mammary epithelial cells by insulin and insulin-like growth factors, *J. Endocrinol.* 123:319.

Zurfluh, L.L., Bolten, S.L., Byatt, J.C., McGrath, M.F., Tou, J.S., Zupec, M.E. and Krivi, G.G., 1990, Isolation of genomic sequence encoding a biologically active bovine TGF-α protein, *Growth Factors* 3:257.

EXPRESSION OF *HOX* GENES IN NORMAL AND NEOPLASTIC MOUSE MAMMARY GLAND

Yael Friedmann and Charles W. Daniel

Department of Biology
Sinsheimer Laboratories
University of California
Santa Cruz, CA 95064, U.S.A.

INTRODUCTION

The striking phenomenon of homeosis, in which one body part is substituted for another, was described and named in *Drosophila* more than a century ago (Bateson, 1894). Homeobox genes have been intensively studied, particularly in recent years, as advances in understanding of the molecular genetic basis for homeosis captured the interest and imagination of developmental biologists. It is now clear that homeotic genes function as master regulators, determining the developmental pathway and cell fate of individual *Drosophila* segments.

Homeobox genes (homeogenes) are highly conserved and homologues are found throughout the animal kingdom (Bieberich *et al.*, 1991; Murtha *et al.*, 1991; Pendleton *et al.*, 1993). They comprise a large collection of developmental regulators whose common structural characteristic is a box motif of about 183 bases, coding for the 61 amino acid homeodomain that is the helix-turn-helix DNA binding region of the functional homeoprotein (Scott *et al.*, 1989). It is generally accepted that the homeoprotein functions as a transcription factor, regulating the expression of downstream genes (Andrew and Scott, 1992). *Drosophila* genes with homeotic potential are included in the Antennapedia and Bithorax complexes.

The closely related murine homologues, the *Hox*-C genes, occur in four clusters that, like their Dipteran homologues, are arranged along their respective chromosomes in a manner that is generally co-linear with their expression sequence (reviewed by Akam, 1989; Kessel and Gruss, 1990; McGinnis and Krumlauf, 1992). Most of the mammalian homeobox genes, including the 38 *Hox*-C in the mouse and human, were discovered by cross hybridization and only later, and with considerable difficulty, have the phenotype of a few of them been determined (McGinnis and Krumlauf, 1992). Because of the shortage of developmental mutations in mammals, information on homeobox gene function has come from reverse genetics - either misexpression through the insertion of homeobox transgenes, or generation of the null phenotype through targeted gene disruption of the mouse germ line (reviewed by McGinnis and Krumlauf, 1992). Transgenic mice in which *Hox*-1.1 (Balling *et al.*, 1989) and *Hox*-4 (Wolgemuth *et al.*, 1989) were overexpressed, for example, were dominant lethals showing severe developmental defects. Misexpression of *Hox* 4.6 in the chick limb bud led to pattern alterations and apparent homeotic transformation of the digit (Morgan *et al.*, 1992).

Why might expression of homeobox genes be expected in the mammary gland, which is not considered a segmental structure? One argument is that the mammary gland is embryonic-like during its phase of ductal development following menarche, even though development occurs in the subadult animal (Sakakura *et al.*, 1979). In addition, there are numerous examples of homeobox gene expression in tissue development that are not considered segmental in nature (review by Sassoon, 1992). Particularly striking examples are the role of the *Hoxd* cluster in limb morphogenesis (Tabin, 1991; Morgan *et al.*, 1992; Tabin, 1992), and in the differentiation of haematopoietic cell lineages, in which many *Hox* genes are switched on or off in blocks, indicating coordinate regulation. Distinct patterns of *Hox* gene activation are found in different lineages, suggesting that these genes may play an important role in cell lineage determination during both normal and leukemic haematopoiesis (Magli *et al.*, 1991; Celetti *et al.*, 1993).

Thus, the mammary gland appears to be a candidate target for homeobox gene action. It represents a variety of developmental interactions and pathway decisions as it undergoes its complex developmental cycle of branching morphogenesis, functional differentiation, secretion, and involution, in which master regulatory genes are likely to play a role. In this chapter we report the expression of several *Hox* genes in the mouse mammary gland, and describe in more detail the activity of members of the *Hoxd* cluster.

RESULTS

Detection of candidate homeobox genes expressed in the mouse mammary gland

cDNA was made from mouse mammary gland RNA and was used to amplify homeoboxes using degenerate oligonucleotide primers from conserved domains within the homeobox. The PCR products were cloned, and the DNA sequences of several individual clones were determined. We found expression of genes from all four major clusters and chose the *Hoxd* cluster for further investigation.

Different *Hoxd* genes are independently regulated and regulation is related to developmental stage

Total cellular RNA was extracted at the onset of puberty (about five weeks after birth), two time points during adulthood of virgin animals (8-9 weeks after birth and 16 weeks after birth), early pregnancy (5-8 days post coitus), late pregnancy (16-18 days p.c.) and lactation. The RNA was poly A$^+$ enriched and northern blots of these RNAs were hybridized consecutively with *Hoxd*-3, *Hoxd*-4, *Hoxd*-8, *Hoxd*-9, *Hoxd*-10, *Hoxd*-11, *Hoxd*-12 and L7 probes. L7 RNA was used as a control for amount of RNA and its integrity. Figure 1 shows the pattern of expression of the RNAs of representatives of the *Hoxd* genes during mammary development. *Hoxd*-9 and *Hoxd*-4 transcripts were readily detected in glands from pubescent and mature virgins and animals in early stages of pregnancy. Transcript levels decreased in glands from animals that were in late stages of pregnancy and could hardly be detected during lactation. During lactation, high levels of transcripts for milk proteins may dilute other mRNAs, as can be seen for the L7 loading control. *Hoxd*-9 and *Hoxd*-4 may therefore be expressed during lactation at higher levels than indicated. *Hoxd*-12 was not detected in any of these stages of development. Table 1 summarizes the mRNA expression pattern of all the *Hoxd* genes that were examined. *Hoxd*-8 and *Hoxd*-10 pattern of expression in the normal gland was very similar to the one seen for *Hoxd*-9 and *Hoxd*-4, while *Hoxd*-3 and *Hoxd*-11 were not expressed in any of the normal stages of the mammary gland development (similar to *Hoxd*-12). These different

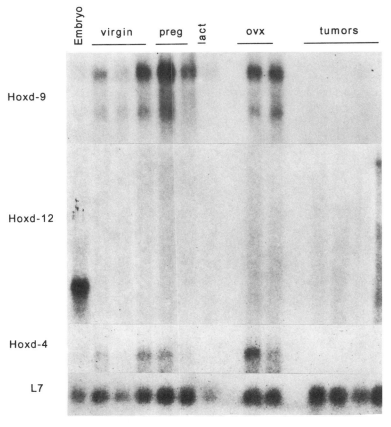

Figure 1. RNA expression of *Hoxd*-9, *Hoxd*-12 and *Hoxd*-4 at different developmental stages of the mouse mammary gland, in response to ovariectomy and in mammary tumors. Poly A$^+$ RNAs were extracted from different stages of mammary gland development, from glands of animals that were ovariectomized and from transformed mammary glands. RNA was subjected to northern blot analysis with random primed, ^{32}P-labeled cDNA probes for *Hoxd*-9, *Hoxd*-12 and *Hoxd*-4. None of the clones contains the homeobox sequence itself. A human ribosomal L7 probe was used as a control for the amount and integrity of the RNA. Each lane contains 5 μg poly A$^+$ enriched RNA. Lanes from left to right: 14 days old embryo, glands from five weeks old animals, glands from 8-9 weeks old animals, glands from 16 weeks old animals, glands from 5-8 days pregnant animals, glands from 15-18 days pregnant animals, glands from lactating animals, glands from animals that were ovariectomized at 5 weeks and allowed to mature for 4 more weeks, glands from animals that were ovariectomized at 12 weeks and allowed to mature for 4 more weeks, tumor D2a, tumor D2d, tumor D1a, 35S Myc tumor.

expression patterns indicate that the *Hoxd* genes are independently regulated in the normal gland and that *Hoxd*-4, *Hoxd*-8, *Hoxd*-9 and *Hoxd*-10 transcript levels are somewhat developmentally regulated during the gland cycle.

Hoxd expression in glands from ovariectomized animals

Mammary gland development is stimulated by the ovarian hormones estrogen and progesterone (Topper and Freeman, 1980; Haslam, 1987; Silberstein *et al.*; 1994). These act directly on the mammary gland (Daniel *et al.*, 1987), probably by regulating growth factors (Dembinski and Shiu, 1987), and may also act indirectly, through mediation of pituitary secretions (Vonderhaar, 1987). The mammary ducts become much smaller in diameter and simpler in pattern of branching in response to ovariectomy, as can be seen in Figure 2 in a whole-mount preparation.

If homeogenes have a significant role in mammary gland development, it is likely that a linkage between them and the mammogenic hormones that are required for growth, morphogenesis, and functional activity would be discovered. To determine if *Hoxd* transcript levels are regulated by ovarian secretions, we isolated RNA from glands of animals that had been ovariectomized at 5 or 12 weeks of age and allowed to mature for 4 weeks after surgery, a time period that was determined adequate for the ovarian steroids to be depleted from the tissues. Figure 1 shows the result of representative northern blot hybridizations with *Hoxd*-9, *Hoxd*-4 and *Hoxd*-12 probes, followed by L7 as a loading control.

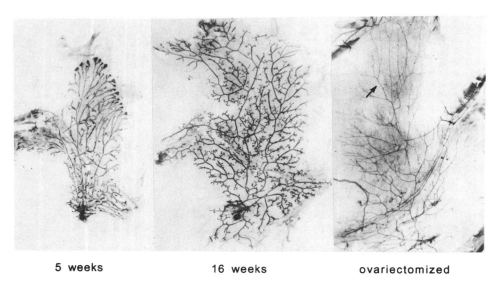

| 5 weeks | 16 weeks | ovariectomized |

Figure 2. Whole mount preparations of glands from 5 weeks old animals, 16 weeks old animals and a gland from a mouse that was ovariectomized at 5 weeks of age and allowed to mature for 4 more weeks. Arrow points to an epithelial duct, showing a decrease in ductal diameter and a simplification of ductal patterning in response to ovariectomy. All pictures were taken at the same magnification.

All genes except for *Hoxd*-4 show the same level of expression in glands from ovariectomized animals as their intact counterparts (glands from subadult and mature virgin animals). *Hoxd*-4, on the other hand, shows a higher level of expression in glands from animals that were ovariectomized at 5 weeks of age and allowed to mature for 4 more weeks than in any normal stage of development. Glands of animals that were ovariectomized when the animal was cycling (12 weeks old when ovariectomized and 16 weeks old when glands were taken) do not show the same effect.

These results suggest that *Hoxd*-4 transcripts are down-regulated by ovarian secretions, at least during some stages of the mammary gland cycle; the candidate hormones are estrogen, progesterone and possibly prolactin, which is estrogen-regulated (Lyons, 1958).

Table 1. *Hoxd* expression during mammary gland development, in response to ovariectomy and in mammary tumors.[1,2]

	5 weeks	8-9 weeks	16 weeks	early pregnant	late pregnant	lactating[3]	ovariectomized at 5 weeks	ovariectomized at 12 weeks	tumor D2a	tumor D2d	tumor D1a	Myc tumor
Hoxd-3	-	-	-	-	-	-	-	-	-	-	+	-
Hoxd-4	++	++	++	++	+	-	++++	++	-	-	-	-
Hoxd-8	+++	+++	+++	+++	+	+	+++	+++	-	-	-	-
Hoxd-9	++	++	++	+++	+	+	++	++	-	-	-	-
Hoxd-10	+	+	++	+++	++	++	++	++	-	-	-	-
Hoxd-11	-	-	-	-	-	-	-	-	-	-	-	-
Hoxd-12	-	-	-	-	-	-	-	-	-	-	-	+++

1. Levels are compared for each gene separately and cannot be compared between different genes.
2. Levels are adjusted according to L7 loading control.
3. During lactation, high levels of transcripts for milk proteins may dilute other mRNAs, (as can be seen for L7 loading control in figure 1). Hox expression may therefore be expressed during lactation at higher levels than indicated.

Hoxd expression in neoplastic mammary glands

Total cellular RNA was extracted from three tumors that arose spontaneously from hyperplastic alveolar nodule lines (Medina, 1973) and from a tumor induced by misexpression of the *myc* proto-oncogene in cultured mammary epithelium, followed by transplantation into mammary gland-free fat pads (Bradbury *et al.*, 1991). RNA was poly A$^+$ enriched and was subjected to a northern blot hybridization analysis using the previously described *Hoxd* probes. As can be seen from Figure 1 and Table 1, two *Hoxd* genes (*Hoxd*-3 and *Hoxd*-12) were expressed in one of the mammary tumors. These two genes were not expressed in any of the normal developmental stages of the mammary gland. *Hoxd*-12 transcript size in the tumor (~ 8.4 kb) is different from its size in the embryo (~ 2.7 kb). *Hoxd*-4, *d*-8, *d*-9 and *d*-10 which showed expression during the normal development of the mammary gland, were not expressed in any of the mammary tumors that were tested.

DISCUSSION

Our initial exploratory studies were aimed at determining whether homeobox genes were expressed in the mammary gland in enough numbers and enough abundance to warrant further investigation. We amplified homeobox sequences from mammary gland cDNA that, after sequence analysis, was found to contain members from each of the four *Hox* clusters. In order to obtain semi-quantitative data we used sequence-specific probes to non-conserved regions 5' to the homeobox. In most cases we were able to obtain a clearly identifiable band on northern blots without enriching for poly A$^+$ (not shown), indicating a reasonably abundant level of transcripts.

In this initial investigation, rather than carry out a large scale survey of the 38 known *Hox* genes and numerous other non-clustered homeobox genes, we chose to examine in more detail the expression of members of the *Hoxd* group. This cluster is of particular interest because of its role in patterning of the vertebrate limb. Beginning with a nondescript cluster of undifferentiated cells, the developing chick limb displays progressively increasing cellular diversity. Using a combination of classical transplantation technology, activation of *Hox* genes by retinoic acid, and molecular techniques, the role of *Hoxa* and *Hoxd* clusters in limb morphogenesis has been elegantly documented (Tabin, 1991; Morgan *et al.*, 1992; Tabin, 1992). Thus, *Hoxd* well illustrates an extensive role for homeobox genes in vertebrate development.

In the normal mammary gland, four out of seven members of the *Hoxd* cluster examined displayed some degree of expression. Differences between individual genes (*Hoxd*-4 vs. *Hoxd*-11, for example) indicate differential regulation. Declining transcript levels in late pregnancy observed in *Hoxd*-4, *Hoxd*-8, *Hoxd*-9 and *Hoxd*-10 indicate a degree of developmental regulation. Steady-state mRNA was also reduced during lactation, but dilution by milk protein mRNA makes this conclusion tentative. In any case, the data show that homeogene expression is not simply uniform or constitutive, and in some cases is related to mammary stage.

Two of the three *Hoxd* genes that were not expressed in the normal gland showed expression in mammary tumors, indicating an association between the altered expression of *Hoxd* genes and mammary cancer. Such a relationship with other tumors and homeobox genes has been documented. In human leukemia, altered regulation of developmental homeobox control has been convincingly demonstrated (Celetti *et al.*, 1993). Also, other cancers display altered expression of homeobox genes (Kennedy *et al.*, 1991; Celetti *et al.*, 1993; De Vita *et al.*, 1993), and misexpression of homeobox genes can transform cells *in vitro* and produce tumors in mice upon transplantation (Perkins *et al.*, 1990; Song *et al.*, 1992).

Hormones of reproduction drive the mammary gland through its cycles of growth, secretory differentiation, lactation, and post-weaning involution. Primary among these hormones are the ovarian steroids, for which receptors exist in both the mammary epithelium and its contiguous stroma (Haslam and Shyamala, 1981; Daniel *et al.*, 1987), and a direct mammogenic role for estradiol has been demonstrated (Silberstein *et al.*, 1994; Haslam, 1988). Genes regulating tissue-specific responses to these hormones should be downstream of the primary steroid response elements, and it is reasonable to suppose that if *Hox* genes play a role in the mammary gland, their expression will be linked to this primary endocrine regulation.

We tested this proposition by ovariectomizing animals and comparing their levels of *Hox* expression with endocrine-intact controls. Expression of *Hoxd*-4 was substantially increased in mice ovariectomized at five weeks, but not at 12 weeks. This suggests that *Hoxd*-4 transcripts may be down-regulated by ovarian steroids in certain stages of development. The age difference in response may indicate that *Hoxd*-4 is regulated by ovarian steroids during the period of active ductal growth but not in the adult, where the gland is relatively quiescent. A small increase in expression during early pregnancy is consistent with this interpretation.

The functional role of homeobox genes in the mammary gland, if any, is of course unknown at this point, but several possibilities come to mind that are consistent with the observed phenotype of these genes in systems that are more amenable to genetic analysis. It is expected that the expression of many homeobox genes will occur in the mammary stroma. Homeobox action in these mesodermal tissues could be reflected in altered production or turnover of mammary-associated matrix components, or in the differentiation of mesenchyme-like cells into fibroblasts. Homeobox genes may also regulate the fate of epithelial cells in the end bud as cells are channeled into various differentiated populations in the subtending ducts, a possibility consistent with the known action of homeotic genes in determining cell fate. A particularly intriguing possibility is that homeobox genes might influence expression of cell-cell or cell-matrix adhesion molecules, thereby directly regulating morphogenesis and pattern formation.

Our results indicate that several *Hox* genes are expressed in the mouse mammary gland. It seems likely that many homeobox genes, both within and outside of the major clusters, will be found to be expressed at some level in one or more stages of the mammary cycle. Working out the functional implications of this gene activity will be a challenging but potentially fruitful task.

SUMMARY

There has been no report of expression of homeobox-containing genes in the breast. We amplified homeobox cDNA from mouse mammary gland and found expression of members from each of the four major *Hox* gene clusters. Northern analysis showed expression of these at significant levels. In order to determine if expression was restricted to only a few representatives, a survey was carried out on the *Hoxd* cluster using sequence-specific cloned probes for northern analysis. Expression in the normal mammary gland was seen in four out of the seven members of the *Hoxd* genes examined; highest levels were found in glands from immature and mature virgin animals and in glands of early pregnant animals, with decreasing transcripts detected during later stages of pregnancy and lactation. Three *Hoxd* genes were not expressed in the normal gland, indicating that the individual genes are independently regulated. In order to determine if expression is linked to mammogenic hormones, *Hox* expression was studied in ovariectomized animals. Most of the *Hoxd* genes were not significantly modulated by reduction in ovarian steroids, but *Hoxd*-4 expression was increased significantly in glands from animals that were ovariectomized at a young age, indicating possible negative regulation by steroid hormones. Looking at mammary tumors, we found two *Hoxd* genes to be expressed, each in a different tumor. Our findings indicate that several but not all *Hoxd* genes are expressed in the normal gland and some are expressed in mammary tumors and not in the normal gland. Expression is differentially regulated and is related to developmental stage. A relationship between mammogenic steroid hormones and *Hox* expression suggests that these genes may play a role in mammary development or function.

REFERENCES

Akam, M., 1989, *Hox* and *HOM*: Homologous gene clusters in insects and vertebrates, *Cell* 57:347.
Andrew, D.J. and Scott, M.P., 1992, Downstream of the homeotic genes, *New Biol.* 4:5.
Balling, R., Mutter, G., Gruss, P. and Kessel, M., 1989, Craniofacial abnormalities induced by ectopic expression of the homeobox gene *Hox*-1.1 in transgenic mice, *Cell.* 58:337.

Bateson, W. (1894). Materials for the Study of Variation. London, MacMillan and Co.

Bieberich, C.J., Ruddle, F.H. and Stenn, K.S., 1991, Differential expression of the *Hox* 3.1 gene in adult mouse skin, *Ann NY Acad Sci.* 642:346.

Bradbury, J.M., Sykes, H. and Edwards, P.A., 1991, Induction of mouse mammary tumours in a transplantation system by the sequential introduction of the *myc* and *ras* oncogenes, *Int J Cancer* 48:908.

Celetti, A., Barba, P., Cillo, C., Rotoli, B., Boncinelli, E. and Magli, M. C., 1993, Characteristic patterns of *Hox* gene expression in different types of human leukemia, *Int J Cancer.* 53:237.

Daniel, C.W., Silberstein, G.B. and Strickland, P., 1987, Direct action of 17 beta-estradiol on mouse mammary ducts analyzed by sustained release implants and steroid autoradiography, *Cancer Res.* 47:6052.

Dembinski, T.C. and Shiu, R.P.C., 1987, Growth factors in mammary gland development and function, *in* "The Mammary Gland: Development, Regulation and Function", M.C. Neville and C.W. Daniel, eds., Plenum Press, London.

De Vita, G., Barba, P., Odartchenko, N., Givel, J.C., Freschi, G., Bucciarelli, G., Magli, M.C., Boncinelli, E. and Cillo, C.G., 1993, Expression of homeobox-containing genes in primary and metastatic colorectal cancer, *Eur. J. Cancer* 6:8887.

Haslam, S.Z., 1987, Role of sex steroid hormones in normal mammary gland function. "The mammary gland: development, regulation, and function" M.C. Neville and C.W. Daniel, eds., New York, in *Plenum Press.* 499.

Haslam, S.Z., 1988, Local Versus Systemically Mediated Effects of Estrogen on Normal Mammary Epithelial Cell Deoxyribonucleic Acid Synthesis, *Endocrinology* 122:860.

Haslam, S.Z. & Shymala, G., 1981, Relative distribution of estrogen and progesterone receptors among the epithelial, adipose, and connective tissue components of the normal mammary gland, *Endocrinology* 108:825.

Kennedy, M.A., Gonzalez-Sarmiento, R., Kees, U.R., Lampert, F., Dear, N., Boehm, T. and Rabbitts, T.H.M, 1991, *Hox*-11, a homeobox-containing T-cell oncogene on human chromosome 10q24, *Proc. Natl. Acad. Sci. USA* 88:8900.

Kessel, M. and Gruss, P., 1990, Murine developmental control genes, *Science* 249:374.

Lyons, W.R., 1958, Hormonal synergism in mammary growth, *Proc. Royal Soc. London Series B* 149:303.

Magli, M.C., Barba, P., Celetti, A., De Vita, G., Cillo, C. and Boncinelli, E., 1991, Coordinate regulation of *Hox* genes in human haematopoietic cells, *Proc. Natl. Acad. Sci. USA.* 88:6348.

Medina, D., 1973, Preneoplastic lesions in mouse mammary tumourigenesis, *Methods Cancer Res.* 7:3.

McGinnis, W. and Krumlauf, R., 1992, Homeobox genes and axial patterning, *Cell.* 68:283.

Morgan, B.A., Izpisua-Belmonte, J.C., Duboule, D. and Tabin, C.J., 1992, Targeted misexpression of *Hox*-4.6 in the avian limb bud causes apparent homeotic transformations, *Nature.* 358:236.

Murtha, M. T., Leckman, J. F. and Ruddle, F.H., 1991, Detection of homeobox genes in development and evolution, *Proc. Natl. Acad. Sci. USA* 88:10711.

Pendleton, J.W., Nagai, B.K., Murtha, M.T. and Ruddle, F.H., 1993, Expansion of the *Hox* gene family and the evolution of chordates, *Proc. Natl. Acad. Sci. USA.* 90:6300.

Perkins, A., Kongsuwan, K., Visvader, J., Adams, J.M. and Cory, S., 1990, Homeobox gene expression plus autocrine growth factor production elicits myeloid leukemia, *Proc. Natl. Acad. Sci. USA* 87:8398.

Sakakura, T., Sakagami, Y. and Nishizuka, Y., 1979, Persistence of responsiveness of adult mouse mammary gland to induction by embryonic mesenchyme, *Dev. Biol.* 72:201.

Sassoon, D., 1992, *Hox* genes: a role for tissue development, *Amer. J. Respir. Cell Mol. Biol.* 7:1.

Scott, M.P., Tamkun, J.W. and Hartzell, G., 1989, The structure and function of the homeodomain, *Biochim. Biophys. Acta* 989:25.

Silberstein, G.B., Van Horn, K., Harris, G.S. and Daniel, C.W., 1994, Essential role of endogenous estrogen in directly stimulating mammary growth demonstrated by implants containing pure antiestrogens, *Endocrinology*, in Press.

Song, K., Wang, Y. and Sassoon, D., 1992, Expression of *Hox*-7.1 in myoblasts inhibits terminal differentiation and induces cell transformation, *Nature* 360:477.

Tabin, C.J., 1991, Retinoids, homeoboxes, and growth factors: toward molecular models for limb development, *Cell* 66:199.

Tabin, C.J., 1992, Why we have (only) five fingers per hand: *Hox* genes and the evolution of paired limbs, *Development* 116:289.

Topper, Y.J. and Freeman, C.S., 1980, Multiple hormone interactions in the development biology of the mammary gland, *Physiol. Rev.* 60:1049.

Vonderhaar, B.K.,1987, Prolactin. Transport, function and receptors in mammary gland development and differentiation, *in* "The Mammary Gland: Development, Regulation and Function", M.C. Neville and C.W. Daniel, eds., Plenum Press, London.

Wolgemuth, D.J., Behringer, R.R., Mostoller, M.P., Brinster, R.L. and Palmiter, R.D., 1989, Transgenic mice overexpressing the mouse homoeobox-containing gene *Hox*-1.4 exhibit abnormal gut development, *Nature.* 337:464.

LOCAL SIGNALS FOR GROWTH CESSATION AND DIFFERENTIATION IN THE MAMMARY GLAND

Richard Grosse

The Gade Institute
University of Bergen
N-5021 Bergen, Norway

INTRODUCTION

The mammary gland provides a unique model to study postnatal processes of growth and differentiation. Development of the mouse mammary gland is a complex multistage process, which begins in the embryo with the mammary *anlagen* giving rise to primary and secondary sprouts. Sparsely branching ducts invade the stroma at puberty, followed by the development of lobuloalveolar structures, and functional differentiation during pregnancy (Topper and Freeman, 1980). By use of endocrine ablation, organ culture systems and various cell culture models, it has been demonstrated that several steroid hormones, prolactin and growth hormone regulate this process (for reviews, see Banerjee and Antoniou, 1985; Streuli and Bissell, 1991). The combined action of aldosterone, prolactin, insulin and cortisol is sufficient to promote lobuloalveolar development and functional differentiation in organ cultures of mammary glands from sexually immature mice, pretreated with estradiol and progesterone (Banerjee *et al.*, 1973). Although the systemic importance of ovarian and pituitary hormones has been well documented, these hormones are virtually incapable of stimulating proliferation or inhibiting growth of mammary epithelial cells (MEC) *in vitro*.

In addition to systemic control, it has become clear that the local microenvironment provides many of the signals necessary for epithelial differentiation (Sakakura, 1991). Formation of the basement membrane by endogenous synthesis or by external provision is sufficient for induction of tissue specific differentiation (Streuli and Bissell, 1990). It has been suggested that the basement membrane induces the expression of genes linked to functional differentiation via a direct biochemical effect involving integrin mediated signals rather than by affecting cell polarity, cell-cell interaction or cytokine production (Stoker *et al.*, 1990). Contrary to the situation found during neonatal and adult phases of mammary gland morphogenesis, for which a number of growth factors have been described (reviewed by Oka *et al.*, 1990), comparatively little is known about factors likely to be induced by systemic hormones and or the basement membrane during differentiation. Loss of differentiation as it occurs in the majority of ductal carcinomas is associated with inactivation of negative control mechanisms (Sporn and Roberts, 1985). Therefore, a search for peptides conferring growth cessation and functional differentiation in the normal gland may help to define pathways abrogated during malignant transformation.

This article summarises data related to a number of unresolved questions concerning regulation of mammary growth and differentiation. Specifically, what are the local signals causing cessation of growth of epithelial cells when they enter the differentiation pathway? What are the local signals inducing and maintaining different phases of structural and functional differentiation? Do they mediate or complement the activities of systemic hormones?

GROWTH INHIBITORS

Identification of growth inhibitors has been hampered by the lack of reliable test systems and by difficulties in enriching activities detected in cell and tissue extracts, serum or conditioned media. Until now, only a few reports of structurally-defined mammary growth inhibitors have been published.

Table 1. Mammary epithelial growth inhibitors

Factor	Epression	Inhibition	Reference
Mammastatin	MEC	MEC	Ervin *et al.*, 1989
TGF-ß	MEC/Stroma	MEC	Silbersten and Daniel, 1987
MDGI	MEC	MEC	Böhmer *et al.*, 1987
NDF	Fibroblasts	MEC	Peles *et al.*, 1992
gP30	Breast carcinoma	MEC	Lupu *et al.*, 1992

Mammastatin was purified from normal human mammary epithelial cells (Ervin *et al.*, 1989). It seems to inhibit specifically DNA synthesis in cultured normal and transformed mammary epithelial cells. Its structural identity and mode of action are unknown.

The transforming growth factor-ß (TGF) isoforms, comprising TGF-$ß_1$, -$ß_2$ and -$ß_3$ are a family of multifunctional cytokines which can both stimulate and inhibit cell proliferation (reviewed by Sporn and Roberts, 1992). TGF-$ß_1$ is the only growth inhibitor described so far that is capable of affecting formation and growth of mammary ductal buds in virgin mice (Silberstein and Daniel, 1987). It was first shown that localized TGF-$ß_1$ treatment rapidly inhibited DNA synthesis in ductal cells of sub-adult mice whereas lobuloalveolar DNA synthesis was unaffected. This effect might be related to localization of endogenous TGF-$ß_1$ in complexes with extracellular matrix molecules surrounding those ductal structures in which budding was inhibited (Silberstein *et al.*, 1990). In all actively growing ductal buds, a selective loss of TGF-$ß_1$ was observed. Alveolar buds which give rise to formation of secretory alveoli and which are not affected by TGF-$ß_1$ did not show changes in ECM-bound TGF-$ß_1$ (Silberstein *et al.*, 1992). From this data it was concluded that the ECM-bound TGF-$ß_1$ functions to prevent chronic lateral budding and to maintain normal patterns of mammary branching during ductal morphogenesis. TGF-ß$_1$ is produced by epithelial and stromal mammary cells (Silbertsten *et al.*, 1990, 1992). Its mRNA expression level does not change in virgin, pregnant or terminally-differentiated states of development (Robinson *et al.*, 1993). The questions which remain to be answered are how TGF-ß$_1$ (or one of the other isoforms) is being mobilized precisely during the ductal growth phase to prevent lateral branching, and which cell types are actually releasing the active cytokine. A role for TGF-ß$_1$ in regulating mesenchymal-epithelial interactions has been discussed (Sporn and Roberts, 1992). Different members of the steroid hormone family and their antagonists such as tamoxifen, glucocorticoids, type-1 and type-2 anti-estrogens, as well as retinoic acid, induce TGF-ß$_1$ expression in breast cancer cell lines (Knobbe *et al.*, 1994). It would be interesting to determine whether mammogenic hormones which, like cortisol, interact with mesenchymhal cells can control expression and release of TGF-ß in the developing mouse mammary gland. However, there are conflicting reports on the role of endogenous estrogen in stimulating mammary growth. In addition to its well known indirect effects on the mammary gland through pituitary hormones such as prolactin and growth hormone, more recent data have shown a direct effect by using slow-release plastic implants containing anti-estrogens which were inserted into the growth region of endocrine-intact mouse mammary glands (Silberstein *et al.*, 1994). The observed ductal growth inhibition raises the question of involvement of growth factor dependent pathways. One possible explanation for the local anti-estrogen mediated growth inhibition is an activation of TGF-ß dependent signalling pathways (Silberstein *et al.*, 1994).

In order to dissect *in vivo* functions for TGF-ß during ductal branching in virgin mice, targetted expression of the TGF-ß$_1$ gene was achieved in transgenic animals under the control of the MMTV enhancer/promoter (Pierce *et al.*, 1993). In accordance with the results obtained from experiments with exogenously administered TGF-ß$_1$ (Silberstein and Daniel, 1987), expression of the transgene led to retardation of ductal development in female transgenics. Reduction in ductal epithelial DNA synthesis was most apparent at 13 weeks when ductal growth normally declines. Interestingly, alveolar development as it occurred during pregnancy was not inhibited in the transgenic animals and no tumour formation was detected in transgenic animals. It remains to be established whether TGF-ß$_1$ expression might suppress tumour development.

The Mammary Derived Growth Inhibitor (MDGI) (for reviews, see Grosse and Langen, 1990; Grosse *et al.*, 1992) has been purified according to its inhibitory activity on Ehrlich ascites tumour cells (Böhmer *et al.*, 1984). The protein has been sequenced (Böhmer *et al.*, 1987) and full-length cDNAs for the bovine (Kurtz *et al.*, 1990) and mouse (Binas *et al.*, 1992) forms have been cloned. A striking homology was evident between the 14.5 kDa protein and members of a large conservative family of long-chain fatty acid-, retinoid- (designated as Cellular Retinoic Acid Binding Proteins, CRABP), and eicosanoid-binding proteins, collectively called Fatty Acid Binding Proteins (FABP) (for reviews, see Glatz *et al.*, 1993; Veerkamp *et al.*, 1993). Members of this multigene family have been found in a variety of tissues such as heart, skeletal muscle, liver, kidney, brain, intestine, adipose tissue, stomach or skin, and also in the midgut of Manduca sexta larvae or in *Schistosoma mansoni*. MDGI shares closest structural homology with the FABPs isolated from heart and brain. For the adipocyte, intestinal and liver FABP, and for CRABP, a relationship between their expression and the differentiated stage of the tissue has been documented (Glatz *et al.*, 1993). Our own *in situ* (Kurtz *et al.*, 1990) and immunohistochemical (Binas *et al.*, 1992; Mueller *et al.*, 1989) studies showed a developmentally-regulated spatial and temporal expression pattern for MDGI in the bovine and mouse mammary gland. The inhibitor is present in the embryonic mammary rudiments, is not found in the virgin tissue and is expressed in alveolar cells during pregnancy, and in both ductal and alveolar cells in lactation. MDGI expression becomes suppressed during involution (Politis *et al.*, 1992). Its synthesis is induced by prolactin and cortisol suggesting some role in functional differentiation (Binas *et al.*, 1992). We found an intense nuclear MDGI-immunostaining over euchromatic regions in sections of mouse and bovine mammary glands (Mueller *et al.*, 1989). The close parallel between local expression of MDGI and its mRNA suggests that transcriptional mechanisms are a major regulator. Those control mechanisms could be induced by lactogenic hormones, such as prolactin, leading to the enhanced MDGI expression seen in early differentiation. From a mouse genomic library a gene encoding MDGI has been isolated and characterised (Treuner *et al.*, 1994). The exon sequences were found to be identical with the previously isolated MDGI cDNA. The protein product differs in only 9 amino acid positions from Heart-FABP. We concluded that MDGI and H-FABP are encoded by the same gene (Treuner *et al.*, 1994). Several putative regulatory elements for binding transcription factors known to direct tissue and hormone dependent gene expression have been detected in the promoter region of the MDGI gene. A putative TGF-ß$_1$ inhibitory element (TIE) might be functionally important, since TGF-ß$_1$ has been found to down-regulate MDGI expression in differentiated primary mouse MEC (T. Müller, unpublished work). For MDGI, suppression of its gene expression during early ductal morphogenesis may be essential in order to allow development of a functional mammary gland. In order to test directly whether MDGI and H-FABP both possess growth inhibitory activity during mammary gland development and whether growth inhibition is specific for epithelial cells, the respective recombinant and wild-type proteins were produced, and compared in activity in primary cultures of MEC (Yang *et al.*, 1994). Wild-type and recombinant MDGI or H-FABP at 10^{-9}M inhibited DNA synthesis in MEC by about 50-60% whereas growth of mammary stromal cells was unaffected. More distant members of the FABP family like Liver- and Intestinal-FABP had no significant growth inhibitory effect. Therefore, growth inhibition seems to be specific for MEC and for MDGI and the heart type of FABP.

To ascertain whether MDGI might affect ductal DNA synthesis during glandular development, whole-organ mammary glands derived from virgin mice were cultured in serum-free medium containing aldosterone, prolactin, insulin and hydrocortisone (APIH medium; Yang *et al.*, 1994). This hormone milieu results in complete lobuloalveolar development of the gland and terminates in functional differentiation. Treatment of organ cultures with MDGI at 10^{-9}M resulted in appearance of smaller ducts and ductules with numerous side branches. Accordingly, strongest inhibition of DNA synthesis was found in ductular MEC (Table 2).

Table 2. Inhibition of ductal DNA synthesis by MDGI and bovine H-FABP

Protein	Inhibition of DNA synthesis (%)			
	Alveoli	Ducts	Ductules	Overall
w.-t. MDGI	9 ± 2	34 ± 4	61 ± 6	45 ± 5
w.-t. H-FABP	13 ± 2	33 ± 4	57 ± 4	46 ± 4

Overall epithelial cell inhibition of DNA synthesis induced by the recombinant forms of MDGI, human H-FABP and human mutated H-FABP which does not bind fatty acids (106 Thr for 106 Arg) amounted to 49 ± 5, 40 ± 4, 46 ± 4 and 53 ± 5%, respectively. Mammary glands were cultured in the APIH medium for 3 days. Growth inhibition was determined by labelling cells with BrdU (100μM) for 4 h before cultures were immunohistochemically evaluated using an anti-BrdU-antibody.

By comparing FABP-sequences encoded by exon-4, a striking homology was detected for a C-terminal stretch of 11 amino acids, designated P108 (Wallakut *et al.*, 1991; Yang *et al.*, 1994). P108 was tested in the organ culture system under conditions as described before for MDGI. Analogous to MDGI, treatment of APIH-cultured mammary glands with 10^{-9}M P108 produced a strong inhibition of ductal DNA synthesis (Yang *et al.*, 1994). Notably, P108 does not bind fatty acids (Wallukat *et al.*, 1991) thus providing direct evidence for a functional domain in the C-terminus of the MDGI sequence that does not require lipid binding.

Another group of emerging polypeptides possessing growth inhibitory activity is represented by the ligands of the proto-oncogene products *erbB-2, erbB-3*, and *erbB-4*. *Neu* differentiation factor (NDF), which was purified from *ras*-transformed fibroblasts, inhibits growth of human breast cancer cells, probably by cell cycle arrest at the late S or the G2/M phases (Peles *et al.*, 1992). *Neu* is expressed in the developing mouse mammary gland (E. Spitzer, personal communication). It is tempting to assume some role for NDF during growth and differentiation, however it remains to be tested if *neu* or any of the other isoforms (at least twelve different isoforms of proNDF exist) can modulate growth of epithelial cells in the normal gland. Another ligand for *erbB-2*, referred to as gp30, has been purified from human breast cancer cells. As with *neu*, gp30 blocks growth of human breast cancer cell lines *in vitro* (Lupu *et al.*, 1992).

Differentiation factors

It is a strikingly common property of the epithelial growth inhibitors described above that they all affect differentiation of the mammary gland. It is not known whether any of the anti-proliferative activities of the growth inhibitors is causally related to its action on differentiation. It is as reasonable to assume that the same molecule is affecting different and independent signalling pathways which might lead to altered growth or differentiation. In order to gain experimental insight to the complex system of mammary gland development, attention has focused on the establishment of tissue culture models that reflect the functions of pregnant and lactating mammmary tissue *in vivo* (Stoker *et al.*, 1990). These models will help to define better the local factors that regulate growth and differentiation during pregnancy.

By using either hormone responsive murine cell lines (Mieth *et al.*, 1990) or organ explant cultures derived from glands of pregnant mice (Robinson *et al.*, 1993) it was shown that TGF-ß$_1$ can inhibit functional differentiation *in vitro*. HC11 mouse MEC growing as a monolayer can undergo a limited functional differentiation in response to dexamethasone and prolactin. TGF-ß inhibits DNA synthesis and suppresses ß-casein expression in HC11 cells (Mieth *et al.*, 1990). Cytosine arabinoside, which also blocks DNA synthesis, did not affect ß-casein expression. Therefore, inhibition of functional differentiation does not seem to be linked to growth inhibition. Robinson *et al.* (1993) tested the effect of TGF-ß in mouse mammary explants and observed a strong inhibition of functional differentiation. Taking both experimental approaches together, the data show that inhibition of casein synthesis by TGF-ß does not require interactions with extracellular matrix components. We suggested two functions for TGF-ß during pregnancy. First, it could promote differentiation by stimulating synthesis of matrix proteins thereby supporting differentiation of adjacent MEC. On the other hand, it may directly inhibit ongoing ß-casein synthesis at the end of lactation (Mieth *et al.*, 1990).

This assumption was strengthened by results obtained with transgenic animals overexpressing the TGF-ß$_1$ gene during pregnancy under the control of a WAP-promoter (Jhappan *et al.*, 1993). Females were unable to lactate due to inhibition of the formation of lobuloalveolar structures and suppression of milk protein synthesis.

In conclusion, TGF-ß$_1$ is one of the locally-produced growth inhibitors activity which is essential for normal ductal growth, alveolar morphogenesis and functional differentiation. It remains to be clarified how its activities are interconnected with the action of mammogenic hormones, which cell types are involved in its various activities and how TGF-ß functionally interacts with other essential growth factors such as EGF or TGF-α to control normal epithelial development.

To ascertain whether MDGI might regulate differentiation, whole-organ mammary glands derived from virgin mice were treated for 5 days with MDGI at 10^{-9}M (Yang *et al.*, 1994). MDGI treatment led to the appearance of bulbous alveolar bud ends. Alveolar end buds represent a developmental pathway which eventually leads to secretory alveoli at functional differentiation. Indeed, histological and immunohistochemical examination of MDGI treated glands revealed the appearance of monolayered active secretory alveoli with enlarged luminal spaces and ß-casein secretion (Yang *et al.*, 1994). Substantiation that MDGI as well as the human and bovine forms of H-FABP might function as differentiation factors in the mammary gland was assessed by quantifying "early" and "late" markers of functional differentiation such as ß-casein and WAP expression, fat droplet accumulation and secretory activity (Table 3).

Table 3. Stimulation of functional differentiation by MDGI and H-FABP

Protein	ß-casein	WAP	MDGI	Fat Droplet Index	Alveolar Secretion
	mRNA Levels (%)			%	%
w.-t. MDGI	295 ± 12	275 ± 10	295 ± 10	179 ± 21	153 ± 18
w.-t. bovine H-FABP	280 ± 10	287 ± 15	267 ± 12	195 ± 19	157 ± 17
rec. human H-FABP	241 ± 4	251 ± 9	255 ± 10	218 ± 13	271 ± 22

Mammary glands were cultured for 5 days in the APIH medium. Levels are expressed as percent of the RNA level in the control contralateral gland. Fat droplet index was defined as the quotient of the number of alveoli consisting of more than 50% of epithelial cells with intra-epithelial fat droplets to the total number of alveoli. Alveolar secretion was defined as the quotient of the number of alveoli with lumina containing secretory products to the total number of alveoli. An experimental group comprised about 500 different structures for the morphometric analysis.

In mammary glands of primed virgin mice the MDGI gene is not expressed. However, if these glands are taken into culture in the presence of APIH, MDGI expression can be detected after 3 days (Binas *et al.*, 1992). Hormonally-induced MDGI expression which reaches a maximum of 8 days follows closely the pattern that has been found during pregnancy. This finding argues in favour of a local role for endogenous MDGI during growth and differentiation and we have therefore tested whether MDGI expression is a prerequisite for normal development. To this end, antisense phosphothioate oligonucleotides complementary to position 1-17 (relative to the translation initiation site) of the murine

MDGI coding region (as-oligonucleotide) were added to organ cultures for 5 days. Treatment with the as-oligonucleotide almost completely blocked MDGI expression in MEC and led to prevention of alveolar end bud formation and to the appearance of glands which seem to be retarded in development (Yang *et al.*, 1994). These findings might also indicate an intracellular mode of action of MDGI.

In contrast to the situation in organ cultures, MDGI cannot be induced in MEC grown as monolayers, including murine cell lines responsive to lactogenic hormones (R. Grosse, unpublished data). Single MEC provided with an exogenous basement membrane which can be induced to express ß-casein also do not express MDGI unless they form multicellular structures with luminal spaces. These data and our present findings strongly argue in favour of a role for MDGI at a distinct state of development which is characterized by a decline of ductal growth and associated with lobuloalveolar morphogenesis and functional differentiation. Analogous to MDGI, treatment of APIH-cultured mammary glands for 5 days with 10^{-9}M P108 affected ductal branching and caused appearance of more bulbous alveolar bud ends (Yang *et al.*, 1994). In accordance with the morphological changes which suggest suppression of ductal growth by P108, we found that DNA synthesis in MEC was inhibited by about 60% and that functional differentiation was clearly stimulated. The 11 C-terminal residues of MDGI are sufficient to fully mimic the activities of MDGI, including stimulation of ß-casein and WAP expression, and auto-stimulation of endogenous MDGI gene expression.

It was shown recently that EGF strongly suppresses MDGI mRNA expression and prevents differentiation in explant cultures derived from virgin or pregnant mouse mammary glands (Binas *et al.*, 1992; Spitzer *et al.*, 1994). It was therefore important to ascertain whether EGF and MDGI might functionally interact during growth and differentiation. To this end, organ cultures were treated for 5 days with both factors in a concentration range of 1-3nM (Yang *et al.*, 1994). EGF exerted mitogenic activities and suppressed the mRNA levels for WAP, ß-casein and MDGI. The effects of EGF on growth and differentiation were suppressed with increasing MDGI concentrations. On the other side, stimulation of functional differentiation by MDGI could be strongly reduced in presence of 3nM EGF. In conclusion, increasing the EGF or the MDGI concentrations was mutually antagonistic.

A further aspect of MDGI concerns its possible role in mediating anti-progestin activities (Li *et al.*, 1994). Onapristone and other antiprogestins proved to possess anti-tumour activity in several hormone-dependent experimental breast cancer models from mouse and rat (reviewed in Horwitz, 1992). These compounds may thus be powerful new tools for the management of metastatic breast cancer. It has been postulated that the anti-tumour action is related to the induction of terminal differentiation and increased apoptosis. Antiprogestins strongly inhibit epithelial DNA synthesis and stimulate functional differentiation in whole mammary gland organ cultures of primed virgin mice (Li *et al.*, 1994). These effects are accompanied by a strong induction of MDGI expression. Both half-maximal inhibition of epithelial DNA synthesis and stimulation of MDGI mRNA expression level were found at about 5ng/ml of the antiprogestin ZK 114043. The data indicate that the growth and differentiation effects of antiprogestins could be locally complemented or mediated by MDGI.

There are no published data about differentiation promoting activities of *erbB-2* ligands in normal epithelium. NDF induces phenotypic differentiation of cultured breast cancer cells, including altered morphology and induction of ß-casein synthesis (Peles *et al.*, 1992). It would be interesting to test for expression of *erbB-2* and possible action of NDF in cultured mammary glands.

CONCLUSION

In comparison to the few reports on growth inhibitors, numerous papers have addressed the role of growth stimulatory factors in mammary gland development. Amongst them, EGF can either stimulate growth or inhibit differentiation. Others like TGF-ß can inhibit growth and differentiation of mammary epithelium.

When mammary epithelial cells undergo differentiation, they ultimately stop growing in order to fulfil their function in lactation. These processes are regulated by systemic hormones and depend on interactions between mesenchyme and parenchyme that might involve the generation of diffusible factors conferring signals for differentiation. If the direct physical contact between basement membrane and epithelium provides an essential basis for onset of differentiation, then this should lead to the induction of activities in the epithelium (growth factors, lipid second messengers, proteases, etc.) that control growth arrest and onset of differentiation in MEC. This might imply mechanisms operating entirely intracellularly. Identification of respective factors and signalling messengers could help us to understand better how cells escape from normal control during malignant transformation.

REFERENCES

Banarjee, M., Wood, B. and Kinder, D., 1973, Whole mammary gland organ culture: selection of appropriate gland, *In Vitro Cell. Dev. Biol.* 9:129.

Banarjee, M.R. and Antoniou, M., 1985, Steroid and polypeptide hormone interaction in milk protein gene expression, *in* "Biochemical actions of hormones", Litwack, G., ed., Academic Press, New York, 237.

Binas, B., Spitzer, E., Zschiesche, W., Erdmann, B., Kurtz, A., Müller, T., Nieman, C., Blenau, W. and Grosse, R., 1992, Hormonal induction of functional differentiation and mammary derived growth inhibitor expression in cultured mouse mammary gland explants, *In Vitro Cell. Dev. Biol.* 28A:625.

Böhmer, F.-D., Lehmann, W., Schmidt, H., Langen, P. and Grosse, R., 1984, Purification of a growth inhibitor for Ehrlich ascites mammary carcinoma cells from bovine mammary gland, *Exp. Cell Res.* 150:466.

Böhmer, F.-D, Kraft, R., Otto, A., Wernstedt, C., Kurtz, A., Müller, T., Rohde, K., Etzold, G., Langen, P., Heldin, C.-H. and Grosse, R., 1987, Identification of a polypeptide growth inhibitor from bovine mammary gland. MDGI sequence homology to fatty acid and retinoid binding proteins, *J. Biol. Chem.* 262:15137.

Ervin, P., Kaminski, M., Cody, R. and Wicha, M., 1989, Production of mammastatin, a tissue specific growth inhibitor by normal human mammary epithelial cells, *Science* 244:1585.

Glatz, J.F.C., Vork, M., Cistola, D. and van der Vusse, G., 1993, Cytoplasmic fatty acid binding proteins: signficance for intracellular transport of fatty acids and putative role on signal transduction pathways, *Prostaglandins Leukotrienes and Essential Fatty Acids* 48:33.

Grosse, R. and Langen, P., 1990, Mammary derived growth inhibitor, *in* "Handbook of Experimental Pharmacol. 95/II", Sporn, M. and Roberts, A., eds., Springer Verlag, Heidelberg 249.

Grosse, R., Böhmer, F.-D., Binas, B., Kurtz, A., Spitzer, E., Müller, T. and Zschiesche, W., 1992, Mammary derived growth inhibitor, *in* "Cancer Treatment and Research Genes Oncogenes and Hormones", Dickson, R.B. and Lippman, M., eds., Kluwer Academic Press, Boston 69.

Horwitz, K., 1992, The molecular biology of Ru 486. Is there a role for antiprogestins in the treatment of breast cancer? *Endoc. Rev.* 13:146.

Jhappan, C., Geiser, A., Kordon, E., Bagheri, D., Hennighausen, L., Roberts, A., Smith, G. and Merlino, G., 1993, Targetting of a transforming growth factor β_1 transgene to the pregnant mammary gland inhibits alveolar development, *EMBO J.* 12:1835.

Knabbe, C., Kopp, A., Jonat, W. and Zugmaier, G., 1994, *J. Cell. Biochem.* 231 (Abstract Y 115).

Kurtz, A., Vogel, F., Funa, K., Heldin, C.-H. and Grosse, R., 1990, Developmental regulation of mammary derived growth inhibitor expression in bovine mammary tissue, *J. Cell Biol.* 110:1779.

Li, M., Spitzer, E., Zschiesche, W., Binas, B., Parczyk, K. and Grosse, R., 1994, Antiprogestins inhibit growth and stimulate differentiation in the normal mammary gland, *J. Cell. Physiol.*, in press.

Lupu, R., Colomer R., Kannan, B. and Lippman, M., 1992, Characterization of a growth factor that binds exclusively to the *erbB-2* receptor and induces cellular responses, *Proc. Natl. Acad. Sci*, USA 89:2287.

Mieth, M., Böhmer, F., Ball, R., Groner, B. and Grosse, R., 1990, Transforming growth factor ß inhibits lactogenic hormone induction of ß-casein expression in HC 11 mouse mammary epithelial cells, *Growth Factors* 4:9.

Mueller, T., Kurtz, A. Vogel, F., Breter, H., Schneider, F., Engstroem, U., Mieth, M., Boehmer, D.-D. and Grosse, R., 1989, A mammary derived growth inhibitor (MDGI) related 70kDa antigen identified in nuclei of mammary epithelial cells, *J. Cell. Physiol.* 138:415.

Oka, T., Yoshimura, M., Lavandero, S., Wada, K. and Ohba, Y., 1990, Control of growth and differentiation of the mammary gland by growth factors, *J. Dairy Sci.* 74:2788.

Peles, E., Bacus, S., Koski, R., Lu, H., Wen, D., Ogden, S., Levy, R. and Yarden, Y., 1992, Isolation of the *neu*/HER-2 stimulatory ligand: a 44kDa glycoprotein that induces differentiation of mammary tumour cells, *Cell* 69:205.

Pierce jr., D.F., Johnson, M., Matsui, Y., Robinson, S., Gold, L., Purchio, A., Daniel, C., Hogan, B. and Moses, H., 1993, Inhibition of mammary duct development but not alveolar outgrowth during pregnancy in transgenic mice expressing active TGF-ß1, *Genes Dev.* 7:2308.

Politis, I., Gorewit, R., Müller, T. and Grosse, R., 1992, Mammary derived growth inhibitor in lactation and involution, *Domestic Animal Endoc.* 9:88.

Robinson, F.D., Roberts, A. and Daniel, C., 1993, TGF-ß suppresses casein synthesis in mouse mammary explants and may play a role in controlling milk levels during pregnancy, *J. Cell Biol.* 120:245.

Robinson, S., Roberts, A. and Daniel, C., 1993, TGF-ß suppresses casein synthesis in mouse mammary explants and may play a role in controlling milk levels during pregnancy, *J. Cell Biol.* 120:245.

Sakakura, T., 1991, New aspects of stroma-parenchyma relations in mammary gland differentiation, *Intern. Rev. Cytol.* 125:165.

Silberstein, G., Horn, K., Shyamala, G. and Daniel, C., 1994, Essential role of endogenous estrogens in directly stimulating mammary growth demonstrated by implants containing pure antiestrogens, *Endocrinology* 134:84.

Silberstein, G., Strickland, P., Coleman, S. and Daniel, C., 1990, Epithelium-dependent extracellular matrix synthesis in transforming growth factor ß$_1$ growth inhibited mouse mammary gland, *J. Cell Biol.* 110:2209.

Silberstein, G. and Daniel, C., 1987, Reversible inhibition of mammary ductal growth by transforming growth factor ß, *Science* 37:291.

Silberstein, G., Flanders, G., Roberts, A. and Daniel, C., 1992, Regulation of mammary morphogenesis: evidence for extracellular matrix-mediated inhibition of ductal budding by TGF-ß$_1$, *J. Cell Biol.* 152:354.

Spitzer, E., Zschiesche, W., Binas, B., Grosse, R. and Erdmann, B., 1994, EGF and TGF-α modulate structural and functional differentiation of the mammary gland from pregnant mice, *J. Cell. Biochem.* in press.

Sporn, M. and Roberts, A., 1992, Transforming growth factor-ß: recent progress and new challenges, *J. Cell Biol.* 119:1017.

Sporn, M. and Roberts, A., 1985, Autocrine growth factors and cancer, *Nature* 313:745.

Stoker, A., Streuli, C., Martins-Green, M. and Bissell, M., 1990, Designer microenvironments for the analysis of cell and tissue function, *Curr. Opin. Cell Biol.* 2:864.

Streuli, C. and Bissell, M., 1990, Expression of extracellular matrix components is regulated by substratum, *J. Cell. Biol.* 110:1405.

Streuli, C. and Bissell, M., 1991, Mammary epithelial cells, extracellular matrix and gene expression, *in* "Regulatory mechanisms in Breast Cancer", Lippman, M. and Dickson, R., eds., Kluwer Academic Publishers, Norwell, MA. 365.

Topper, Y. and Freeman, C., 1980, Multiple hormone interactions in the development biology of the mammary gland, *Physiol. Rev.* 60:1049.

Treuner, M., Kozak, C., Gallhan, D., Spitzer, E., Grosse, R. and Mueller, T., 1994, Cloning and characterization of the mouse gene encoding mammary-derived growth inhibitor heart-fatty acid binding protein, *Gene*, in press.

Veerkamp, J., van Kuppevelt, T., Maatman, R. and Prinsen, C., 1993, Structural and functional aspects of cytosolic fatty acid binding proteins, *Prostaglandins, Leukotrienes and Essential Fatty Acids* 49:887.

Wallukat, G., Böhmer, F.-D., Engstroem, U., Langen, P., Hollenberg, M., Behlke, J., Kuehn, H. and Grosse, R., 1991, Modulation of the ß-adrenergic response in cultured rat heart cells. II. Dissociation from lipid-binding activity of MDGI, *Mol. Cell. Biochem.* 102:49.

Yang, Y., Spitzer, E., Kenney, N., Zschiesche, W., Li, M., Kromminga, A., Müller, T., Spener, F., Lezius, A., Veerlamp, J., Smith, G., Salomon, D. and Grosse, R., 1994, Members of the fatty acid binding protein family are differentiation factors for the mammary gland, *J. Cell Biol.*, in press.

APOPTOSIS IN MAMMARY GLAND INVOLUTION: ISOLATION AND CHARACTERIZATION OF APOPTOSIS-SPECIFIC GENES

Wolfgang Bielke[1], Guo Ke[1], Robert Strange[2] and Robert Friis[1]

[1]Laboratory for Clinical and Experimental Research
University of Bern
Tiefenaustrasse 120
CH-3004 Berne, Switzerland
[2]AMC Cancer Research Center
1600 Pierce St.
Denver, CO 80214, USA

INTRODUCTION

The massive cell death of secretory epithelium in the mammary gland following each cycle of pregnancy/lactation offers an especially convenient model for studies on the biochemistry of programmed cell death. Programmed cell death in the mammary gland takes the form of an apoptosis (Walker *et al.*, 1989; Strange *et al.*, 1992). Few examples of *in vivo* apoptosis occur so synchronously, involve such a large cell mass, or are so approachable for study. At the same time apoptosis in the mammary gland is important not merely for academic reasons. Nullparity is a positive risk factor for women, while pregnancy and lactation correlate with a lowered probability of developing mammary carcinoma (Yoo *et al.*, 1992; Yang *et al*, 1993). One can speculate that in the course of the mammary involution which follows on pregnancy/lactation, the apoptotic death of premalignant cells, cells bearing initial mutations conferring an elevated potential for neoplasia, is a desirable, but unspecific side-effect of the massive restructuring of the gland. The suggestion that genes like *bcl-2* which interfere with apoptotic cell death can act as oncogenes because they tilt the balance normally existing between cell proliferation and death (Raff, 1992), lends credence to this speculation. Furthermore, when tumors develop and the oncologist resorts to DNA damaging radiation or chemotherapy, again it is the death of tumor cells by apoptosis which can free the patient of disease.

Apoptosis is an extremely active field of biological research owing to its relevence for carcinogenesis.   Investigators suffer presently from an increasing awareness of the heterogeneity of the apoptotic process, both in terms of its induction, and in terms of the properties exhibited by the dying cells (Clarke, 1990; Schulte-Hermann *et al.*, 1990; Schwartz *et al.*, 1990; Schwartz *et al.*, 1993). Whether there exists a general apoptotic pathway which can be triggered by many inducers, or whether many pathways lead to apoptosis, the resulting apoptotic cells exhibit important common features such as physical stability, membrane integrity, and retention of cytoplasmic and nuclear contents. Only in this way is an extensive cell death without inflammation possible.

At present no estimate of the number of gene products participating in an apoptotic pathway can be made. Since this form of cell death is in the highest sense "conservative", aiming to avoid waste, many subsidiary subprogrammes can be anticipated. Because we concluded that an understanding of apoptosis, and more important, the capability to manipulate apoptosis will only be realized when a broader view of the pathway can be achieved, we have employed a mammary involution model to clone genes differentially expressed at this time of extensive apoptotic cell death. We soon discovered, however, that in a whole organ, it is difficult to distinguish directly between parallel programs of gene expression in a complex biological process. Thus, in the involuting mammary gland, tissue remodelling, basement membrane reorganization, cell death and stromal responses to these events are coupled, interactive processes (Li *et al.*, 1994a). In this communication, we will describe the use of a "coincident-expression" approach to allow detection of "apoptosis-associated" genes. This approach combines the advantages of the recently described Differential Display method (Liang and Pardee, 1992; Liang *et al.*, 1992) which allows the visualization of thousands of individual polymerase chain reaction (PCR) fragments, with the side-by-side comparison of a set of RNAs obtained from mammary glands at different stages in development, and from ventral prostate glands, undergoing a similar involution subsequent to castration.

METHODS

Animals

Forced weaning of Sprague Dawley female rats was carried out to induce mammary involution as described previously for mice (Strange *et al.*, 1992). Sprague Dawley male rats (500 g) were castrated under Nembutol anaesthetic through the scrotum, removing both testes and epididymus.

Isolation of RNA and northern blots

RNA was prepared from inguinal mammary glands of rats at different developmental stages or from the ventral prostate glands of rats before, and at different intervals after castration, according to the method of Chomczynski and Sacchi (1987). RNA was also prepared from various tissues of normal mice and rats, using the same procedure and mindful of the importance of rapid dissections. Agarose gel elecrophoresis of total RNA and polyA-enriched RNA was performed as previously described (Strange *et al.*, 1992; Li *et al.*, 1994a).

In Situ DNA fragmentation assay

Mammary gland tissue samples were fixed in freshly prepared (from paraformaldehyde) 4% formaldehyde solution in phosphate buffered saline at pH 7.4, 0°C for 4 hrs., dehydrated, and embedded in paraffin. 4 μm sections were prepared and digested for 30 min at room temperature with proteinase K (10 μg/ml) in Tris buffer (Tris, 20 mM, pH 8.0; ethylene diamine tetra-acetic acid, 1 mM), after which an *in situ* terminal transferase reaction was performed essentially according to Gavrieli *et al.* (1992), but using as substrate ^{35}S-thiophosphate dATP (Amersham) as described (Li *et al.*, 1994a). Detection was with autoradiography using NTB-2 film (Kodak).

Differential display

The "RNAmap" kits (primer sets A and B) (GenHunter Corp., Brookline, MA) were employed to amplify 0.2 μg DNase I-treated total RNA. The reactions were performed according to the instructions provided with the kit and as described (Bielke *et al.*, 1994). PCR reactions were analyzed on 6% denaturing polyacrylamide/urea sequencing gels. Bands of interest were removed from the dried gel, re-amplified using the same primers, and subcloned using blunt-end ligation in the *Sma-1* site of the Bluescript plasmid (Stratagene). Sequencing was performed in both directions using the dideoxynucleotide chain termination method (Sequenase 2.0 kit; United States Biochemicals, Cleveland, OH). Sequence comparisons against the EMBO Gene Bank were carried out using the Sequence Analysis Software Package of the Genetics Computer Group (FASTA program - version 6.0; University of Wisconsin, Madison, WI).

RESULTS AND DISCUSSION

The histological sections presented in Figure 1 illustrate the massive changes in the rodent's mammary gland organization which are carried out within a period of less than 2 weeks after weaning. The endonucleolytic DNA fragmentation typical of apoptosis (Wyllie, 1980) has been demonstrated for the involuting mammary gland (Strange *et al.*, 1992). The vast pool of free 3' ends thus created allows a thousand-fold excess of terminal transferase labelling as compared to normal cells. In the tissue sections shown in Figure 1 an *in situ* terminal transferase reaction as described by Gavrieli *et al* (1992) has been carried out with ^{35}S-thiophosphate dATP as substrate, detected using autoradiography. The intensely black deposits above the cells are the silver grains from the autoradiography. Although this method produces rather unattractive pictures, it has the major advantage over other detection systems such as alkaline phosphatase or peroxidase, that a stoichiometry assures proportionality between reaction and intensity of detection. This means in practice, that one can very accurately observe which cell types show reaction, and which are negative. Three important facts emerge from a study of Figure 1. Firstly, the onset of DNA fragmentation is very sharp; in 5 separate experiments, positive cells were never observed at 48 hours after weaning, but were always observed by 60 - 72 hours. Secondly, it was secretory epithelium in the collapsed mammary alveoli which showed positive reaction. No reaction was ever observed in the clearly distinguishable stroma, fat cells, or blood vessels. Unfortunately, myoepithelial cells could not be clearly distinguished with this autoradiographic method, but based on independent experiments utilizing alkaline phosphatase detection, myoepithelial cells are also negative. Thirdly, the intense terminal transferase positive reaction was transient; at 11 days post-lactation no positive cells could be detected, though other staining procedures (data not shown) allow us still to recognize numerous apoptotic bodies. This latter finding presumably means that the condensed nuclear structure becomes impenetrable to the enzyme, probably as a result of transglutaminase crosslinking of proteins.

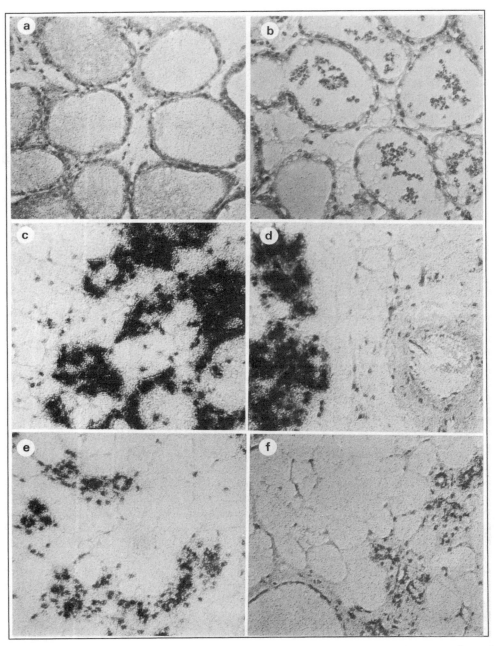

Figure 1. An *in situ* terminal transferase reaction was done as described in Methods. Sections of lactating (a) and involuting mammary glands 2 days (b), 3 days (c), 4 days (d), 6 days (e) and 11 days (f) after litter removal are shown after autoradiography and eosin staining (x 480).

Isolation of genes regulated in mammary involution

Differential screening of bacteriophage Lambda cDNA mammary involution libraries allowed Li *et al.* (1994a) to isolate a large number of clones exhibiting down-regulation or up-regulation within a few days post-lactation. In the former category, i.e. down-regulated, she found 5 members of the casein family, whey acidic protein, and GlyCAM-1 (Dowbenko *et al.*, 1993). The category of genes found to be up-regulated included some surprises. Table 1 summarizes the properties of some of the more interesting examples which, on the basis of partial sequences, could be identified in the EMBO Data Bank. Several highly promising, but unknown sequences are under investigation, but remain functionally undefined.

Table 1. Genes upregulated in mammary involution

Known gene	Mammary expression	Mammary localization	Function
stromelysin	peak 4-6 days	myoepithelium	tissue remodelling
protective protein	peak 4-6 days	?	"
24p3/NGAL	peak 2 days	secret. epithelium	"
WDNM 1	peak 2 days	intra-alveolar stroma	"
TIMP	peak 2-3 days	intra-alveolar stroma	"
Adipocyte differentiation related protein	peak 2-3 days	stroma/fat	lipid storage
stearoyl CoA desaturase	peak 4-6 days	stroma/fat	"
leucine zipper protein	peak 2-4 days	?	transcription factor
transglutaminase	peak 4-6 days	secret. epithelium	physical stabilization of apoptotic cells
gas-1	peak 3-6 days	secret. epithelium	?

From Table 1 it is evident that not "death"-associated genes, but more probably tissue remodelling genes were the main harvest of this initial differential screening approach. It is, of course, presently impossible to distinguish clearly between tissue remodelling and apoptosis. To some degree these processes must always overlap. Stromelysin, a metalloproteinase strongly upregulated in mammary involution was expressed in myoepithelial cells according to immunohistochemistry (Li *et al.*, 1994b). The "mouse protective protein" associates with lysosomal enzymes and restores ß-galactosidase and neuoamidase activities in galactosialidosis cells (Galjart *et al.*, 1988). Its predicted amino acid sequence bears strong homology to the yeast carboxypeptidase Y, suggesting that "protective protein", too, bears proteolytic activity.

24p3 (Hraba-Renevey *et al.*, 1989) was also found to be up-regulated in involuting mammary secretory epithelium according to northern blotting and *in situ* hybridization (Li *et al.*, 1994a). 24p3 has been recently discovered to be identical to NGAL, a protein capable of complexing with the 92 kDa gelatinase in neutrophils, and presumably capable of constitutively activating the proteolytic activity (Kjeldsen *et al.*, 1993). Another modulator/inhibitor of proteolytic activity, the tissue inhibitor of metalloproteinases (TIMP-1) is up-regulated (Strange *et al.*, 1992) in the intra-alveolar stroma (Li *et al.*, 1994b) following weaning. The WDNM1 gene also detected as a gene up-regulated in mammary involution, bears sequence homology with protease inhibitors (Dear and Kefford, 1991), and may be an additional modulator of proteolysis. WDNM1 expression seems to be in stroma surrounding involuting alveoli, according to *in situ* hybridization (data not shown).

Proteolytic activities and their regulators should not automatically be qualified as tissue remodelling functions, however, since the recent functional identification of the C. elegans *ced-3* gene, one of two recognized, cell-autonomous genes essential for programmed cell death in that system, as a cysteine protease (Yuan *et al.*, 1993). Interestingly, *ced-3* is the homolog to the mammalian interleukin-1 beta-converting enzyme (ICE), which overexpressed in fibroblasts, induces programmed cell death (Miura *et al.*, 1993). Until more is learned about how the various proteases and their modulators act to initiate, carry through, and clean up after programmed cell death, no conclusions can be forthcoming about primary and secondary roles in programmed cell death.

Several genes were also isolated with presumptive functions related to the reactivation of lipid storage after lactation: adipocyte differentiation-related protein and stearoyl CoA desaturase. Finally, a leucine zipper protein, and transglutaminase, two probable apoptosis-associated genes, were detected (Li *et al.*, 1994a).

A Coincident Expression Assay for cloning apoptosis-associated genes

Tissue remodelling is especially drastic in the mammary gland. The organ functions cyclically in the normal life of the mouse, and is restored within two week after weaning to a state not very different from the gland of an adult virgin. We speculated that if a comparison could be done between clones expressed in the involuting mammary gland and a separate involuting system, one lacking significant tissue remodelling, then organ-specific and remodelling-specific isolates could be allowed to cancel themselves out. This selection protocol we call a "Coincident Expression Assay".

The rat ventral prostate was an obvious candidate for this alternative system. Castration made possible a synchronous hormone ablation equivalent to that induced in the mammary gland by forced weaning, and an extensive literature documents the involution by apoptosis in the ventral prostate (English *et al.*, 1989; Colombel *et al.*, 1992). Figure 2 illustrates the relative expression of the tissue remodelling enzyme stromelysin, and its inhibitor, TIMP-1 observed in involuting mammary glands and prostate glands, respectively. Stromelysin was undetectable in prostate and TIMP was detected transiently as a faint expression. In consequence, we decided to compare the genes shown in Table 1 for their expression at different stages of mammary gland and prostate involution. Not surprisingly, most of the clones previously identified in mammary gland involution failed to show expression in the involuting prostate (data not shown).

A new strategy for cloning apoptosis-associated genes was required if efficient comparisons between involuting mammary and prostate glands was to be feasible. For this reason the recently described Differential Display method of Liang and Pardee (1992) was adopted. This method is a polymerase chain reaction (PCR) cloning procedure, in which parallel RNA samples, for example RNAs from different stages of mammary and prostate gland involution, are first employed as templates for an oligo dT-primed, reverse transcriptase-dependent reaction. After the second strand synthesis takes place, 40 cycles of PCR are carried out using many parallel sets of reactions, each set with an individual, short second oligonucleotide primer representing a distinct random sequence. The reactions are performed using radioactively-labeled dATP, and assayed on denaturing polyacrylamide sequencing gels. Figure 3 illustrates the results of such an experiment for one primer pair (20 pairs in all were evaluated; Bielke *et al.*, 1994):

first strand primer: 5' -TTTTTTTTTTTMC
(where **M** is degenerate; A, G or C)
second strand primer: 5' -GTTGCGATCC

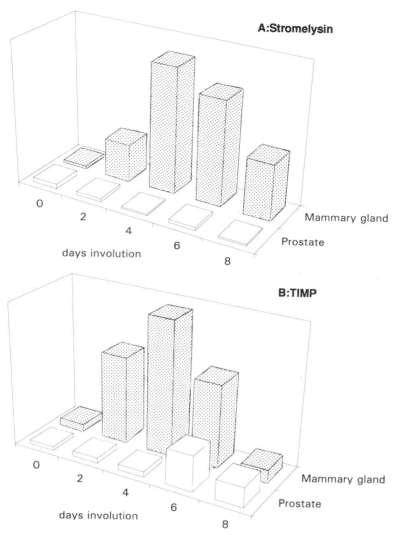

Figure 2: A schematic representation of northern blot data for stromelysin (A) and TIMP (B) expression at different times during involution is shown for mammary gland and ventral prostate.

Figure 3 shows a coincident expression fragment, indicated with arrows from the right and left sides, and an example of a lactation-specific clone, designated with a star.

Gas-1, an example of successful selection for an apoptosis-associated gene?

Of 12 clones isolated after reamplification of Differential Display bands, 5 could be detected on northern blots and proved to be genuinely up-regulated in a coincident manner in both mammary glands and prostate undergoing involution (Bielke *et al.*, 1994). Figure 4 illustrates the northern blot expression pattern of a rat Differential Display Coincident clone which shares approximately 90% homology with the 3' end of the mouse *gas-1* gene (Del Sal *et al.*, 1992). A transcript of 2.9 kb is detected in rat polyA-enriched RNAs from different stages of mammary gland and prostate involution. The expression in the mammary

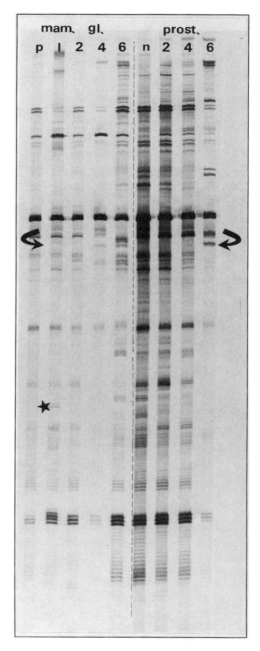

Figure 3. For illustrative purposes, the Differential Display for a single primer pair is presented.

gland is reproducibly bimodal, expression high at 2 days, peaking at 3 days, dropping thereafter, only to rise again at 6 days before ultimately falling. Prostate expression is similar. Figure 4 contains a final track, in which RNA from a rat receiving a testosterone depot on the fifth day, one day before sacrifice, was electrophoresed. Animals receiving this treatment re-express, for example sulfated glycoprotein-2 (also called testosterone repressed prostate message; Buttyan *et al.*, 1989) within 24 hours after receiving a testosterone depot (data not shown). The implication is that *gas-1* expression is part of a process which is irreversible by 5 days. Figure 5 shows the expression pattern of the rat *gas-1* related isolate in different normal organs of the rat. Note that several organs quiescent for proliferation such as brain, heart or skeletal muscle express, while others such as ovary and intestine which are not likely to be quiescent, also express strongly. It seems likely that the latter category may be involved with apoptosis, which is an established part of the normal growth of both organs.

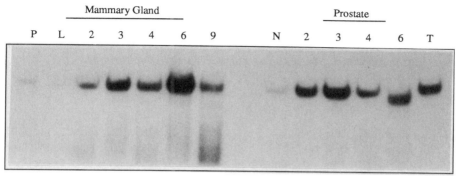

Figure 4. A northern blot shows the expression of *gas-1* at different stages of mammary and prostate gland involution. The transcript is at 2.9 kb.

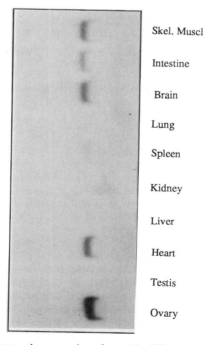

Figure 5. A northern blot illustrates the expression of *gas-1* in different tissues. The transcript is at 2.9 kb.

Gas-1 is a growth arrest gene (Schneider *et al.*, 1988; Del Sal *et al.*, 1992). It is a unique gene with no known relatives. We were able to use our 3' homologous sequences to isolate clones which were 100% homologous to the *gas-1* gene from a mouse mammary involution library which extended through the coding sequences. Therefore, it seems likely that the rat isolate is the genuine homolog of the mouse gene.

Gas-1 is a putative transmembrane protein exhibiting an RGD-motif; hence, it seems a likely candidate as an integrin binding protein (Del Sal *et al.*, 1992). It is expressed in mouse fibroblasts reaching confluence, or after serum starvation. Re-feeding with serum or transfer with trypsin results in a rapid loss of detectable *gas-1* message. Overexpression of *gas-1* in mouse fibroblasts was capable of inducing growth inhibition (Del Sal *et al.*, 1992). In short, many features of the *gas-1* gene make it a likely candidate for an effector in an apoptosis pathway. Based on a variety of data, we can speculate that apoptosis may be an option for a cell at the G_1/S border, or perhaps even later in cell cycle.

What remains to be discovered is whether *gas-1* can also act there. Present data leave this question unanswered. In any case, in *gas-1* the Coincident Expression Assay has brought out a promising clone, one which is conceivable as an apoptosis regulator.

REFERENCES

Bielke, W., Guo, K., Saurer, S. and Friis, R.R., 1994, Apoptosis in the rat mammary gland and ventral prostate: detection of "death-associated genes using a coincident-expression cloning approach, Submitted for publication.

Buttyan, R., Olsson, C.A., Pintar, J., Chang, C., Bandyk, M., Ng, P.-Y. and Sawczuk, I.S., 1989, Induction of the TRPM-2 gene in cells undergoing programmed death, *Mol. Cell Biol.* 9:3473.

Chomczynski, P. and Sacchi, N., 1987, Single-step method of RNA isolation by guanidinium thiocyanate-phenol-chloroform extraction, *Analyt. Biochem.* 162:156.

Clarke, P.G.H., 1990, Developmental cell death: morphological diversity and multiple mechanisms, *Anat. Embryol.* 181:195.

Colombel, M., Olsson, C.A., Ng, P.-Y., and Buttyan, R., 1992, Hormone-regulated apoptosis results from reentry of differentiated prostate cells onto a defective cell cycle, *Cancer Res.* 52:4313.

Dear, T.N., and Kefford, R.F., 1991, The WDNM1 gene product is a novel member of the 'four-disulphide core' family of proteins, *Biochem. Biophys. Res. Commun.* 176:247.

Del Sal, G., Ruaro, M.E., Philipson, L., and Schneider, C., 1992, The growth arrest-specific gene *gas1*, is involved in growth suppression, *Cell* 70:595.

Dowbenko, D., Kikuta, A., Fennie, C., Gillett, N., and Lasky, L.A., 1993, Glycosylation-dependent cell adhesion molecule 1 (GlyCAM 1) mucin is expressed by lactating mammary gland epithelial cells and is present in milk, *J. Clin. Invest.* 92:952.

English, H.F., Kyprianou, N., and Isaacs, J.T., 1989, Relationship between DNA fragmentation and apoptosis in the programmed cell death in the rat prostate following castration, *The Prostate* 15:233.

Galjart, N.J., Gillemans, N., Harris, A., van der Horst, G.T.J., Verheijen, F.W., Galjaard, H., and d'Azzo, A., 1988, Expression of cDNA encoding the human "protective protein" associated with lysosomal ß-galactosidase and neuraminidase: homology to yeast proteases, *Cell* 54:755.

Gavrieli, Y., Sherman, Y., and Ben-Sasson, S.A., 1992, Identification of programmed cell death *in situ* via specific labelling of nuclear DNA fragmentation, *J. Cell Biol.* 119:493.

Hraba-Renevey, S., Türler, H., Kress, M., Salomon, C., and Weil. R., 1989, SV40-induced expression of mouse gene 24p3 involves a post-transcriptional mechanism, *Oncogene* 4:601.

Kjeldsen, L., Johnsen, A.H., Sengeløv, H., and Borreegaard, N., 1993, Isolation and primary structure of NGAL, a novel protein associated with human neutrophil gelatinase, *J. Biol. Chem.* 268:10425.

Kyprianou, N.,English, H.F., and Isaacs, J.T., 1990, Programmed cell death during regression of PC-82 human prostate cancer following androgen ablation, *Cancer Res.* 50:3748.

Li, F. Bielke, W., Guo, K., Andres, A.-C., Jaggi, R., Friis, R.R., Niemann, H., Bernis, L., Geske, F.J., and Strange, R., 1994a, Isolation of cell death associated cDNAs from involuting mouse mammary epithelium, submitted for publication.

Li, F., Strange, R., Friis, R.R., Djonov, V., Altermatt, H.-J., Saurer, S., Niemann, H., and Andres, A.-C., 1994, Expression of stromelysin-1 and TIMP-1 in the involuting mammary gland and in early invasive tumors of the mouse, *Int. J. Cancer,* in press.

Liang, P., and Pardee, A.B., 1992, Differential display of eukaryotic messenger RNA by means of the polymerase chain reaction, *Science* 257:967.

Liang, P., Averboukh, L., Keyomarsi, K., Sager, R., and Pardee, A.B., 1992, Differential display and cloning of messenger RNAs from human breast cancer *versus* mammary epithelial cells, *Cancer Res.* 52:6966.

Miura, M., Zhu, H., Rotello, R., Hartwieg, E., and Yuan, J., 1993, Induction of apoptosis in fibroblasts by IL-1ß-converting enzyme, a mammalian homolog of the C. elegans cell death gene *ced-3, Cell* 75:653.

Raff, M.C., 1993, Social controls on cell survival and cell death, *Nature* 356:397.

Schneider, C., King, R.M., and Philipson, L., 1988, Genes specifically expressed at growth arrest of mammalian cells, *Cell* 54:787.

Schulte-Hermann, R., Timmermann-Rosiener, I., Barthel, G., and Bursch, W., 1990, DNA synthesis, apoptosis and phenotypic expression as determinants of growth of altered foci in rat liver during phenobarbital promotion, *Cancer Res.* 50:5127.

Schwartz, L.M., Kosz, L., and Kay, B.K., 1990, Gene activation is required for developmentally programmed cell death, *Proc. Natl. Acad. Sci. USA* 87:

Schwartz, L.M., Smith, S.W., Jones, M.E.E., and Osborne, B.A., 1993, Do all programmed cell deaths occur via apoptosis? *Proc. Natl. Acad. Sci. USA* 90:980.

Strange, R., Li, F., Saurer, S., Burkhardt, A., and Friis, R.R., 1992, Apoptotic cell death and tissue remodelling during mouse mammary gland involution, *Development* 115:49.

Walker, N.I., Bennett, R.E., and Kerr, J.F.R., 1989, Cell death by apoptosis during involution of the lactating breast in mice and rats, *Amer. J. Anat.* 185:19.

Wyllie, A.H., 1980, Glucocorticoid-induced thymocyte apoptosis is associated with endogenous endonuclease activation, *Nature* 284:555.

Yang, C.P., Weiss, N.S., Band, P.R., Gallagher, R.P., White, E., and Daling, J.R., 1993, History of lactation and breast cancer risk, *Amer. J. Epidemiol.* 136:1050.

Yoo, K.-Y., Tajima, K., Kurioshi, T., Hirose, K., Yoshida, M., Miura, S., and Murai, H., 1992, Independent protective effect of lactation against breast cancer: a case-control study in Japan, *Amer. J. Epidemiol.* 135:726.

Yuan, J., Shaham, S., Ledoux, S., Ellis, H.M., and Horvitz, H.R., 1993, The C.elegans cell death gene *ced-3* encodes a protein similar to mammalian interleukin-1ß-converting enzyme, *Cell* 75:641.

THE ROLE OF *ERBB*-FAMILY GENES AND *WNT* GENES IN NORMAL AND PRENEOPLASTIC MAMMARY EPITHELIUM, STUDIED BY TISSUE RECONSTITUTION

Paul A.W. Edwards[1], Clare Abram[1], Susan E Hiby[1], Christina Niemeyer[2], Trevor C. Dale[2], and Jane M Bradbury[1]

[1] Division of Cell and Genetic Pathology,
Department of Pathology, University of Cambridge,
Tennis Court Road, Cambridge, CB2 1QP, U.K.
[2] Haddow Laboratories, Institute of Cancer Research,
15 Cotswold Road, Sutton, Surrey, SM2 5NG, U.K.

INTRODUCTION

One approach to understanding the control of growth and three-dimensional organisation of mammary epithelium is to identify genes that may be involved and then perturb their expression. We have developed a method for expressing genes specifically in mouse mammary epithelium by reconstituting mammary epithelium from genetically-manipulated cells (Edwards *et al.*, 1988; Edwards 1993). We summarise here recent results we have obtained with genes of the *erbB* and *Wnt* families.

There are currently two main routes to the identification of genes that play a role in the control of mammary growth. The more familiar is the study of mammary tumours, which has led, for example, to interest in the *erbB* family of receptor tyrosine kinases. An alternative is to look for the mammalian homologues of genes that control the development of simpler organisms such as Drosophila, which has led to interest in *hox* genes (see Friedman and Daniel; this volume). The *Wnt* genes were identified by a combination of the two approaches. The first mammalian *Wnt*, *Wnt-1*, was first identified as *int-1*, an oncogene activated in mouse mammary tumours, but it subsequently turned out to be a mammalian homologue of the Drosophila segment polarity gene *wingless*.

The *erbB* family

The *erbB* family of genes are thought to play an important role in the development of tumours of mammary epithelium. The family currently consists of *erbB* (the EGF receptor), *erbB*2, *erbB*3 and *erbB*4. *erbB*2, *erbB*3 and *erbB*4 appear to be receptor tyrosine kinases like EGF-R. Several ligands for EGF-R are known, including EGF and TGF-α, but they appear not to bind to the other family members. There has been confusion over ligands for *erbB*2: the heregulins, also known as NDFs, were identified as molecules that activated *erbB*2 tyrosine kinase, but it now appears that they are ligands for *erbB*4, and that the activation of *erbB*2 occurs by heterodimerisation with the activated *erbB*4 (Culouscou *et al.*, 1993; Plowman, personal communication). Meanwhile, Matsuda *et al.*, 1993 have reported activation of *erbB*2 by estradiol, suggesting that estradiol or a related molecule may be the true ligand for *erbB*2. Like *erbB* and *erbB*2, *erbB*3 (Lemoine *et al.*, 1992) and *erbB*4 may be overexpressed in some breast tumours.

*erbB*2

There is a good case that *erbB*2 is acting as an oncogene in breast cancer. *erbB*2 is expressed at relatively high levels in at least 20% of breast carcinomas, and this is usually, but not always, associated with amplification of the gene (e.g. Berger *et al.*, 1988). The protein appears to be normal, at least the known transforming mutations of *erbB*2 have not been found in human tumours (for references see Varley and Walker, 1993). A high level of expression of normal protein could be mitogenic even in the absence of ligand, as high expression of normal *erbB*2 can transform fibroblasts *in vitro* in the absence of any (identified) ligand (DiFiore *et al.*, 1987). Most important, an oncogenic effect has been demonstrated *in vivo* - ectopic expression of *erbB*2 in mammary epithelium causes tumours. Expressing rat *erbB*2, in the overactive mutant form *neu*, in mammary epithelium either in transgenic mice or by inoculation into the mammary ducts of rats, caused carcinoma-*in-situ* and frank tumours (Muller *et al.*, 1988; Bouchard *et al.*, 1989; Wang *et al.*, 1991). Subsequently Guy *et al.* (1992) showed that the normal form of rat *erbB*2 was also effective, and that some of the tumours metastasised.

We recently described the results of an initial study of expressing *erbB*2 in mammary epithelium using our reconstitution system (Bradbury *et al.*, 1993). We were curious to know what the earliest events in tumour progression were. Our hypothesis was that there would be milder lesions in mammary epithelium induced by *erbB*2 overactivity, preceding the carcinomas-*in-situ* and full carcinomas observed by other groups. A major advantage of our reconstitution model (discussed in more detail below) is that only a few of the cells in the epithelium express the inserted gene, which is a more realistic model of tumour development than expressing the gene in all the cells, as occurs in transgenic mice expressing a gene from a tissue-specific promoter. Furthermore, mild lesions that arise are not overgrown by more aggressive lesions and so should be easier to find. We did indeed find a spectrum of milder lesions, and we present here examples of the simplest we have identified.

erbB

erbB/EGFR is considered to be over-expressed in perhaps 40% of breast tumours, and these are mostly estrogen receptor-negative ones, suggesting that high EGFR may substitute for high estrogen receptor (for references see Varley and Walker, 1993). This is intriguing because EGF can act as a local mediator of estrogen action at least in uterus (Nelson *et al.*, 1991). However, high expression of a protein in tumours is not by itself a strong argument that the protein is a major player in the development of the tumours. In particular it has not yet been shown that overactivity of *erbB* has a biological effect in mammary epithelium.

We therefore set out to see if overactivity of *erbB*/EGF-R in mammary epithelium can contribute to tumour development, and if its effect on mammary epithelium is the same as that of over-active *erbB*2. If *erbB* is an oncogene in mammary epithelium, then expression of an overactive form of the gene in the epithelium should produce an abnormal pattern of growth, as shown for other oncogenes (Edwards, 1993). The lesions should also progress to tumours, either on extended growth *in vivo* or with the addition of a further oncogene. We describe elsewhere in this volume (Abram *et al.*, 1994) the results of an exploratory study in which we have introduced into mammary epithelium the original oncogene form of *erbB*, v-*erbB* from the chicken erythroblastosis retrovirus. The growth pattern of the epithelium was substantially affected, and although we have not yet allowed any of the lesions to progress to full tumours, they resemble preneoplastic lesions and are consistent with overactivity of *erbB* contributing to neoplastic development.

Wnt genes

The prototype gene *Wnt*-1 was first identified as *int-1*, a site of consistent provirus integration in tumours induced by mouse mammary tumour virus (MMTV), but the *Wnt* genes (not to be confused with the other *int* genes, which are distinct) are perhaps better thought of as mammalian homologues of Drosophila *wingless*. *Wingless* is required throughout development but best known is its role in the cell-cell signalling that organises segments. It appears that Wingless and Wnt proteins signal from one cell to neighbouring cells, at most a few cell diameters away (reviewed in Nusse and Varmus, 1992). The *Wnt* gene family includes at least 10, perhaps 20 closely-related genes in the mouse. Most of the known ones were cloned by the polymerase chain reaction (PCR) exploiting conserved motifs (Gavin *et al.*, 1990). At least 6 are normally expressed in mammary gland, the level of expression varying with development of the gland, suggesting that they are involved in development (Gavin and McMahon, 1992; Buhler *et al.*, 1993; Weber-Hall *et al.*, 1994).

Work on the effects of *Wnt* gene expression in mammary epithelium began with the expression of *Wnt*-1 in mammary epithelium of transgenic mice by Tsukamoto *et al.* (1988). They found that *Wnt*-1 induced a uniform hyperplasia resembling to some extent early lobulo-alveolar development. We obtained similar hyperplasia when we expressed *Wnt*-1 in our reconstituted mammary epithelium (Edwards *et al.*, 1992). As it was emerging that other members of the *Wnt* family were expressed in normal mammary gland, and we were struck by the similarity of *Wnt*-1 hyperplasia to mid-pregnant growth, we suggested that *Wnt*-1 was acting by mimicking another member of the *Wnt* family that normally signalled the development of side-branch growth in early pregnancy. It followed that one or more members of the family should be involved in directing side-branch growth in the normal gland, and that expressing it/them in the virgin gland should produce a hyperplasia.

We have therefore prepared retroviruses that express various *Wnt*s, and have examined in more detail the expression of *Wnt*s at the message level in normal mammary epithelium (Weber-Hall *et al.*, 1994). During the development of the retroviruses we made the unexpected observation that some of them could mildly 'transform' fibroblasts (Bradbury *et al.*, 1994), an unexpected finding as previously the only cell known to respond to *Wnt* genes *in vitro* were the mammary epithelial cell lines C57MG and RAC311C (Brown *et al.*, 1986; Rijsewijk *et al.*, 1987) and the phaeochromocytoma line PC12 (Shackleford *et al.*, 1993). This observation is important because it suggests that *Wnt*s may signal to mesenchymal cells as well as to epithelial cells. We report here the results of expressing *Wnt*-4 in mammary epithelium, which strongly supports our initial hypothesis that at least one *Wnt* plays a key role in development of side-branches in pregnancy.

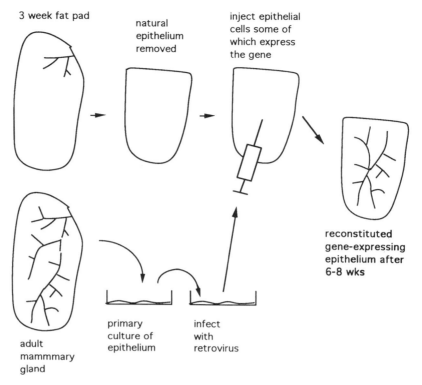

3 week fat pad

natural epithelium removed

inject epithelial cells some of which express the gene

reconstituted gene-expressing epithelium after 6-8 wks

adult mammmary gland

primary culture of epithelium

infect with retrovirus

Figure 1. Tissue reconstitution method for expressing genes in mammary epithelium

Introducing genes into mammary epithelium by tissue reconstitution

We have developed a 'tissue reconstitution' method for introducing genes into mammary epithelium *in vivo*, which permits us to express genes ectopically (Edwards *et al.*, 1988; Edwards, 1993), illustrated in Figure 1. Normal mammary epithelial cells are isolated from one mouse and very briefly put into primary culture for introduction of a gene using replication-defective, helper-free retroviruses, which insert genes relatively efficiently without spreading from cell to cell (reviewed by Miller and Rosman, 1989). The retrovirus directs expression of the gene from a constitutive promoter such as the ß-actin promoter. The genetically-manipulated cells are transplanted into the 'cleared' mammary fat pad of a host mouse, i.e. a fat pad from which the endogenous epithelium has been removed (deOme *et al.*, 1959). The transplanted cells grow to reconstitute a 'tree' of glandular epithelium in which some cells express the introduced gene (Figure 3). Apart from the effects of the gene, the epithelium is like the natural epithelium except that it is not connected to a nipple, and so does not maintain lactation after parturition as it is not suckled.

This tissue reconstitution approach has advantages over the more familiar one of expressing oncogenes in transgenic mice from mammary-specific promoters (e.g. Tsukamoto *et al.*, 1988; Andres *et al.*, 1988): only individual clones of cells express the oncogene, as in natural tumourigenesis, and hormone-insensitive promoters can be used (Bradbury *et al.*, 1991). We also avoid lethal effects of unwanted expression of the introduced gene during mouse development, as observed for example by Stocklin *et al.* (1993).

The procedure works well and is usually very informative. We have expressed a variety of genes, including the mutant oncogenes *v-myc* (Edwards *et al.*, 1988); *v-Ha-ras* (Bradbury *et al.*, 1991); *neu/erbB2* (Bradbury *et al.*, 1993) and the normal genes *Wnt-1* (Edwards *et al.*, 1992), *c-myc* and FGF-4 (Edwards, 1993). All produced alterations of growth pattern, which were all different and characteristic - for example alterations to branch frequency, multilayering of epithelium, or inappropriate alveolus formation (Edwards, 1993).

MATERIALS AND METHODS

Retroviruses

The *erbB2* retrovirus INA-*neu* (Bradbury *et al.*, 1993) contains the transforming *neu* oncogene in the retrovirus vector INA, a derivative of the vector pgagneoSRV in which the rat ß-actin promoter replaces the SV40 promoter (Morgenstern and Land, 1990). To construct a retrovirus expressing *Wnt-4*, the *Wnt-4* coding region was first amplified from a mouse mammary gland cDNA library, with addition of a Kozak consensus sequence (T.C. Dale, unpublished). After cloning into the Bluescript vector and sequencing, it was subcloned into the retrovirus vector pINA. The resulting virus INA-*Wnt-4* expresses the *Wnt-4* gene from transcripts initiated by the internal ß-actin promoter. Retrovirus-producing cells were made by transfection followed by infection of packaging cells as described (Edwards *et al.*, 1992; Bradbury and Edwards, 1988), except that the packaging line GP+E86 was used (Markovitz *et al.*, 1988). The control retrovirus INA was made by packaging the empty vector pINA (Edwards *et al.*, 1992). To test the ability of the retrovirus to transfer *Wnt-4*, cell lines 10T1/2 and C57MG were infected and selected for neo expression with G418. *Wnt-4* mRNA expression was confirmed by RT-PCR, using one primer specific for *Wnt-4* and the other primer specific for retrovirus sequences, so that only retrovirus-mediated expression would be detected. The identity of the PCR products was confirmed by Southern blotting with a *Wnt-4* probe.

Transplant methods

Transplants of mammary epithelium were set up essentially as previously described, using female Balb/c mice (Edwards *et al.*, 1988; Bradbury *et al.*, 1991, 1993). Primary cultures of adult normal mammary epithelium were prepared by collagenase digestion and cryopreserved (Edwards *et al.*, 1988). On day 0 epithelium from 1 to 2 mice was plated per 5-cm dish, in a 1:1 mixture of Glasgow modification of Eagle's Medium and F12 medium with 10% foetal calf serum, 5 mg/ml bovine insulin, 5 μg/ml cholera toxin, 5 μg/ml hydrocortisone and 10 ng/ml EGF. On day 2 the cultures were trypsinised and cells from each 5-cm dish replated in a 75-cm^2 flask with a feeder layer of 0.5-1.0 x 10^7 X-irradiated or mitomycinC-treated retrovirus producer cells, in the presence of 4 μg/ml polybrene. On day 4 cultures were harvested by trypsinisation and washed in serum-free, Hepes-buffered medium. Cleared inguinal ([#]4) fat pads were prepared in 22-day-old mice of less than 13g weight (DeOme *et al..*, 1959; Edwards *et al.*, 1988) and injected with one-tenth to one-twentieth of the cells from one flask (estimated to be 2 - 10 x 10^4 mammary epithelial cells) in 5 μl. After growth, for 9 to 12 weeks, transplants were whole mounted and stained with alum carmine. Representative areas of interest in such whole mounts were subsequently sectioned.

RESULTS AND DISCUSSION

*erbB*2

   We have previously described a series of transplants that expressed *erbB*2, in the form
of the overactive mutant *erbB*2, *neu* (Bradbury *et al.*, 1993). We have examined in more
detail the whole mounts and histology of these transplants to look for the 'mildest'
disturbances of growth pattern, in other words the alterations of growth pattern that seem
to be the minimum recognisable deviation from normal, as such lesions may precede the
more complex abnormalities. In most (11/15) of the transplants in animals that were not
mated, at least a few of the ducts had very small alveolus-like structures scattered along
them in rows of single alveoli and/or in small clusters. In hosts that had undergone a cycle
of pregnancy, lactation and at least 14 days involution, many transplants (13/25) showed
foci where involution seemed not to have occurred, i.e. substantial clusters of alveoli were
retained along ducts. It seems clear that these are a consequence of expression of the virus
since (i) clusters like this were not seen in the control transplants in the contralateral fat

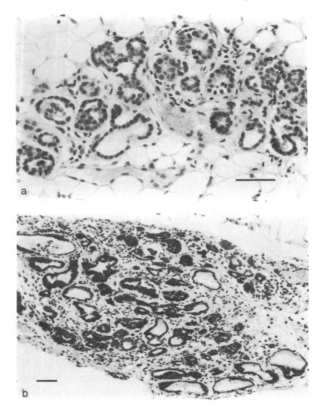

Figure 2. Histology of mild lesions induced by expressing *neu* in reconstituted mammary epithelium. (a)
Mild lesion resembling a cluster of alveoli in involuted mammary epithelium, 2 weeks after cessation of
lactation. Some of the 'alveoli' have extra epithelial cells forming papillae or bridges in the lumens.
Haematoxylin and eosin stain (b) More extensive lesion showing less regular acini and with some
epithelium thin and generally monolayered, interspersed with patches of multilayered epithelium.
Immunostained for *neu* protein with light nuclear counterstain, as this emphasises the arrangement of the
epithelial component. Scale bars 50 μm.

pads which were infected with the control retrovirus and (ii) in sections they stained for *neu* protein, with the anti-*neu* antibody 21N as described previously (Bradbury *et al.*, 1993). The whole mounts also showed more obviously abnormal areas, often appearing as solid swellings of ducts with no visible lumen (see photographs in Bradbury *et al.*, 1993). When representative examples of lesions were sectioned, we found that, as expected, several were little more than clusters of acini, like a little focus of alveolar development. Typical examples are shown in Figure 2. Figure 2a shows a cluster of epithelial structures resembling alveoli, but with a slightly disturbed arrangement of the epithelium, some having epithelial cells projecting into or crossing the lumen. In Figure 2b there is quite extensive multilayering of the epithelium, the epithelial units are of rather variable size, and the stroma is quite cellular. These lesions appear closely related to the more abnormal lesions we described previously, where either the stroma is very dense and eosinophilic, and the epithelium appears compressed into flattened sacs, or the lumens are filled to give solid rods of carcinoma-*in-situ* (see Bradbury *et al.*, 1993). Thus there seems to be a continuous spectrum of lesions, suggesting either progression to more abnormal structure or a range of severity determined by the level of expression of the *neu* protein.

Wnt-4

32 transplants were made from cells infected with the INA-*Wnt*-4 retrovirus. In each host mouse the contralateral gland contained a control transplant, infected with the control retrovirus vector INA. 9 of the *Wnt*-4 transplants showed a characteristic change in growth pattern, different from anything we have seen before (Figure 3). The epithelium was strikingly over-branched compared to control transplants or normal virgin mammary epithelium. The resulting hyperplastic pattern was somewhat like a normal early to mid-pregnant epithelium, and also quite like transplants expressing *Wnt*-1 (Edwards *et al.*, 1992). There were however, subtle, characteristic differences, which were seen more clearly by cutting small fragments from whole mounts and examining them at higher power on a conventional microscope (Figure 3 c, d). The termini of the small ducts in the *Wnt*-4 epithelium formed a characteristic finger-like structure, with some suggestion that it was made up of alveolar-like units. These do not seem to be exactly the same as structures seen in mid pregnancy or the termini of *Wnt*-1-expressing glands (not shown). In addition, we found that where transplants had not completely filled the fat pad, and so had terminal end buds at the growing tips of the ducts, the structure of the end buds was abnormal (Figure 3 e, f). The buds had a thin wall of epithelium and instead of the wall being thickened at the distal, growing end, as in the normal end bud, it was thickened at the base where it was joined to the duct.

Our hypothesis, on the basis of work with *Wnt*-1 (Edwards *et al.*, 1992), had been that (1) there would be at least one member of the *Wnt* family that was normally expressed in mammary gland that would give a hyperplasia, and (2) this *Wnt* would normally be involved in directing side-branch growth during early to mid pregnancy. *Wnt*-4 does in fact give a hyperplasia similar at a superficial level to that given by *Wnt*-1, and it gives a pattern of small branches close to that of mid-pregnancy. *Wnt*-4 expression in the normal gland does increase in early pregnancy, but there is some expression in the virgin gland (Gavin and McMahon, 1992; Weber-Hall *et al.*, 1994). Overall, one very attractive way to explain our observations would be that indeed *Wnt*-4 expression is normally switched on in early pregnancy and directs side-branch development. The modest difference between the pattern of growth of our *Wnt*-4-expressing epithelium and normal mid-pregnant epithelium could be due to the absence of other mid-pregnancy signals or proper control of *Wnt*-4 expression;

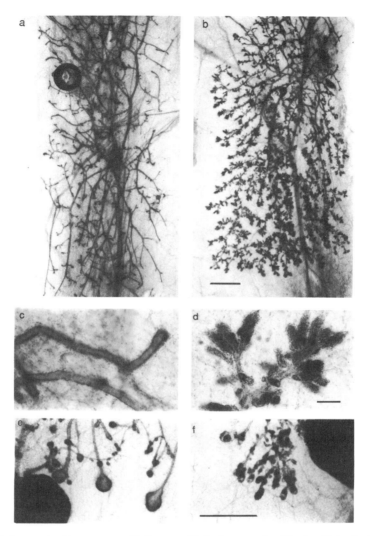

Figure 3. Whole mounts of mammary epithelium constitutively expressing *Wnt*-4. Stained with the nuclear stain carmine. Epithelium appears as dark lines. The dark oval bodies in **e** and **f** are lymph nodes. **a, c, e** show control transplants infected with control retrovirus. **b, d, f** show epithelium expressing the *Wnt*-4 gene product, resulting in proliferation of small side-branches of epithelium somewhat resembling those formed in early pregnancy. **a** and **b** show almost complete fat pads. Dark circle in **a** is an air bubble. Scale bar 1 mm. **c** and **d** are corresponding higher power views comparing detailed morphology of the ends of the branches at the periphery of the transplants. Scale bar 100 μm. **e** and **f** medium-power views of partly-grown transplants to show the morphology of the tips of the growing epithelium, the terminal end buds. Those in the control transplant **e** are typical of normal growing ducts, those in **f**, carrying the *Wnt*-4 retrovirus, are abnormal - smaller, with a thin distal wall and thickened proximal wall. Scale bar 1 mm.

the low level of expression of *Wnt*-4 in the virgin might be associated with the small amount of side-branching seen in most virgin mice. To evaluate this hypothesis we need to know more about the normal distribution of *Wnt*-4 expression and to make a detailed comparison between normal pregnant and *Wnt*-4 expressing epithelium.

Clearly there are other possible explanations of our observations. For example, the change in epithelial growth pattern induced by *Wnt*-4 in our transplants could be unphysiological. It is likely that we are inducing expression of *Wnt*-4 throughout both the basal and luminal epithelial cell layers, and this may be quite different from the normal pattern of expression, which is not yet known. *Wnt*-4 may cause a hyperplasia like that produced by *Wnt*-1, because, like *Wnt*-1, it mimics or interferes with some other, unknown signal.

CONCLUSION

We and others have now expressed a wide variety of genes in mammary epithelium, and perhaps the most striking observation is the diversity of effects on growth pattern. From the perspective of the cancer biologist, this underlines the diversity of effects that tumour mutations can have on the three-dimensional growth of cells, far exceeding in complexity what *in vitro* analysis can show. From the developmental point of view, trying to interpret the effects of inappropriate expression of *Wnt*-4 or *erbB*2 on the three-dimensional pattern of mammary epithelium presents a formidable challenge.

ACKNOWLEDGEMENTS

We thank Sallie McKenna for technical assistance; Barry Potter for histology; the Cancer Research Campaign, Agriculture and Food Research Council, Association for International Cancer Research and Breast Cancer Research Trust for support.

REFERENCES

Andres, A.C., van der Walk, M.A., Schonenberger, C.A., Fluckiger, F., LeMeur, M., Gerlinger, P. and Groner, B., 1988, Ha-*ras* and c-*myc* oncogene expression interferes with morphological and functional differentiation of mammary epithelial cells in single and double transgenic mice. *Genes Develop.* 2:1486.

Berger, M.S., Locher, G.W., Saurer, S., Gullick, W.J., Waterfield, M.D., Groner, B. and Hynes, N.E., 1988, Correlation of c-*erb*-B2 gene amplification and protein expression in human breast carcinoma with nodal status and nuclear grading, *Cancer Res.* 48:1238.

Bouchard, L., Lamarre, L., Tremblay, P.J. and Jolicoeur, P., 1989, Stochastic Appearance of Mammary Tumors in Transgenic Mice Carrying the MMTV/c-*neu* oncogene, *Cell* 57:931.

Bradbury, J.M., Arno, J. and Edwards, P.A.W., 1993, Induction of epithelial abnormalities that resemble human breast lesions by the expression of the *neu*/*erbB*2 oncogene in reconstituted mouse mammary gland, *Oncogene* 8:1551.

Bradbury, J.M., Niemeyer, C.C., Dale, T.C. and Edwards, P.A.W., 1994, Alterations of the growth characteristics of the fibroblast cell line C3H10T1/2 by members of the Wnt gene family, *Oncogene* 9:2597.

Bradbury, J.M., Sykes, H. and Edwards, P.A.W., 1991, Induction of mouse mamary tumours in a transplantation system by the sequential introduction of the *myc* and *ras* oncogenes, *Int. J. Cancer* 48:908.

Brown, A.M.C., Wildin, R.S., Prendergast, T.J. and Varmus, H.E., 1986, A retrovirus vector expressing the putative mammary oncogene *int*-1 causes partial transformation of a mammary epithelial cell line, *Cell* 46:1001.

Buhler, T.A., Dale, T.C., Kieback, C., Humphreys, R.C. and Rosen, J.M., 1993, Localisation and quantification of *Wnt*-2 gene expression in mouse mammary development, *Dev. Biol.* 155:87.

Culouscou, J-M, Plowman, G.D., Carlton, G.W., Green, J.M. and Shoyab, M., 1993, Characterisation of a breast cancer cell differentiation factor that specifically activates the HER4/p180*erbB*4 receptor, *J. Biol. Chem.* 268:18407.

De Ome, K.B., Faulkin, L.J. Jr., Bern, H.A. and Blair, P.B., 1959, Development of mammary tumors from hyperplastic alveolar nodules transplanted into gland-free mammary fat pads of female C3H mice, *Cancer Res.* 19:515.

DiFiore, P.P., Pierce, J., Kraus, M.H., Segatto, O., King, C.R. and Aaronson, S.A., 1987, erbB2 is a potent oncogene when overexpressed in NIH3T3 cells, *Science* 237:178.

Edwards, P.A.W., 1993, Tissue reconstitution models of breast cancer, *Cancer Surveys* 16:79.

Edwards, P.A.W., Ward, J.L. and Bradbury, J.M., 1988, Alteration of morphogenesis by the v-*myc* oncogene in transplants of mouse mammary gland, *Oncogene* 2:407.

Edwards, P.A.W., Hiby, S.E., Papkoff, J. and Bradbury, J.M., 1992, Hyperplasia of mouse mammary epithelium induced by expression of the *Wnt*-1 (*int*-1) oncogene in reconstituted mammary gland, *Oncogene* 7:2041.

Gavin, B.J., McMahon, J.A. and McMahon, A.P., 1990, Expression of multiple novel *Wnt*-1/*int*-1 related genes during fetal and adult mouse development, *Genes Develop.* 4:2319.

Gavin, B.J. and McMahon, A.P., 1992, Differential regulation of the *Wnt* gene family during pregnancy and lactation suggests a role in post natal development of the mammary gland, *Mol. Cell Biol.* 12:2418.

Gullick, W.J., Bottomley, A.C., Lofts, F.J., Doak, D.G., Mulvey, D., Newman, R.H., Crumpton, M.J., Sternberg, M.J.E. and Campbell, I.D., 1992, Three dimensional structure of the transmembrane region of the proto-oncogenic and oncogenic forms of the *neu* protein, *EMBO J.* 11:43.

Guy, C.T., Webster, M.A., Schaller, M., Parsons, T.J., Cardiff, R.D. and Muller, W.J., 1992, Expression of the *neu* protooncogene in mammary epithelium of transgenic mice induces metastatic disease, *Proc. Natl. Acad. Sci. USA*, 89:10578.

Lemoine, N.R. Barnes, D.M., Hollywood, D.P., Hughes, C.M., Smith, P., Dublin, E., Prigent, S.A., Gullick, W.J. and Hurst, H.C., 1992, Expression of the erbB3 gene product in breast cancer, *Brit. J. Cancer* 66:1116.

Markovitz, D., Goff, S. and Bank, A., 1988, A safe packaging line for gene transfer: separating viral genes on two different plasmids, *J. Virol.* 62:1120.

Matsuda, S., Kadowaki, Y., Ichino, M., Akiyama, T., Toyoshima, K. and Yamamoto, T., 1993, 17ß-estradiol mimics ligand activity of the c-*erbB2* protooncogene product, *Proc. Natl. Acad. Sci. USA* 90:10803.

Miller, A.D. and Rosman, G.J., 1989, Improved retroviral vectors for gene transfer and expression, *BioTechniques* 7:980.

Morgenstern, J. and Land, H., 1990, Advanced mammalian gene transfer: high titre retroviral vectors with multiple drug selection markers and a complementary helper-free packaging cell line, *Nucl. Acids Res.* 18:3587.

Muller, W.J., Sinn, E., Pattengale, P.K., Wallace, R. and Leder, P., 1988, Single-step induction of mammary adenocarcinoma in transgenic mice bearing the (activated) T-*neu* oncogene, *Cell* 54:105.

Nelson, K., Takahashi, T., Bossert, N.L., Walmer, D.K. and McLachlan, J.A., 1991, Epidermal growth factor replaces estrogen in the stimulation of female genital-tract growth and differentiation, *Proc. Natl. Acad. Sci. USA* 88:21.

Nijsewijk, F., van Deemter, L., Wagenaar, E., Sonnenberg, A. and Nusse, R., 1987, Transfection of the *int*-1 mammary oncogene in cuboidal RAC mammary cell line results in morphological transformation and tumorigenicity, *EMBO J.* 6:127.

Nusse, R. and Varmus, H.E., 1992, *Wnt* genes, *Cell* 69:1073.

Shackleford, G.M., Willert, K., Wang, J. and Varmus, H.E., 1993, The *Wnt*-1 protooncogene induces changes in morphology, gene expression and growth factor responsiveness in PC12 cells, *Cell* 11:865.

Stocklin, E., Botteri, F. and Groner, B., 1993, An activated allele of the c-*erbB*-2 oncogene impairs kidney and lung function and causes early death of transgenic mice, *J. Cell Biol.* 122:199.

Tsukamoto, A.S., Grosschedl, R., Guzman, R.C., Parslow, T. and Varmus, H.E., 1988, Expression of the *int*-1 gene in transgenic mice is associated with mammary gland hyperplasia and adenocarcinomas in male and female mice, *Cell* 55:619.

Varley, J.M. and Walker, R.A., 1993, The molecular pathology of human breast cancer, *Cancer Surveys* 16:31.

Wang, B., Kennan, W.S., Yasukawa-Barnes, J., Lindstrom, M.J. and Gould, M.N., 1991, Frequent induction of mammary carcinomas following *neu* oncogene transfer into *in situ* mammary epithelial cells of susceptible and resistant rat strains, *Cancer Res.* 51:5649.

Weber-Hall, S.J., Phippard, D.J., Niemeyer, C.C. and Dale, T.C., 1994, Developmental and hormonal regulation of *Wnt* gene expression in the mouse mammary gland, *Differentiation*, in press.

66

*V-ERB*B INDUCES ABNORMAL PATTERNS OF GROWTH IN MAMMARY EPITHELIUM

Clare L. Abram[1], Jane M. Bradbury[1],
Martin J. Page[2] and Paul A.W. Edwards[1]

[1]Department of Pathology, University of Cambridge,
Tennis Court Road, Cambridge, CB2 1QP, U.K.
[2]Wellcome Foundation Ltd, Langley Court,
South Eden Park Road, Beckenham, Kent, BR3 3BS, U.K.

Members of the *erb*B family of growth factor tyrosine kinase receptors have been implicated in breast cancer (Walker and Varley, 1993). The epidermal growth factor receptor (EGFR) or *c-erb*B is overexpressed in around 35% of breast tumours (Sainsbury *et al.*, 1985) but there is no evidence that this affects mammary epithelial development *in vivo*. We have expressed a constitutively activated form of the EGFR, the *v-erb*B oncogene from the chicken erythroblastosis retrovirus, in mammary epithelium using our novel tissue reconstitution system (Edwards *et al.*, 1988). In previous experiments using this system, the introduction of activated rat *neu/c-erb*B2 into the mammary gland produced abnormalities in epithelial growth resembling those seen in human breast cancer (Bradbury *et al.*, 1993).

The NTK *v-erb*B retrovirus was obtained as a producer cell line (von Ruden *et al.*, 1992) expressing the *v-erb*B oncogene under the control of the thymidine kinase promoter. Control retrovirus contained a retroviral vector backbone with no oncogene. The transplantation procedure was carried out as described previously (Edwards *et al.*, 1988; this volume). Primary mammary epithelial cells prepared by collagenase digestion of glands from female balb/c adult mice were infected with retrovirus by plating onto a feeder layer of X-irradiated retrovirus producer cells for 48h then ~10^4 cells were injected into the pre-cleared mammary fat pad of three week old female balb/c mice. The contralateral gland of each mouse was injected with cells infected with a control retrovirus. After 10-15 weeks, the transplanted glands were wholemounted and stained with carmine. Areas of abnormality were sectioned and stained with haematoxylin/eosin.

A total of 69 mice were transplanted with *v-erb*B containing epithelium and with control epithelium contralaterally. 26/69 glands showed growth abnormalities under the dissecting microscope whilst none of the control glands showed any abnormality. The abnormalities included enlarged epithelial ducts, enlarged ill-defined alveoli, and macrophages and/or loose epithelial cells in the lumen of some ducts (Figure 1). Several

Intercellular Signalling in the Mammary Gland
Edited by C.J. Wilde *et al.*, Plenum Press, New York, 1995

of the glands were sectioned and stained with haematoxylin/eosin, and the histology showed a higher density of epithelial ducts than in the control glands. In particular, enlarged epithelial cells were present in the lumens and the duct epithelium was multilayered and irregular.

These results show that *v-erb*B disrupts mammary epithelial cell growth. This suggests that activated EGFR may contribute to breast tumour development.

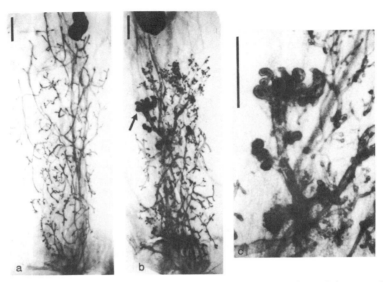

Figure 1. Wholemounts of carmine stained mouse mammary glands. **a,** gland containing control retrovirus and showing no growth abnormalities. **b,** gland containing the *v-erb*B retrovirus with abnormal growth indicated by the arrow. **c,** abnormality at a higher magnification. Scale bars = 2mm.

ACKNOWLEDGEMENTS

We thank Barry Potter for histology, Thomas von Rüden for NTK *v-erb*B retrovirus; Philip Starling for photography and MRC and the Wellcome Foundation plc for a studentship for CA.

REFERENCES

Bradbury, J.M., Arno, J. and Edwards, P.A.W., 1993, Induction of epithelial abnormalities that resemble human breast lesions by the expression of the *neu/erb*B-2 oncogene in reconstituted mouse mammary gland, *Oncogene* 8:1551.

Edwards, P.A.W., Ward, J.L. and Bradbury, J.M., 1988, Alteration of morphogenesis by the *v-myc* oncogene in transplants of mammary gland, *Oncogene* 2:407.

von Rüden, T., Kandels, S., Radaszkiewicz, T., Ullrich, A. and Wagner, E., 1992, Development of a lethal mast cell disease in mice reconstituted with bone marrow cells expressing the *v-erb*B oncogene, *Blood* 12:3145.

Sainsbury, J.R.C., Malcolm, A.J., Appleton, D.R., Farndon, J.R. and Harris, A.L., 1985, Presence of epidermal growth factor receptor as an indicator of poor prognosis in patients with breast cancer, *J. Clin. Pathol.* 38:1225.

Walker, R.A. and Varley, J.M., 1993, The molecular pathology of human breast cancer, *Cancer Surveys* 16:31.

INTERPLAY BETWEEN EPIDERMAL GROWTH FACTOR RECEPTOR AND ESTROGEN RECEPTOR IN PROGRESSION TO HORMONE INDEPENDENCE OF HUMAN BREAST CANCER

Ton van Agthoven and Lambert C.J. Dorssers

Department of Molecular Biology, Dr Daniel den Hoed
Cancer Center, P.O. Box 5201, 3008 AE Rotterdam,
The Netherlands

Breast growth and development is regulated by interactions of steroid hormones and polypeptide growth factors with their specific cellular receptors (Clarke *et al.*, 1992). Estrogen receptors (ER) have been detected in more than 50% of primary human breast tumours. Presence of ER in primary tumours identifies patients with a lower risk of relapse and prolonged survival. Therapeutic strategies using antiestrogens are based on the dependence of breast cancer cells on estrogens for proliferation. These antagonists can bind to the ER, but are unable to stimulate cell proliferation. Approximately 50% of patients with ER-positive primary tumours will have an objective response to endocrine therapy. However, the duration of response is limited and the majority of patients will experience a relapse due to development of hormone-independent tumours.

Epidermal growth factor (EGF) and transforming growth factor α are important polypeptide growth factors for normal and malignant breast cells. They exert their signals through the EGF receptor (EGFR), which is over-expressed in approximately 40-60% of breast tumours. EGFR expression is inversely related to ER expression in primary breast tumours. Moreover, EGFR is expressed in tumours unresponsive to endocrine therapy (Klijn *et al.*, 1992). However, about 50% of ER positive tumours contain EGFRs. Most assays of ER and EGFRs have been performed in tumour and non-malignant breast tissue extracts, which do not discriminate between the cellular origin of these receptors. Therefore, it is not known whether the receptors are expressed in the same or in different cells. To study the mechanisms underlying the development of antiestrogen resistance, we have investigated the consequences of EGFR expression in altering the estrogen-dependent phenotype. Experiments were conducted in which estrogen-dependent, EGFR-negative breast cancer cells were converted into EGFR-positive cells. We have also determined the expression patterns of ER and EGFRs at the single cell level in a series of primary breast tumours and non-malignant breast tissues using dual-staining immunohistochemistry.

RESULTS AND DISCUSSION

Proliferation of ZR-75-1 human breast cancer cells is estrogen-dependent, and fully inhibited in the presence of antiestrogens. The parental cells were infected with a retrovirus carrying the EGFR complementary DNA. The derived ZR/HERc cells were ER and EGFR-positive and acquired a proliferative response to EGF. Treatment with antiestrogens in the presence of EGF resulted in bypassing of estrogen-dependent growth. Furthermore, in contrast to the parental cells, prolonged antiestrogen treatment of ZR/HERc cells resulted in progression to estrogen independence and down-regulation of ER. Interference of ER and EGFR signal transduction pathways in ZR/HERc cells was observed when both were simultaneously activated (Van Agthoven et al., 1992).

Dual-staining immunohistochemistry for ER and EGFRs was performed on 28 cytosolic ER positive breast carcinomas. ER positive tumour cells were detected in 26 (93%) and EGFR positive tumour cells were detected in 7 carcinomas (25%). In 5 tumours both ER and EGFRs were detected, but localized in distinct tumour cells. Most tumour biopsy specimens contained both ER and EGFRs. However, in the majority of the tumours containing both ER and EGFRs, these were inversely expressed in different subsets of tumour cells or in normal components present in the specimen. Coexpression in individual tumour cells was a rare phenomenon. However, in non-malignant luminal epithelial cells, adjacent to the tumour cells, we frequently observed coexpression of ER and EGFRs in individual cells. Expression of ER and EGFRs was heterogeneous with respect to tumour biopsy compound and thus this study further documents the heterogeneous nature of primary breast tumours. In normal breast tissues and in benign breast lesions, ER expression was exclusively restricted to actin-negative cells of the epithelial lineage, but absent in actin-positive myoepithelial breast cells. EGFR expression was not only detected in luminal and myoepithelial cells but also in stromal cells. Double-positive cells were frequently observed in luminal epithelial cells in fibroadenomas and occasionally in luminal epithelial cells of normal breast tissue and mastopathies (Van Agthoven et al., 1994). Demonstration of ER and EGFRs in individual normal luminal epithelial cells shows that expression is not mutually exclusive and suggests a role in normal development. It is tempting to speculate the existence of a direct control mechanism of ER and EGFR gene expression, which may explain their uncoupling during malignant progression in breast cancer.

REFERENCES

van Agthoven, T., van Agthoven, T.L.A., Portengen, H., Foekens, J.A. and Dorssers, L.C.J., 1992, Ectopic expression of epidermal growth factor receptors induces hormone independence in ZR-75-1 human breast cancer cells, *Cancer Res.* 52:5082.

van Agthoven, T., Timmermans, M., Foekens, J.A., Dorssers, L.C.J. and Henzen-Logmans, S.C., 1994, Differential expression of estrogen, progesterone and epidermal growth factor receptors in normal, benign and malignant breast tissues using dual-staining immunohistochemistry, *Amer. J. Pathol.* 144:1238.

Clarke, R., Dickson, R.B. and Lippman, M.E., 1992, Hormonal aspects of breast cancer. Growth factors, drugs and stromal interactions, *Crit. Rev. Oncol. Hematol.* 12:1.

Klijn, J.G.M., Berns, P.M.J.J., Schmitz, P.I.M. and Foekens, J.A., 1992, The clinical significance of epidermal growth factor receptor (EGF-R) in human breast cancer: a review on 5232 patients, *Endocr. Rev.* 13:3.

EFFECT OF PLANE OF NUTRITION ON MAMMARY GLAND DEVELOPMENT IN PREPUBERTAL GOATS

Christine E. Bowden, Karen Plaut and Rhonda L. Maple

Department of Animal and Food Science
University of Vermont, Burlington, VT 05405, USA

A high plane of nutrition fed prior to puberty has been shown to negatively affect mammary gland development in heifers by decreasing secretory tissue and increasing adipose tissue (Sejrsen et al., 1982; Stelwagen and Grieve, 1990). Heifers reared on a high plane of nutrition have been shown to have depressed milk yields throughout their productive lives (Peri et al., 1992). Presently, it is not known why a high plane of nutrition inhibits mammary parenchymal development. Hormones and local growth factors play an important role in mammary gland development (Plaut, 1993). It is possible that a high plane of nutrition alters levels of hormones or growth factors that are needed for parenchymal development. Because heifers take up to 10 months to reach puberty, research in this area has been limited. Goats reach puberty 3-5 months sooner than heifers, therefore a similar study of caprine mammary development could be carried out in almost half the time. This study was conducted to test the usefulness of goats for such studies, and to test if a high plane of nutrition affects prepubertal mammary development.

METHODS

Ten French Alpine goats, age 37 ± 4 days, were paired on the basis of weight and randomly assigned to an *ad libitum* or a restricted diet of Jersey milk. Restricted animals received 70% of their pair mate's milk intake for four weeks, then 50% for nine weeks. All animals had free access to hay and water. At 130 ± 6 days of age, goats were slaughtered. Mammary glands were removed, frozen in liquid N_2 and stored at -80°C. Frozen glands were separated into parenchyma and adipose tissue by colour (Sejrsen et al., 1982) and analysed for protein, DNA, lipid, and hydroxyproline. Protein was analyzed by Bio-Rad Microassay (Hercules, CA). DNA was quantified in a fluorescence assay using Hoechst reagent (Labarca and Paigen, 1980). Lipid content was determined by extracting freeze dried samples with petroleum ether for 24 h. Hydroxyproline was measured colorimetrically (Woessner, 1961). Data were analysed using Students paired t-test.

RESULTS AND DISCUSSION

Table 1. Effect of plane of nutrition on mammary composition in prepubertal goats.

	Ad Libitum		Restricted	
	Mean	% of Gland	Mean	% of Gland
Gland Weight, g	69.2*		44.5	
Parenchyma, g	16.3	24.5	13.2	31.3
Adipose, g	52.9*	75.5	31.3	68.7
Protein, g	5.5	8.1*	6.2	14.4
DNA, mg	58.7	0.06*	88.5	0.09
Lipid, g	45.7[+]	65.4	28.8	62.5
Hydroxyproline, g	334.3*	0.5	206.7	0.5

*$P < 0.05$ [+]$P < 0.12$

Goats fed on a high plane of nutrition have larger mammary glands than goats fed a restricted diet. Increased gland size in goats on a high plane of nutrition is due to increased adipose deposition. Despite a smaller gland size, restricted animals have more secretory tissue than *ad libitum* animals based on protein and DNA analysis. These results are consistent with those found in heifers (Sejrsen *et al.*, 1982; Stelwagen and Grieve, 1990), indicating that goats would serve as a good model for studying the effect of plane of nutrition on hormones and growth factors in the mammary gland.

ACKNOWLEDGEMENTS

We would like to thank Wanda Calor and the William H. Miner Institute, Chazy, NY, for their input in designing the experiment and for donating the goats.

REFERENCES

Labarca, C. and Paigen, K., 1990, A simple rapid, and sensitive DNA assay procedure, *Analyt. Biochem.* 102:34.

Peri, I., Gertler, A., Bruckental, I. and Barash, H., 1992, The effect of manipulation in energy allowance during the rearing period of heifers on hormone concentrations and milk production in first lactation cows, *J. Dairy Sci.* 76:742.

Plaut, K., 1993, Role of epidermal growth factor and transforming growth factors in mammary development and lactation, *J. Dairy Sci.* 76:1526.

Sejrsen, K., Huber, J.T., Tucker, H.A. and Akers, R.M., 1982, Influence of nutrition on mammary development in pre- and postpubertal heifers, *J. Dairy Sci.* 65:793.

Stelwagen, K. and Grieve, D.G., 1990, Effect of plane of nutrition on growth and mammary gland development in Holstein heifers, *J. Dairy Sci.* 73:2333.

Woessner Jr., J.F., 1961, The determination of hydroxyproline in tissue and protein samples containing small proportions of this imino acid, *Arch Biochem Biophys.* 93:440.

ACTIVATION OF BASIC FIBROBLAST GROWTH FACTOR (bFGF) BY HEPARAN SULPHATE (HS)

Hai-Lan Chen, Philip S. Rudland, John A. Smith
and David G. Fernig

Cancer and Polio Research Fund Laboratories
Department of Biochemistry
University of Liverpool, Liverpool L69 3BX, U.K.

Fibroblast growth factors (FGFs) deliver their signals to cells by interacting with cellular receptors. There are two types of receptors for FGFs, namely heparan sulphate proteoglycans (HSPGs) and receptor tyrosine kinases (FGFRs). The functions of the HSPG receptors include: (i) the activation of FGFs which enables them to bind to the FGFRs with high-affinity; (ii) the extracellular storage of bFGF and hence its sequestration from the cell-surface tyrosine kinase receptors. bFGF is a normal product of mammary myoepithelial, but not epithelial, cells (Rudland et al., 1993). The rat mammary cells that are stimulated to proliferate by bFGF are stroma-derived fibroblasts and epithelium-derived intermediate cells of the terminal end buds and myoepithelial cells. These cell types are stimulated to proliferate as a consequence of their possessing receptors for bFGF while epithelial cells from normal mammary gland neither respond to bFGF nor possess detectable receptors for bFGF (Fernig et al., 1990, 1993). However, malignant mammary epithelial cells possess FGFRs, though not detectable HSPG receptors for bFGF, and they are responsive to exogenous bFGF (Fernig et al., 1993).

As a start to defining the structures within heparan sulphate (HS) that are responsible for binding and activating bFGF in mammary cells, HS has been purified from a series of rat mammary (Rama) cell lines that represent the different cellular compartments of the mammary gland: Rama 27, fibroblastic, derived from the stroma of the normal mammary gland; Rama 37, epithelial, derived from a benign tumour; Rama 401, myoepithelial, derived from the normal mammary gland; Rama 800, malignant epithelial, derived from a moderately metastatic tumour. HS was purified from the culture medium (secreted/shed HS), from a trypsin sensitive fraction (plasma membrane HS) and from a Triton-X100/urea fraction (extracellular matrix HS). The activating function of the purified HS was then investigated using Rama 27 fibroblasts grown in a sulphur-free culture medium containing 10 mM chlorate, which inhibits the synthesis of endogenous HS (Rapraeger et al., 1991).

The HS-depleted Rama 27 cells were rendered quiescent by serum deprivation. The ability of the purified HS to restore bFGF-dependent DNA synthesis was then determined. DNA synthesis in HS-depleted Rama 27 fibroblasts, in the absence of exogenously-added heparin, is 100-fold less sensitive to bFGF than in the presence of heparin, although the stimulation of DNA synthesis by epidermal growth factor is unaffected. Concentrations of heparin as low as 30 ng/ml restore the DNA stimulatory activity of bFGF.

HS purified from mammary cell lines possessed different bFGF-activating abilities. The plasma membrane HS of Rama 27 fibroblasts, Rama 401 myoepithelial cells and Rama 800 malignant cells could restore the activity of bFGF when assayed on HS-depleted Rama 27 fibroblasts. HS purified from the extracellular matrix of Rama 27 fibroblasts, Rama 37 epithelial cells and Rama 401 myoepithelial cells was also able to restore the activity of bFGF. Similarly HS purified from the culture medium of Rama 37 epithelial cells, Rama 401 myoepithelial cells and Rama 800 malignant cells restored the activity of bFGF. However, HS isolated from the culture medium of Rama 27 fibroblasts, from the plasma membrane of Rama 37 epithelial cells and from the Triton-X100/urea fraction of the Rama 800 malignant epithelial cells was unable to activate bFGF.

As Rama 27 fibroblasts and Rama 401 myoepithelial cells possess HSPG receptors for bFGF the presence of bFGF-activating HS in these cells is expected. bFGF-activating HS is secreted into the medium or localised to the extracellular matrix in Rama 37 cells, in agreement with the absence of bFGF binding sites on these cells. Similarly, the absence of bFGF-activating HS from the extracellular matrix fraction of the malignant Rama 800 cells is not surprising since malignant cells produce little or no extracellular matrix (Fernig et al., 1993). The Rama 800 cells do not possess detectable HSPG binding sites for bFGF, which may be accounted for by these cells producing 40-fold less HS than, for example, Rama 401 myoepithelial cells. However, the small amount of plasma membrane-associated HS produced by the Rama 800 malignant cells is able to activate bFGF as efficiently as that from the Rama 401 cells. These results suggest that the bFGF-activation and bFGF-storage functions of the HS produced by the cells of the mammary gland may be due to different structures within the HS. Regulation of the expression of bFGF-binding and bFGF-activating HS by mammary cells may therefore be a key control point in the modulation of the bFGF-growth stimulatory circuit in the developing mammary gland. The presence of bFGF-activating HS and the concomitant absence of bFGF-binding HS in malignant Rama 800 cells suggests that in malignant tumours the bFGF growth-stimulatory circuit is unregulated.

REFERENCES

Fernig, D.G., Barraclough, B.R., Rudland, P.S., Wilkinson, M.C. and Smith, J.A., 1993, Synthesis of basic fibroblast growth factor and its receptor by mammary epithelial cells derived from malignant tumours but not from normal mammary gland, *Int. J. Cancer*, 54:629.

Fernig, D.G., Smith, J.A. and Rudland, P.S., 1990, Appearance of basic fibroblast growth factor receptors upon differentiation of rat mammary epithelial to myoepithelial-like cells in culture, *J. Cell. Physiol.*, 142:108.

Rapraeger, A.-C., Krufka, A. and Olwin, B.B., 1991, Requirement of heparan sulphate for bFGF-mediated fibroblast growth and myoblast differentiation, *Science*, 252:1705.

Rudland, P.S., Platt-Higgins, A.M., Wilkinson, M.C. and Fernig, D.G., 1993, Immunocytochemical identification of basic fibroblast growth factor in the developing rat mammary gland, *J. Histochem. Cytochem.* 41:887.

HIGH DENSITY CULTURE OF IMMUNO-MAGNETICALLY SEPARATED HUMAN MAMMARY LUMINAL CELLS

Catherine Clarke, Paul Monaghan, and Michael J. O'Hare

Institute of Cancer Research
Haddow Laboratories
15 Cotswold Road
Sutton, Surrey SM2 5NG, U.K.

When the two epithelial cell types of the human breast, myoepithelial and luminal, are grown together *in vitro*, the myoepithelial cells come to dominate cultures because of their greater proliferative potential. However, it is the luminal cells which are of interest in many studies since they are the source of most breast cancers. In order to study luminal cells in culture, the two cell types must be separated. This has previously been achieved using flow cytometry (O'Hare *et al.*, 1991) with numbers limited to about 10^5 cells of each type. For the current study of differentiation of luminal cells, large numbers of cells were required and an immunomagnetic separation technique was developed using the MACS system (Miltenyi *et al.*, 1990; Clarke *et al.*, 1994). Such purified populations of luminal cells have been grown on various substrates at high density, and their structural differentiation examined by light and electron microscopy.

METHODS

Primary cultures of human breast epithelium were incubated overnight with calcium-free medium resulting in detachment of a luminally-enriched population and leaving an adherent myoepithelially-enriched fraction. These were labelled with antibodies to epithelial membrane antigen (ICR2) and CALLA/CD10 (PHM6) respectively, followed by anti-rat or anti-mouse antibody coated MACS microbeads, and separated in the MACS magnetic separation system. MACS-separated populations of luminal cells were grown in culture at high density (10^6 cells/cm^2) in RPMI 1640 medium containing 10% FCS, hydrocortisone (5 μg/ml), insulin (5 μg/ml) and cholera toxin (100 ng/ml) on either plastic coverslips (Thermanox), porous membranes (Cyclopore), dermal collagen I gels (Vitrogen) or EHS basement membrane complex (Matrigel). Cultures were monitored by phase contrast microscopy and fixed and embedded for electron microscopy after 7-10 days. Sections from the resulting structures were examined by light and electron microscopy.

RESULTS AND DISCUSSION

The purity of the MACS-separated luminal cells was shown both by flow cytometric analysis of CALLA and EMA, and by triple immunofluorescent labelling of cytokeratin markers, to be >95%. Luminal cells grown on plastic showed a typical attenuated morphology with some desmosomes but no junctional complexes even when maintained at high density. On Cyclopore membranes, the cells were more cuboidal, and occasional tight junctions, but no organised junctional complexes were seen. On both released collagen I gels and EHS matrix, the cells formed clusters which appeared similar by phase contrast microscopy. Sections through the clusters on collagen I showed cell polarisation, with junctional complexes between the cells at the upper surface (i.e. not in contact with the gel) and those cells surrounding small lumina. The clusters of cells on EHS matrix were surrounded by the gel, and formed large lumina containing necrotic debris. The surrounding cells had their apical surfaces facing the lumen, and formed a continuous sheet sealed by tight junctions. The overall appearance of the differentiated luminal cells on this substrate was virtually indistinguishable from that of the corresponding cells *in vivo*.

The MACS system has previously been used for lymphocyte separations but this is, to our knowledge, the first use of the system for epithelial cell separations. The large number ($>10^7$) of highly purified luminal and myoepithelial cells produced by this method can also be used for cellular and molecular studies of growth and differentiation with the advantage that the 50 nm beads do not interfere with cell-cell interactions.

In this study we have demonstrated that the substrate on which luminal cells are grown influences their differentiated state, with the highest degree of structural differentiation seen when the cells are grown on a basement membrane complex. Previous studies on rodent luminal cells indicate that these conditions may also allow functional differentiation with lactation-associated markers being expressed under appropriate hormonal conditions (Li *et al.*, 1987). It is now possible, using the MACS separated cells, to undertake equivalent experiments using human luminal cells.

ACKNOWLEDGEMENTS

This work is funded by the Cancer Research Campaign.

REFERENCES

Clarke, C., Titley, J., Davies S. and O'Hare, M.J., 1994, An immunomagnetic separation method using superparamagnetic (MACS) beads for large-scale purification of human mammary luminal and myoepithelial cells, *Epithelial Cell Biol.*, 3:38.

Li, M.L., Aggeler, J., Farson, D., Hatier, C., Hassel J. and Bissell, M.J., 1987, Influence of a reconstituted basement membrane and its components on casein gene expression and secretion in mouse mammary epithelial cells, *Proc. Natl. Acad. Sci. USA.* 84:136.

Miltenyi, S., Muller, W., Weischel W. and Radbruch, A., 1990, High gradient magnetic cell separation with MACS, *Cytometry* 11:231.

O'Hare, M.J., Ormerod, M.G., Monaghan, P., Lane E.B. and Gusterson, B.A., 1991, Characterisation *in vitro* of luminal and myoepithelial cells isolated from the human mammary gland by cell sorting, *Differentiation* 46:209.

MOLECULAR ANALYSIS OF THE EFFECTS OF OESTRADIOL IN AN *IN VIVO* MODEL OF NORMAL HUMAN BREAST PROLIFERATION

Robert B. Clarke[1], Ian J. Laidlaw[2], Anthony Howell[2] and Elizabeth Anderson[1]

Departments of [1]Clinical Research and [2]Medical Oncology, Christie Hospital NHS Trust, Manchester M20 9BX, UK

We report here the results of experiments investigating the effects of oestradiol (E_2) and progesterone (Pg) alone or in combination upon human breast epithelial cell proliferation together with steroid receptor, growth factor and growth factor receptor expression.

Pieces of histologically normal human breast were implanted subcutaneously on the dorsal aspect of athymic nude mice. After 14 days, silastic pellets containing 0.5, 1, 2 or 6 mg E_2 or 4 mg progesterone were inserted s.c. Another group of mice received a combination of a 2 mg E_2 implant for 21 days supplemented by a 4 mg Pg implant after the first 7 days. Cell proliferation was assessed by [^{3}H]-thymidine labelling, histo-autoradiography and scoring labelled cell nuclei as a percentage of the total cells counted (TLI index). Expression of oestrogen receptor (ER), progesterone receptor (PR) and transforming growth factor β (TGFβ) was assessed immunocytochemically in adjacent frozen sections using a kit (ER-ICA and PR-ICA, Abbott Diagnostics) and a rabbit polyclonal antibody against TGFβ_{1-3}. Expression of insulin-like growth factor receptor type 1 (IGF-R1) mRNA was determined by reverse-transcriptase polymerase chain reaction (RT-PCR) performed on total RNA extracted from samples.

The proliferative effects of steroid hormones on normal human breast epithelium are summarised in Figure 1. The median TLI was reduced from a pre-implantation level of 1.5% to 0.49% after 14 days implantation and remained at this level in untreated control mice (Figure 1a). Insertion of a 2 mg E_2 silastic pellet which produced human mid-cycle levels of serum E_2 significantly increased TLI at both 7 and 14 days treatment (Figure 1b; $p < 0.0001$, Mann-Whitney U-test). Figure 1c shows that implantation of a 4 mg Pg pellet alone did not alter the TLI from that seen before treatment even after 14 days of treatment. Finally, Pg combined with E_2 administration in a manner that approximated the pattern of the human menstrual cycle had no significant effect over the effects of E_2 alone (Figure 1d). In dose-response experiments, a significant relationship was demonstrated between the

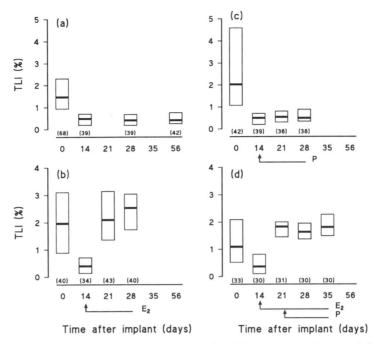

Figure 1. Proliferative activity of human breast tissue implanted into nude mice that were left untreated (a); that received a 2 mg E_2 pellet (b); that received a 4 mg Pg pellet (c) or that received both an E_2 and Pg pellet (d). TLI = thymidine labelling index; the columns indicate the IQ range of each measurement whilst the bars represent the medians. The numbers in parentheses above the abscissa are the number of observations in each group whilst the arrows indicate the duration of steroid treatment.

amount of E_2 administered in a pellet or the mouse serum E_2 levels and the change in TLI over a 14 day treatment period (data not shown). Immunocytochemical analysis of ER and PR indicated no effect of treatment on ER levels, but E_2 treatment led to a 15-20 fold increase in the number of cells expressing PR. Immunocytochemical analysis of TGFβ content revealed an apparent increase in stromal but not epithelial expression of this factor in E_2-treated samples. IGF-R1 mRNA levels measured by RT-PCR showed a significant increase after 7 and 14 days E_2 treatment compared to 4 mg Pg treatment or untreated controls.

We conclude that E_2, but not Pg alone or in combination with E_2, stimulates proliferation of normal human breast tissue implanted into athymic nude mice in a dose-dependent manner. Since E_2 induces high levels of expression of PR in this model yet progesterone has no additive proliferative effects, it is likely that oestrogen is the major mitogenic stimulus in the normal human breast *in vivo*. Measurement of the time-course of the increased proliferation caused by E_2 (data not shown) suggests that the maximal breast cell proliferation seen in the second half of the menstrual cycle is in response to the mid-cycle peak of E_2 secretion and not to the high levels of Pg seen at this time. Further analysis of the effects of oestrogen in this model have shown that growth factor and growth factor receptor expression is altered. Paradoxically, TGFβ expression in the stromal compartment was increased by E_2 treatment. IGF-RI mRNA has been shown to be increased by E_2, which may be significant since synergistic effects of E_2 and IGFs on proliferation have been shown to occur in human breast cancer cells *in vitro*.

JUXTACRINE GROWTH STIMULATION OF MOUSE MAMMARY CELLS IN CULTURE

Ursula K. Ehmann, Michael S.C. Chen and Aldrin A. Adamos

Pathology Research, Veterans Affairs Medical Center,
3801 Miranda Avenue, Palo Alto, CA 94304, U.S.A.

Mouse mammary epithelial cells (MMEC) from normal glands readily proliferate in culture when plated with lethally irradiated cells of the LA7 rat mammary tumour line (Ehmann *et al.*, 1984). These cells can be sub-cultured at least 12 times (32 cell generations, if fresh feeders are added to the new cultures in numbers large enough to achieve confluency. Cells cultured in this manner appear normal with respect to morphology and dome formation. Their normality is also apparent when introduced into mouse mammary fat pads where they proliferate to form normal mammary ducts (Ehmann *et al.*, 1987).

The nature of the growth stimulus provided by the LA7 feeder cells has been the subject of our investigation. In several types of experiments culture medium conditioned by LA7 cells did not stimulate MMEC proliferation at all or supported very little compared to that induced by the LA7 feeder cells themselves (Ehmann, 1992). This result suggested that growth stimulation by LA7 cells takes place at very close range. We are studying the possibility that the feeder cells stimulate MMEC by a growth factor delivered fairly directly from one cell to another either by secretion or through gap junctions.

RESULTS AND DISCUSSION

That the LA7 feeder cells supply the only growth stimulus to the MMEC was evident. Growth without feeders was minimal, but MMEC proliferated profusely in the presence of feeders even without serum or other growth factors in the medium. An intimate relationship between the feeder LA7 and MMEC developed in culture. Their membranes juxtaposed each other with MMEC membranes extending themselves for long distances along feeder cell membranes, sometimes even wrapping themselves around the feeders. They also formed gap junctions with each other. The gap junctions were detected by injection of lucifer yellow into single LA7 cells, which, within minutes, transmitted this fluorescent dye to their immediate MMEC neighbours. The identity of the two cell types was confirmed after microinjection by an antiserum specific to mouse cells, i.e., the MMEC.

We are continuing to pursue the possibility that the growth signal is delivered to the MMEC through the gap junctions. However, in order to completely rule out transmission of a soluble growth factor from the LA7 feeder cells we have measured the growth rates of MMEC plated with a concentrate of LA7-conditioned medium. In this medium, in which molecules >3000 m.w. were concentrated by a factor of 20, the MMEC proliferated well (Figure 1). The control cells, cultured without conditioned medium, decreased in numbers

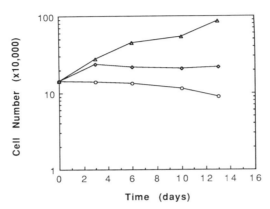

Figure 1. Proliferation of MMEC in passage 5 with or without medium conditioned by LA7 cells. Base medium for all samples was DMEM/F12 supplemented with 0.5% newborn calf serum, 10 μg/ml insulin, and the trace elements Cu, Se, Mn, Si, Mo, V, Ni, Sn, and Zn. o, basic control medium; Δ, LA7-conditioned medium with molecules M_r >3000 concentrated x 20; ◊, control medium plus 10 ng/ml EGF.

during this time. Cells fed with control medium containing epidermal growth factor (EGF) increased only slightly in cell numbers at first, then remained steady.

Normal MMEC are stimulated to proliferate in culture by cells of the LA7 rat mammary tumour line. It is possible that the growth stimulus is delivered by gap junctions which form between the two cell types. Now, also, a factor produced by the LA7 feeders has been shown to stimulate MMEC proliferation. That proliferation was seen only when the growth factor was concentrated suggests that it is delivered fairly directly to the MMEC in coculture. Whether this involves the extracellular matrix, whether the factor is usually secreted, or whether it is simply delivered from membrane to membrane is not clear. At the present time we are trying to determine by Western blotting techniques whether this growth factor is one of those commonly known as autocrine factors or known to be secreted by or responded to by mammary cells. These include transforming growth factor α, EGF, and basic fibroblast growth factor.

REFERENCES

Ehmann, U.K., 1992, Feeder cells impart a growth stimulus to recipient epithelial cells by direct contact, *Exp. Cell Res.* 199:314.
Ehmann, U.K., Guzman, R.C., Osborn, R.C., *et al.*, 1987, Cultured mouse mammary epithelial cells: normal phenotype after implantation, *J. Nat. Cancer Inst.* 78:751.
Ehmann, U.K., Peterson, W.D., Jr., Misfeldt, D.S., 1984, To grow mouse mammary epithelial cells in culture, *J. Cell Biol.* 98:1026.

FIBROBLAST GROWTH FACTORS IN MAMMARY DEVELOPMENT AND CANCER

David G. Fernig, Roger Barraclough, Youqiang Ke,
John A. Smith, Angela Platt-Higgins, Mark C. Wilkinson,
John H.R. Winstanley and Philip S. Rudland

Cancer and Polio Research Fund Laboratories
Department of Biochemistry
University of Liverpool, PO Box 147,
Liverpool, UK.

Rat mammary myoepithelial-like cell lines possess both specific high-affinity receptors for bFGF (K_d 30 pM to 280 pM) and specific low-affinity receptors (K_d 5-300 nM) (Fernig *et al.*, 1990a, 1993). On the basis of their sensitivity to competition by heparin and digestion by heparinase, the low-affinity receptors have been identified as heparan sulphate (HS), one half to two thirds of which may be associated with the extracellular matrix (Fernig *et al.*, 1992). In addition, whilst the HS receptors for bFGF on the myoepithelial-like Rama 401 cells have a single K_d of about 20 nM, they possess three distinct dissociation rate constants, suggesting that these receptors have heterogenous binding characteristics (Fernig *et al.*, 1992). In marked contrast to the myoepithelial-like cells, the parental rat epithelial cell lines derived from the normal mammary gland or benign tumours do not express cell-surface bFGF-receptors, presumably the cause of the failure of bFGF to stimulate their growth (Fernig *et al.*, 1990a,b). A series of rat cell lines intermediate in characteristics between the epithelial and myoepithelial cells are representative of the cells found in terminal end buds (TEBs). *In vivo* it is the cells of the TEBs, not the ducts, that are dividing and differentiating (Fernig *et al.*, 1990b). Both high- and HS low-affinity receptors for bFGF are first expressed at the cell surface at an early step in the differentiation pathway of epithelial cells to a myoepithelial-like cell. Thereafter, the number of both classes of receptors for bFGF increases, in the same order as the cells stage in differentiation to the myoepithelial phenotype, up to a maximum value for the myoepithelial cells (Fernig *et al.*, 1990a).

Myoepithelial cells derived from the normal rat mammary gland and from benign rat mammary tumours contain both bFGF mRNA and 7 ng bFGF/10^6 cells. However, bFGF mRNA and activity are virtually absent from the epithelial cell lines derived from the normal mammary gland, and from benign mammary tumours (Barraclough *et al.*, 1990; Fernig *et al.*, 1993). Furthermore, in the cell lines intermediate in characteristics between epithelial and myoepithelial cells the levels of bFGF mRNAs and activity increase as the myoepithelial characteristics of the cells increase (Barraclough *et al.*, 1990). Extracts of virgin rat mammary gland contain 50 ng/g tissue of bFGF-like activity (Barraclough *et al.*, 1990), and immunochemical analysis of these extracts indicates that all four translation products of the bFGF mRNA are present (Rudland *et al.*, 1993). bFGF mRNA has been detected in normal human mammary gland and in benign lesions including neoplasms (Anandappa *et al.*, 1994). Immunocytochemical localisation of bFGF in the developing rat mammary gland indicates that, as in cell cultures, bFGF is associated with the myoepithelial cells, with the intermediate cells of the TEBs and with the basement membrane, but not with epithelial cells of the ducts or alveoli (Rudland *et al.*, 1993). Moreover, large numbers of HS storage receptors for bFGF are associated with the myoepithelial cells of the resting ducts and basement membrane, but are absent from TEBs (Rudland *et al.*, 1993).

Thus it would appear likely that bFGF is involved in controlling the growth and differentiation of the mammary gland. The physiological role of bFGF may include not only the growth and differentiation of the intermediate cells of TEBs, but also its abstraction by HS storage receptors for bFGF may be associated with quiescent myoepithelial cells of the ducts and alveoli.

Malignant rat mammary epithelial cells possess high-affinity receptors for bFGF, but not HS low-affinity storage receptors for bFGF. Rat mammary malignant epithelial cells produce bFGF and bFGF mRNA is present in 25% of human malignant mammary cells (Fernig *et al.*, 1993; Anandappa *et al.*, 1994). Approximately 25% of human mammary carcinomas express bFGF mRNA at levels which are equivalent to, or higher than, those of the benign tissue despite the loss of bFGF-producing myoepithelial cells in invasive carcinomas (Anandappa *et al.*, 1994). Thus, the absence of HS storage receptors for bFGF on the malignant rat and human mammary epithelial cells with the greatest malignant potential suggests that alterations to the HSPG receptors for bFGF may also play a key role in the contribution to growth and dissemination of malignant mammary tumours by bFGF.

REFERENCES

Anandappa, S.Y., Winstanley, J.H.R., Leinster, S., Green, B., Rudland, P.S. and Barraclough, R., 1994, Comparative expression of fibroblast growth factor mRNAs in benign and malignant breast disease. *Br. J. Cancer*, 69:772.

Barraclough, R., Fernig, D.G., Rudland, P.S., and Smith, J.A., 1990, Synthesis of basic fibroblast growth factor upon differentiation of rat mammary epithelial to myoepithelial-like cells in culture, *J. Cell. Physiol.*, 144:333.

Fernig, D.G., Smith, J.A., and Rudland, P.S., 1990a, Appearance of basic fibroblast growth factor receptors upon differentiation of rat mammary epithelial to myoepithelial-like cells in culture, *J. Cell. Physiol.*, 142:108.

Fernig, D.G., Smith, J.A., and Rudland, P.S., 1990b, Relationship of growth factors and differentiation in normal and neoplastic development of the mammary gland, *in* "Breast Cancer: Cellular and Molecular Biology" Vol 2, Lippman, M.E. and Dickson, R. eds, Kluwer Nijhoff, Norwell MA.

Fernig, D.G., Rudland, P.S., and Smith, J.A., 1992, Modulation of bFGF action by low-affinity receptors in rat mammary myoepithelial-like cells in culture, *Growth Factors*, 7:27.

Fernig, D.G., Barraclough, R., Ke, Y., Wilkinson, M.C., Rudland, P.S. and Smith, J.A., 1993, Synthesis of basic fibroblast growth factor and its receptor by mammary epithelial cells derived from malignant tumours but not from normal mammary gland, *Int. J. Cancer*, 54:629.

Rudland, P.S., Platt-Higgins, A.M., Wilkinson, M.C., and Fernig, D.G., 1993, Immunocytochemical identification of basic fibroblast growth factor in the developing rat mammary gland, *J. Histochem. Cytochem.*, 41:887.

EVIDENCE OF GROWTH FACTOR PRODUCTION BY SHEEP MYOEPITHELIAL AND ALVEOLAR EPITHELIAL CELLS: POTENTIAL FOR AUTOCRINE/PARACRINE INTERACTIONS

Isabel A. Forsyth, James A. Taylor,
Maggy Villa and R. Stewart Gilmour

Cellular Physiology Department
Babraham Institute, Babraham,
Cambridge CB2 4AT, U.K.

INTRODUCTION

The major endocrine factors which control the development of the mammary gland in pregnancy are well known. However, exposure of ruminant mammary epithelial cells *in vitro* to oestradiol, progesterone, glucocorticoids, prolactin, growth hormone and placental lactogen has little or no effect on DNA synthesis (Winder *et al.*, 1989). Hormone action *in vivo* may be indirect through stimulation of or interaction with growth factors, some of which may be produced locally and act in an autocrine or paracrine manner.

Within a basement membrane, mammary gland alveoli contain two cell types: (1) alveolar epithelial cells (AEC) which differentiate to produce milk and (2) myoepithelial cells (MEC) which contract in response to oxytocin in the lactating gland to force secreted milk into the ducts. Clearly, to achieve normal development of the gland, the growth of these two cells needs to be co-ordinated. We have cultured AEC and MEC from the mammary glands of sheep and present evidence that AEC are the source of a growth factor of the epidermal growth factor (EGF/TGF-α) family, while MEC contain messenger RNA for insulin-like growth factor-II (IGF-II).

METHODS

AEC were prepared from sheep mammary tissue by collagenase-hyaluronidase digestion and cultured on hydrated gels of rat tail collagen as described by Winder *et al.* (1992). Cultures of MEC were obtained by explant outgrowth on plastic (Forsyth and Villa, 1994).

Intercellular Signalling in the Mammary Gland
Edited by C.J. Wilde *et al.*, Plenum Press, New York, 1995

DNA synthesis was measured for both cell types by incorporation of [^{3}H]-methyl thymidine. Total RNA was isolated from confluent monolayers of sheep MEC and IGF-II messenger RNA analysed by RNase protection assay (Li *et al.*, 1993).

RESULTS

Identification of cell types

AEC were identified on the basis of their polygonal shape, reaction with a monoclonal antibody to cytokeratin and ability to respond to lactogenic hormones with production of α-lactalbumin (Winder *et al.*, 1992). MEC were stellate, reacted with antibodies to vimentin, α-smooth muscle actin and α-actinin and responded to oxytocin (Forsyth and Villa, 1994). Cultures of AEC on collagen gels contained a small subpopulation identified as MEC (Winder *et al.*, 1992) but after the second or third subculture MEC were essentially free of AEC.

Evidence of growth factor production

DNA synthesis in AEC was inhibited in a dose-dependent manner by heparin, with a maximum effect of 5 μg/ml. In AEC from pregnant sheep (n=5), DNA synthesis was reduced ($P < 0.02$) to $21.2 \pm 2.0\%$ (mean $\pm$ SEM) of control values on day 4 of culture. Inhibition could be at least partially reversed ($P < 0.02$) by TGF-α (10 ng/ml) to $65.3 \pm 9.4\%$ of control. Heparin did not significantly inhibit DNA synthesis in MEC.

RNase protection analysis showed the presence in MEC of a region complementary to an IGF-II riboprobe.

CONCLUSIONS

The results suggest that sheep AEC, but not MEC, are the source of a heparin-inhibited growth factor acting through the EGF receptor and that MEC synthesize IGF-II. Since both AEC (McGrath *et al.*, 1991; Winder *et al.*, 1992) and MEC (Forsyth and Villa, 1994) respond to TGF-α and IGFs by synthesizing DNA, the possibility exists of co-ordinate development through paracrine and autocrine interactions.

REFERENCES

Forsyth, I.A., and Villa, M., 1994, Stimulation of DNA synthesis in myoepithelial cells derived from sheep mammary gland by insulin-like growth factor-I (IGF-I) and transforming growth factor-α (TGF-α), *J. Endocr.* 140 supplement:201.

Li, J., Saunders, J.C., Gilmour, R.S., Silver, M., Fowden, A.L., 1992, Insulin-like growth factor-II messenger ribonucleic acid expression in fetal tissues of the sheep during late gestation: effects of cortisol, *Endocrinology* 132:2083.

McGrath, M.F., Collier, R.J., Clemmons, D.R., Busby, W.H., Sweeny, C.A., and Krivi, G.G., 1991, The direct *in vitro* effect of insulin-like growth factors (IGFs) on normal bovine mammary cell proliferation and production of IGF binding proteins, *Endocrinology* 129:671.

Winder, S.J., Turvey, A., and Forsyth, I.A., 1989, Stimulation of DNA synthesis in cultures of ovine mammary epithelial cells by insulin and insulin-like growth factors, *J. Endocr.* 123:319.

Winder, S.J., Turvey, A., and Forsyth, I.A., 1992, Characteristics of ruminant mammary epithelial cells grown in primary culture in serum-free medium, *J. Dairy Res.* 59:491.

DERIVATION OF CONDITIONALLY IMMORTAL MAMMARY CELL LINES

Katrina E. Gordon, Bert Binas, A. John Clark
and Christine J. Watson

BBSRC Roslin Institute
Roslin, Midlothian EH25 9PS, U.K.

Conditionally immortal mammary cell lines are being derived to provide an *in vitro* model system for studying mammary gland growth, differentiation and apoptosis. The approach being adopted utilises transgenic mice harbouring an enhancerless temperature sensitive SV40 T-antigen (tsA58) gene construct driven by the milk protein promoter ß-lactoglobulin (BLG) as the source of mammary cells. The expectation was that expression of the immortalising agent T antigen would be limited to the secretory epithelial cells of the mammary gland.

Some lines of transgenic mice generated developed tumours in a number of unexpected anatomical sites due to ectopic expression from the BLG promoter. The tumour phenotype appears to be copy-number related.

The establishment, maintenance and characterisation of cell lines derived from these mice is currently underway. In parallel, mammary cells from immortomouse (tsA58 driven by the H-2K promoter) are being cultured and compared to BLG/tsA58 cells.

EFFECT OF RELAXIN ON PRIMARY CULTURES OF GOAT MAMMARY EPITHELIAL CELLS

Sylvia Kondo[1], Colin J. Wilde[2], Gillian D. Bryant-Greenwood[1], Eleanor Taylor[2], Malcolm Peaker[2] and Lynn M. Boddy-Finch[2]

[1]Department of Anatomy and Reproductive Biology,
University of Hawaii at Manoa, Honolulu,
Hawaii 96822, USA
[2]Hannah Research Institute, Ayr, KA6 5HL, U.K.

Relaxin (RLX) is a peptide hormone produced in the female reproductive tract. However, its site of synthesis, pattern of secretion and structure can vary between species (reviewed by Bryant-Greenwood and Schwabe, 1994). Recent studies have demonstrated synthesis of RLX by the mammary gland in the guinea-pig (Peaker *et al.*, 1989) and its immunolocalisation in the normal, cystic and neoplastic human breast (Mazoujian and Bryant-Greenwood, 1990). These findings imply a possible role for relaxin in influencing the growth of mammary epithelial cells, directly as a mitogen or indirectly through remodelling of the extracellular matrix, in addition to its main functions which are to promote uterine quiescence and softening of the uterine cervix during pregnancy. Indeed, there are reports of RLX stimulating mammary cell growth *in vivo* in mice (Bani *et al.*, 1986), rats (Wahab and Anderson, 1989) and pigs (Hurley *et al.*, 1991). Furthermore, RLX is reported to promote the growth of mammary cell lines *in vitro* (Bigazzi *et al.*, 1992). The present study investigated the effect of RLX on proliferation of goat mammary epithelial cells in primary culture.

Mammary cells were isolated from late-pregnant (d 108) British Saanen goats by collagenase/hyaluronidase digestion and enriched for epithelial cells by discontinuous centrifugation using a Percoll gradient. Cells were cultured on tissue culture plastic in M199/Ham's F-12 (1:1) medium with combinations of relaxin, 10 ng ml^{-1} EGF, 5 μg ml^{-1} insulin and at high (20% v/v horse serum, 5% v/v foetal calf serum [FCS]) or low (5% v/v FCS) concentrations of serum. Cell proliferation was measured by incorporation of [^{3}H]thymidine. DNA synthesis was assayed using a fluorimetric method. Purified porcine relaxin was prepared from pregnant sow ovaries (Sherwood and O'Byrne, 1974).

A wide range of concentrations of relaxin (between 10^{-5} and 10^{-11} M) had no effect on the proliferation of mammary cells isolated from pregnant goats and cultured with high serum medium for 4 d. Similar results were obtained when cells were cultured in low serum medium and when serum was replaced with 1% (w/v) BSA, confirming that relaxin-antagonising activity was not present in serum. Porcine RLX has high structural homology with insulin (James *et al.*, 1977), a known mitogen for mammary tissue. However, the addition of 10^{-6} M RLX could not compensate for the absence of insulin in primary cell cultures. Moreover, RLX showed no synergistic effect on cell proliferation in the presence of insulin and/or EGF over a 48 h period.

Culture of pregnant goat mammary epithelial cells on an extracellular basement membrane extracted from the Engelbreth-Holm-Swarm (EHS) murine tumour promotes cell differentiation. Multicellular aggregates form lobular-alveolar structures (mammospheres) capable of milk protein secretion (L.M. Boddy & C.J. Wilde, unpublished work). In one preliminary experiment, addition of RLX (10^{-6} M) to these differentiated cultures resulted in inhibition of protein secretion by 26% and total protein synthesis to a lesser degree (approximately 13%) over a 4 h period. However, further experiments are required to confirm an autocrine effect of RLX on the mammary gland.

In conclusion, RLX over a wide range of concentrations had no effect on cell proliferation in goat mammary cell cultures. Therefore, although porcine RLX has high structural homology with insulin (James *et al.*, 1977), it did not mimic its mitogenic effects. This contrasts with studies using mammary cell lines *in vitro*, but is consistent with a report that RLX did not stimulate cellular proliferation in mammary tissue explants from pregnant or lactating pigs (Buttle and Lin, 1991). The lack of response to RLX may be due to absence in cell culture of paracrine factors which synergise with RLX, since experiments *in vivo* have shown a stimulatory effect of RLX on mammary cell growth in mice (Bani *et al.*, 1986), rats (Wahab and Anderson, 1989) and pigs (Hurley *et al.*, 1991).

ACKNOWLEDGEMENTS

This work was funded by the Scottish Office Agriculture and Fisheries department. SK was the recipient of an NIH Fellowship award HD07421.

REFERENCES

Bani, G., Bigazzi, M. and Bani, D., 1986, The effects of relaxin on the mouse mammary gland. II The epithelium, *J. Endocrinol. Invest.* 9:145.

Bigazzi, M., Brandi, M.L., Bani, G. and Bani Sacchi, T., 1992, Relaxin influences the growth of MCF-7 breast cancer cells, *Cancer* 70:639.

Bryant-Greenwood, G.D. and Schwabe, C., 1994, Human relaxins: chemistry and biology, *Endocr. Rev.*, 15:5.

Buttle H.L. and Lin, C.L., 1991, The effect of insulin and relaxin upon mitosis (*in vitro*) in mammary tissue from pregnant and lactating pigs, *Domestic Animal Endocrinol.* 8:565.

Hurley, W.L., Doane, R.M., O'Day-Bowman, M.B., Winn, R.J., Mojonnier, L.E. and Sherwood, O.D., 1991, Effect of relaxin on mammary development in ovariectomzed pregnant gilts, *Endocrinology* 128:1285.

James, R., Niall, H., Kwok, S. and Bryant-Greenwood, G.D., 1977, Primary structure of porcine relaxin: homology with insulin and other growth factors, *Nature* 267:544.

Mazoujian G. and Bryant-Greenwood, G.D., Relaxin in breast tissue, *The Lancet* 335:298.

Peaker, M., Taylor, E., Tashima, L., Redman, T.L., Greenwood, F.C. and Bryant-Greenwood, G.D., 1989, Relaxin detected by immunocytochemistry and northern analysis in the mammary gland of the guinea-pig, *Endocrinology* 125:693.

Sherwood O.D. and O'Byrne, E.M., 1974, Purification and characterization of porcine relaxin, *Arch. Biochem. Biophys.* 160:185.

Wahab, I.M. and Anderson, R.R., 1989, Physiologic role of relaxin on mammary gland growth in rats, *Proc. Soc. Exp. Biol. Med.* 192:285.

DIFFERENTIAL EFFECTS OF GROWTH FACTORS ACTING BY AUTOCRINE AND PARACRINE PATHWAYS IN BREAST CANCER CELLS

Anna Perachiotti and Philippa D. Darbre

Department of Biochemistry and Physiology
University of Reading, Whiteknights, Reading RG6 2AJ, UK

INTRODUCTION

The extent and nature of the interaction between oestrogen and insulin-like growth factors (IGFI, IGFII) in growth regulation of breast cancer cells has been extensively debated in recent years (Yee, 1992). However, one question which has received little attention concerns the relevance of the source of the IGF to breast cancer cell growth. IGFI is considered to be a paracrine regulator of breast cancer cells whilst IGFII may act by both autocrine and paracrine mechanisms (Yee, 1992). It remains a possibility that there could be different consequences to breast cancer cells when a growth factor acts by autocrine or paracrine pathways. Even in cell culture experiments, the implicit assumption seems to be that up-regulation of endogenous growth factor gene expression is synonymous with exogeneous administration of growth factor. Our recent data using a cell culture model has suggested that this assumption may not be correct and that the consequences to breast cancer cells of enhanced endogenous IGFII gene expression acting in a true autocrine loop can be different from administration of an exogenous source of IGFII.

CELL CULTURE MODEL SYSTEM USED

In previous studies Daly *et al.* (1991), developed an expression vector which would allow IGFII expression to be regulated in transfected cells by metal ions. The vector was constructed by linking part of the metallothionein IIA promoter to the coding sequence of the IGFII gene. Using this vector, one stable transfected clone of MCF7 human breast cancer cells (entitled MI7 cells) showed zinc-inducible expression of IGFII mRNA and inducible secretion of IGFII protein from the cells. Administration of zinc resulted in a markedly enhanced growth rate which could be inhibited by the monoclonal antibody α-IR3, which blocks the binding site of the IGFI receptor (IGFIR). MI7 cells thus provide a model system in which expression of IGFII from a transfected expression vector can function in a true autocrine mechanism and, furthermore, in which this IGFII expression can be regulated by zinc ions. We have studied several molecular consequences of enhanced IGFII gene expression in these MI7 cells and the results have been compared with effects of administration of an exogenous source of IGFII to parental untransfected MCF7 cells.

RESULTS

Administration of IGFII by the two routes revealed striking differences in response of several molecular parameters. 1) Enhanced autocrine production of IGFII in MI7 cells resulted in a serum-inducible increase in TGFß$_1$ mRNA (by northern blotting) which could not be reproduced in untransfected parental MCF7 cells. 2) pS2 is a well-established oestrogen-inducible gene in MCF7 human breast cancer cells but we found that enhanced autocrine production of IGFII could result in increased pS2 mRNA in the absence of oestrogen. However, this induction of pS2 mRNA could not be reproduced by exogenous addition of IGFII to parental untransfected MCF7 cells. 3) Autocrine production of IGFII resulted in reduced IGFIR in MI7 cells but this could not be mimicked by adding exogenous IGFII to untransfected parental cells. 4) By contrast, studies of IGFII regulation of the secretion of IGFBP revealed that effects were dependent on the IGFII being administered as an exogenous source. In particular, exogenous addition of IGFII to untransfected MCF7 cells resulted in an increase in IGFBP4 secretion which was not found following enhanced autocrine production of IGFII.

CONCLUSIONS

These results demonstrate that the molecular consequences of enhanced endogenous IGFII gene expression acting in an autocrine pathway can be different from those following administration of an exogenous source of IGFII. Whilst exogenous administration of a growth factor may be more equivalent to a paracrine than an autocrine situation, growth factors may well have yet different effects again in a true paracrine situation where cell-to-cell contacts may be involved in transmission of the growth factor to its target cell. However, these data provide a cautionary tale to any scientist tempted to equate autocrine production with exogenous administration and suggest that different pathways of growth factor action may have different mechanisms of action on their target cells.

ACKNOWLEDGEMENTS

This work was supported financially by the Association for International Cancer Research. Financial help to attend this symposium was provided by the University of Reading Research Travel Fund. We are grateful to many scientists who provided various components including cDNAs.

REFERENCES

Daly, R.J., Harris, W.H., Wang, D.Y. and Darbre, P.D., 1991, Autocrine production of an insulin-like growth factor II using an inducible expression system results in reduced oestrogen sensitivity of MCF-7 human breast cancer cells, *Cell Growth and Differentiation* 2:457.

Yee, D., 1992, Insulin-like growth factors in breast cancer, *Breast Cancer Res. Treat.* 22:1.

VALIDATION OF TRANSFORMING GROWTH FACTOR ß-1 BINDING ASSAY FOR BOVINE MAMMARY TISSUE

Karen Plaut[1], Rhonda L. Maple[1],
Anthony V. Capuco[2], and Alan W. Bell[3]

[1]Department of Animal and Food Sciences,
University of Vermont, Burlington, VT 05405, USA
[2]Milk Secretion and Mastitis Laboratory
USDA-ARS, Beltsville, MD, USA
[3]Department of Animal Sciences
Cornell University, Ithaca, NY 14853, USA

Active transforming growth factor-beta-1 (TGF-ß1) implanted in the mammary gland prior to puberty inhibits DNA synthesis in the end buds (Daniel *et al.*, 1989). However, implantation of the same pellet during pregnancy does not influence lobulo-alveolar development (Daniel *et al.*, 1989). This inability of the mammary gland to respond to active TGF-ß1 may be regulated at the level of the receptor for TGF-ß1.

Receptors for TGF-ß1 have been actively studied in many different cell types *in vitro* (Wakefield *et al.*, 1987). Up to five receptor types have been identified in epithelial cells (Cheifetz *et al.*, 1990; Massague *et al.*, 1992; O'Grady *et al.*, 1991). However, receptors for TGF-ß1 have not been characterized in mammary tissue. The objective of this study was to develop an assay to measure binding of TGF-ß1 to mammary membranes in bovine mammary tissue.

METHODS

Mammary tissue was homogenized and centrifuged at 10,000 g for 20 min in 25 mM Tris, 0.25 M sucrose, 10 mM EDTA with 1 mM phenylmethanesulphonyl fluoride, pH 7.4. The supernatant was filtered through cheesecloth and centrifuged at 100,000 g. Crude microsomal membranes were resuspended in 25 mM Tris, 0.02% sodium azide, pH 7.4 and frozen at -80°C. Dissociation of endogenously bound TGF-ß from its receptors was accomplished by incubation of 150 μg of membrane protein for 2-5 minutes with 3 M MgCl$_2$ followed by centrifugation at 5,000 g for 30 min. The membranes were resuspended

Intercellular Signalling in the Mammary Gland
Edited by C.J. Wilde *et al.*, Plenum Press, New York, 1995

in 25 mM Tris, 10 mM $CaCl_2$, 1.0% BSA and 0.02% sodium azide, pH 7.4. Tubes were incubated with 65,000 dpm of ^{125}I-TGF-ß1 (Amersham, Arlington Heights, IL; specific activity 800-1000 $\mu Ci/\mu g$) in the presence or absence of 0-10 ng unlabelled TGF-ß1. After 2 hr at 4°C, buffer was added and membranes centrifuged at 5,000 g for 30 min. Buffer was decanted and pellet counted in a gamma counter. Specific binding was calculated as the difference between binding in the presence (10 ng) versus absence of unlabeled TGF-ß1.

Conditions for the binding assay were optimized by incubating membranes at 4, 23 and 37°C from 0-24 hr. Membrane protein was varied from 140 - 1020 μg of membrane protein. Specificity was determined by competing ^{125}I-TGF-ß1 with 0 -1 μg of transforming growth factor-α, epidermal growth factor, insulin-like growth factor I, TGF-ß1, TGF-ß2 and TGF-ß3. Each assay was performed on at least 2 cows in at least 3 separate assays.

RESULTS AND DISCUSSION

TGF-ß1 specifically binds to bovine mammary membranes indicating that TGF-ß1 may be a physiological regulator of mammary development in the cow. Non-specific binding was high when temperatures of either 23° or 37°C were used. Therefore, 4°C was chosen as the optimum temperature at which to perform the binding reaction. The binding reaction equilibrated in 2 hours at 4°C. This rapid equilibration has been observed previously at 4°C in binding assays on cultured epithelial cells (Kalter and Brody, 1991). Specific binding of TGF-ß1 increased linearly as membrane protein was increased from 140 to 1020 μg membrane protein. Insulin-like growth factor-I, epidermal growth factor and transforming growth factor α were not able to compete in a concentration dependent manner for binding to the TGF-ß1 receptor. TGF-ß1, TGF-ß2 and TGF-ß3 were able to compete with ^{125}I-TGF-ß1 for binding to the receptors. Using Scatchard analysis (1949), it was estimated that the number of receptors were 543 pmoles/mg membrane protein and the affinity was 1.3 x 10^{-10} M. These studies have allowed us to develop a binding assay to study the developmental regulation of TGF-ß receptors in bovine mammary tissue.

REFERENCES

Cheifetz, S., Hernandez, H., Laiho, M., ten Dijke, P., Iwata, K.K. and Massague, J., 1990, Distinct transforming growth factor-ß (TGF-ß) receptor subsets as determinants of cellular responsiveness to three TGF-ß isoforms, *J. Biol. Chem.* 265:20533.

Daniel, C.W., Silberstein, G.B., Van Horn, K., Strickland, P. and Robinson, S., 1989 TGFß1-induced inhibition of mouse mammary ductal growth: developmental specificity and characterization, *Dev. Biol.* 135:20.

Kalter, V.G. and Brody, A.R., 1991, Receptors for transforming growth factor-ß (TGF-ß) on rat lung fibroblasts have higher affinity for TGF-ß1 and for TGF-ß2, *Amer. J. Respir. Cell Mol. Biol.* 4:397.

Massague, J., Cheifetz, S., Laiho, M., Ralph, D.A., Weis, F.M.B. and Zentella, A., 1992, Transforming growth factor-ß, *in* "Cancer Surveys Volume 12, Tumor Supressor Genes, the Cell Cycle and Cancer", Imperial Cancer Research Fund, London.

O'Grady, P., Huang, S.S. and Huang, J.S., 1991, Expression of a new type high molecular weight receptor (Type V receptor) of transforming growth factor ß in normal and transformed cells, *Biochem. Biophys. Res. Comm.* 179:378.

Scatchard, G., 1949, The attraction of proteins for small molecules and ions, *Ann. NY Acad. Sci.* 51:600.

Wakefield, L.M., Smith, D.M., Masui, T., Harris, C.C. and Sporn, M.B., 1987, Distribution and modulation of the cellular receptor for transforming growth factor-beta, *J. Cell. Biol.* 105:965.

ENDOCRINE EFFECT OF IGF-I ON MAMMARY GROWTH IN PREPUBERTAL HEIFERS

Stig Purup[1], Yael Sandowski[2] and Kris Sejrsen[1]

[1]National Institute of Animal Science,
Department of Research in Cattle and Sheep
Research Centre Foulum, DK-8830 Tjele, Denmark
[2]Department of Biochemistry, Food Science and Nutrition,
The Hebrew University of Jerusalem, Rehovot 76100, Israel

Pubertal mammary growth in heifers is stimulated by exogenous GH (Sejrsen et al., 1986; Purup et al., 1993). The mechanism of action of GH on the mammary gland, however, is unclear, but evidence favours an indirect action via IGF-I. IGF-I is produced mainly in the liver, but also locally in other tissues, including the mammary gland. Therefore the in vivo effect of GH may be mediated by systemic as well as locally produced IGF-I. To elucidate this question we investigated the effect of sera from GH-treated and control (placebo-treated) heifers on mammary cell proliferation in vitro.

Mammary tissue parenchyma obtained from prepubertal heifers (BW ~ 200 kg) was used for explant cultures. Cultures, set up as described previously (McFadden et al., 1989), were conducted with tissues from 3 heifers. Pooled sera from control (n=4) and GH-treated (n=4) heifers were added to explant cultures in a concentration of 20% of the growth medium. Proliferation of mammary cells, measured as incorporated [^{3}H]-thymidine into explant DNA showed that growth medium supplemented with serum from GH-treated heifers significantly ($P < 0.001$) increased proliferation in explants compared with that of control serum (Table 1). Addition of IGF-I (300 μg/l) together with sera, showed a stimulatory effect of IGF-I on proliferation only in growth medium supplemented with sera from control heifers but not in medium containing sera from GH-treated heifers.

Mammary epithelial cells isolated from prepubertal heifers were cultured in three-dimensional collagen gels as described previously (Shamay et al., 1988). Sera from control (n=8) and GH-treated (n=8) heifers were added to the growth medium in concentrations of 1, 5 and 10%. Proliferation of epithelial cells was measured as [^{3}H]-thymidine incorporation per well. Cell proliferation was stimulated in a dose-dependent manner by sera from both control and GH-treated heifers (Table 2). Furthermore, growth medium containing sera from GH-treated heifers showed a significant ($P < 0.001$) stimulation of cell proliferation compared with that obtained with control sera.

Table 1. Mammary cell proliferation in mammary explants cultured in the presence of 20% serum from control (C-serum) and GH-treated (GH-serum) heifers.

Exp.	C-serum	C-serum + IGF-I[†]	GH-serum	GH-serum + IGF-I[†]	SEM
1	244	-	1225[***]	-	90
2	475	-	1200[***]	-	96
3	188	-	604[***]	-	67
3a	141	315[*]	645[***]	657	55

Values (DPM/μg DNA) are least square means; N=32 wells (Exp. 1-3) or N=88 wells (Exp. 3a). [*]$P < 0.05$, [***]$P < 0.001$ for significance of treatments compared with C-serum (ANOVA). [†] 300 μg/l.

Table 2. Mammary cell proliferation in mammary epithelial cells cultured with increasing concentrations of sera from control (C; n=8) and GH-treated (GH; n=8) heifers.

	1% serum		5% serum		10% serum		
Exp.	C	GH	C	GH	C	GH	SEM
1[†]	1950[a]	2002[a]	8574[ab]	14993[b]	24289[c]	37759[d]	2526
2[‡]	60844[a]	76217[a]	118838[b]	152767[b]	192715[c]	274919[d]	13302

Values (DPM/well) are least square means; N=144 wells per experiment. Means with different letters within the same row are significantly ($P < 0.05$) different (ANOVA). [†] Specific activity of [^{3}H]-thymidine was 27 Ci/mmol. [‡] Specific activity of [^{3}H]-thymidine was 85 Ci/mmol.

The results demonstrate that growth medium containing serum from GH-treated heifers stimulates proliferation of mammary cells *in vitro* more than serum from control heifers. The results therefore indicate that circulating IGF-I may be the factor responsible for the *in vivo* effect of GH on mammary cell proliferation. Furthermore, preliminary results of *in vitro* experiments indicate that the stimulatory effect of GH-serum and IGF-I can be blocked by IGF-I-antibody (Purup, Sandowski and Sejrsen, unpublished). The observed stimulatory effect of sera from GH-treated heifers in primary cultures of epithelial cells argues against a paracrine effect of IGF-I.

REFERENCES

McFadden, T.B., Akers, R.M. and Beal, W.E., 1989, Influence of breed and hormones on production of milk proteins by mammary explants from prepubertal heifers, *J. Dairy Sci.* 72:1754.

Purup, S., Sejrsen, K., Foldager, J. and Akers, R.M., 1993, Effect of exogenous bovine growth hormone and ovariectomy on prepubertal mammary growth, serum hormones and acute *in vitro* proliferative response of mammary explants from Holstein heifers, *J. Endocrinol.* 139:19.

Sejrsen, K., Foldager, J., Sørensen, M.T., Akers, R.M. and Bauman, D.E., 1986, Effect of exogenous bovine somatotropin on pubertal mammary development in heifers, *J. Dairy Sci.* 69:1321.

Shamay, A., Cohen, N., Niwa, M. and Gertler, A., 1988, Effect of insulin-like growth factor I on deoxyribonucleic acid synthesis and galactopoiesis in bovine undifferentiated and lactating mammary tissue *in vitro*, *Endocrinology* 123:804.

LOCAL CONTROL OF MAMMARY APOPTOSIS BY MILK STASIS

Lynda H. Quarrie, Caroline V.P. Addey and Colin J. Wilde

Hannah Research Institute, Ayr, KA6 5HL, UK.

Mammary involution after lactation is associated with extensive loss of secretory epithelial cells. Laddering of genomic DNA (Wyllie *et al.,* 1980) and altered gene expression indicate that this cell loss occurs by apoptosis. Neither the mechanism nor the regulation of mammary cell death is well-characterised, but both circulating mammogenic hormones and locally-active factors may be involved. Apoptosis induced in lactating mouse mammary gland by litter removal was prevented by prolactin treatment (Sheffield and Kotolski, 1992), whereas non-milking of one gland of lactating goats increased DNA laddering in that gland (Quarrie *et al.*, 1994). In this study, we present evidence for local induction of apoptosis in glands of lactating mice after 24 h of milk stasis.

The litter size of Tuck's No. 1 strain mice were adjusted to 10 pups on day 1 of lactation. On day 10, Vetseal was applied to all right-sided glands to seal these teats, and litter size was reduced to 5 pups. Mammary tissue from sealed and unsealed glands was snap-frozen in liquid N_2 and used to prepare genomic DNA (Sambrook *et al.*, 1989). DNA ($10\mu g$) was end-labelled with $[\alpha^{32}P]dCTP$ (0.3 nCi/$20\mu l$) by incubation with Klenow DNA polymerase (5mU; 10 min, room temperature). Labelled DNA (1 μg) was resolved by electrophoresis in 1.8% (w/v) agarose gels, fixed in 7% (w/v) TCA for 20 min, and exposed to X-ray film for 72 h.

Histological examination of sealed glands showed milk accumulation and changes characteristic of involution not observed in unsealed glands. DNA electrophoresis consistently identified ladders of 180-200 base pairs characteristic of apoptosis in tissue from both sealed and unsealed mammary glands. Laddering in sealed mammary glands was more pronounced after 48 h of milk stasis, and at each time point was consistently greater than in contralateral unsealed glands. The increase in DNA laddering with time after cessation of milk removal, and its preferential (but not exclusive) induction in sealed glands, suggests that mammary apoptosis in mice is regulated by a local mechanism sensitive to milk removal. Local regulation of apoptosis by milk stasis was also suggested by a recent study in lactating goats, but there the response was slower: DNA laddering was not detected in a gland unmilked for 48h or 1 week, but was observed after 2-3 weeks of unilateral treatment (Quarrie *et al.*, 1994). Induction of laddering by milk stasis was also more rapid than that observed after complete litter removal (Strange *et al.*, 1992).

Intercellular Signalling in the Mammary Gland
Edited by C.J. Wilde *et al.*, Plenum Press, New York, 1995

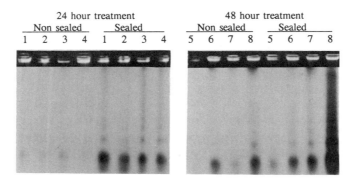

Figure 1. DNA fragmentation in mouse mammary glands sealed or unsealed for 24 or 48 h at peak lactation. The top panel shows DNA stained by ethidium bromide to confirm equal loading of the gel. The lower panel shows the DNA laddering pattern in DNA samples end labelled with [α^{32}P]dCTP. Samples from 8 animals are labelled 1-8.

This may reflect the greater sensitivity afforded by end-labelling of cleaved DNA compared with ethidium bromide staining. It may also explain the detection of DNA laddering in contralateral unsealed glands, which suggests that programmed cell death may occur during the course of lactation. Indeed, end radiolabelling has detected DNA laddering in mammary tissue of untreated day 10-lactating mice (L.H. Quarrie, unpublished work). However, since laddering increased in unsealed glands after the second 24 h period of milk stasis, the results do not exclude the possibility that systemic influences may have contributed bilaterally to the induction of apoptosis. Gland sealing and the reduction in litter size may have decreased galactopoietic hormone levels. Alternatively, a factor released by involuting mammary tissue may have acted systemically to induce apoptosis in the contralateral unsealed gland.

ACKNOWLEDGEMENTS

This work was funded by the Scottish Office Agriculture and Fisheries Department. LHQ is the recipient of a BBSRC studentship.

REFERENCES

Quarrie, L.H., Addey, C.V.P. and Wilde, C.J., 1994, Local regulation of mammary apoptosis in the lactating goat, *Biochem. Soc. Trans.* 22:178S.
Sambrook, J., Fritsch, E.F. and Maniatis, T., 1989, Isolation of high molecular-weight DNA from mammalian cells, *in*: "Molecular Cloning- A Laboratory Manual," 2nd ed., Cold Spring Harbor Laboratory Press, New York.
Sheffield, L.G. and Kotolski, L.C., 1992, Prolactin inhibits programmed cell death during mammary gland involution, *FASEB J.* 6:A1184.
Strange, R. Li, F., Saurer, S., Burkhardt, A. and Friis, R.R., 1992, Apoptotic cell death and tissue remodelling during mouse mammary gland involution, *Development* 115:49.
Wyllie, A.H., 1980, Glucocorticoid induced thymocyte apoptosis is asociated with endogenous endonuclease activation, *Nature* 284:555.

IMMUNOLOCALISATION OF INSULIN-LIKE GROWTH FACTORS AND BASIC FIBROBLAST GROWTH FACTOR IN THE BOVINE MAMMARY GLAND

Dieter Schams[1], Werner Amselgruber[2] and Fred Sinowatz[2]

[1]Institute of Physiology
Technical University of Munich
Freising Weihenstephan, Germany
[2]Institute of Veterinary Anatomy II
Histology and Embryology
University of Munich, Germany

Preparation of the mammary gland for lactation takes place throughout foetal and post-natal life and results from many complex interactions. The cyclical events of mammary growth and differentiation that occur during pregnancy and lactation are controlled by various polypeptide and steroid hormones. However, following the discovery of various growth factors increasing evidence has accumulated indicating that these growth factors may play an important role in the control of mammary growth and differentiation by complementing or mediating the hormonal action. Here, we studied the localisation of insulin-like growth factors (IGF's) and fibroblast growth factor (FGF) in mammary tissue of cows.

Mammary tissue from Brown Swiss cows (mid-lactation stage, n = 3) was collected after slaughter. Using conventional procedures, the tissue was dissected into 0.5 cm thick tissue slices, fixed in Bouin's solution or in methanol and glacial acid (ratio 1:1, w/v) for 24 h, dehydrated in a graded series of ethanol, cleared in xylene and embedded in paraffin. Sixty serial sections of 4-6 μm thickness per tissue sample were cut on a Leitz microtome and every twentieth section was subjected to immunohistochemical testing. The IGF-I rabbit polyclonal antibody (kindly supplied by Kabi Pharmacia, Stockholm, Sweden) was directed to the C-terminal portion and showed no cross-reactivity in a RIA with IGF-II ($< 1\%$) or other related peptides. IGF-II immunoreactivity was tested using a monoclonal antibody purified from mouse ascites (Amano Corp., Japan). The polyclonal rabbit antibody against recombinant human basic FGF (bFGF) showed no cross reaction against acidic FGF. Antigen localisation was achieved using the avidin-biotin immunoperoxidase technique. Potential endogenous peroxide activity was suppressed by incubation with methanolic hydrogen peroxide. The sections were placed in a moist chamber and covered initially with normal goat serum (1:10 diluted) for 30 min to minimize non-specific staining. Overnight incubation at 4°C with 1:1000 (IGF-I), 1:1200 (bFGF) and 10 ng/ml (IGF-II) dilutions of antisera followed. After washing in PBS the sections were incubated for 30 min with biotinylated goat anti-rabbit IgG (IGF-I, bFGF) or with goat anti-mouse IgG (IGF-II). The sections were then reacted with avidin-biotin-peroxidase reagent from a commercial kit (Vector Laboratories, Burlingame, CA, USA). The bound complex was visualised by reaction with 0.05% 3,3'-diaminobenzidine hydrochloride and 0.0006% H_2O_2 in 0.1 M PBS. The specificity of the immunocytochemical reactions was assessed on the basis of absorption tests involving the respective antigens, by serial dilutions of the primary antigen, by replacement of the primary antibodies with buffer, and their substitution with non-immune, protein A-Sepharose-purified goat IgG or with diaminobenzidine reagent alone.

Strong positive staining for IGF-I was observed in the mammary secretion, epithelial cells and stromal cells. By contrast, no positive staining was visible for IGF-II. For bFGF no staining was observed in the epithelial or stromal tissue, but strong, specific labelling was seen in vascular cells and myoepithelial cells. At the cellular level, the reaction product was exclusively localized in the cytoplasm of capillary endothelial cells and pericytes. Cells of lymphatic vessels were negative.

DNA synthesis and/or increase in cell number is strongly stimulated by IGF-I at physiological concentrations in cultures from mouse, rat, bovine and ovine mammary epithelial cells (Forsyth, 1991). The origin of the localized IGF-I can not be determined in our study. From the literature there exists clear evidence for local production as well as for bloodborne IGF-I. Both may be important for the function of mammary gland. Glimm et al. (1988) detected IGF-I in the stroma of bovine mammary gland. The cytoplasm of mammary epithelial cells also showed a strong positive reaction only after treatment *in vivo* with growth hormone. We assume from the negative result for IGF-II that most of the IGF-II secreted into milk is bloodborne and that IGF-II may be less important as a local autocrine/paracrine factor. Basic FGF is known to be an important regulator of angiogenesis in many tissues. The localization of bFGF in mammary blood capillaries and myoepithelial cells suggests a role in vascularisation and maybe differentiation of the mammary gland. We assume from the localisation study that IGF-I and bFGF may be of importance for the local regulation of bovine mammary gland.

REFERENCES

Forsyth, I.A., 1991, The mammary gland, *Baillierés Clin. Endocrinol. Metabol.* 5:809.
Glimm, D.R., Baracos, V.E., Kennelly, J.J., 1988, Effect of bovine somatotropin on the distribution of immunoreactive insulin-like growth factor in lactating bovine mammary tissue, *J. Dairy Sci.* 71:2923.

FLOW SORTING AND CLONAL ANALYSIS OF MOUSE MAMMARY EPITHELIUM

Matthew Smalley, Jenny Titley and Michael J. O'Hare

Institute of Cancer Research
Haddow Laboratories, 15 Cotswold Road,
Sutton, Surrey SM2 5NG, U.K.

The segregation of the adult mammary parenchyme into two major cell types, the luminal and myoepithelial cells, indicates pathways of development and differentiation which should be amenable to lineage analysis if pure cell populations can be prepared. In this study, conditions for the direct clonal culture of the mouse parenchymal cells have been established, and antibodies identified with which different mouse epithelial cell types can be flow sorted. The phenotype of such clones and sorted cells has been analyzed *in vitro* using cell-type specific cytostructural markers, permitting a direct comparison to be made with both flow-sorted human (O'Hare *et al.*, 1991) and rat mammary luminal epithelial and myoepithelial cells (Dundas *et al.*, 1991).

METHODS

A rat monoclonal antibody, 33A10, specific for a cell surface antigen on mouse luminal epithelial cells (Sonnenberg *et al.*, 1986) and a rat anti-CD10 (CALLA/NEP 24.11) antibody, AL2 (specific for mouse myoepithelial cells), were utilized for fluorescence-activated cell-sorting. From a range of mouse monoclonal antibodies against specific cytokeratin polypeptides a set which specifically stained luminal or myoepithelial cells *in vivo* was also identified. Freshly isolated mammary parenchymal cell suspensions from adult (10 week old) virgin animals were sorted, cloned and their *in vitro* phenotype analyzed using the cytostructural markers.

RESULTS

A systematic analysis of the culture conditions demonstrated that virgin adult mouse mammary cells, unlike their human and rodent counterparts, clone efficiently only in a low (<5% v/v) oxygen environment. Analysis of clones derived from both unsorted cells and sorted luminal epithelial cells showed that almost all rapidly came to express the myoepithelial marker cytokeratin 14, in addition to the luminal markers cytokeratin 18 and cytokeratin 19 which typify these cells *in vivo*. However, when cloned mouse luminal cells were grown as 'mammospheres' on Engelbreth-Holm-Swarm murine sarcoma matrix ('Matrigel'), they could be seen, using confocal microscopy, to segregate into an inner layer of cells that expressed cytokeratin 18 surrounded by an outer layer of cells that also expressed cytokeratin 14.

When sorted mouse myoepithelial cells, separated either by positive sorting using the AL2 antibody or as the negative arm of a 33A10 sort, were cloned under the same conditions, no unique phenotype was seen and overall cloning efficiency was very low. Clones obtained in this manner showed the 'mixed' luminal/myoepithelial phenotype, although rare cytokeratin 14-only positive clones could be seen in unsorted cultures.

DISCUSSION

The methods and results described here show that using conventional (i.e. monolayer) culture systems, the phenotype of mouse mammary luminal cells is rapidly modulated *in vitro*, but that a differentiated phenotype is retained, or re-expressed to some degree, on more physiological substrates. These results demonstrate that mouse mammary cells show a lability of phenotype not observed in their human counterparts (O'Hare *et al.*, 1991) and only to a much lesser degree in rat mammary cells (Dundas *et al.*, 1991). The failure to successfully clone a pure myoepithelial cell type from the mouse mammary epithelium also contrasts sharply with rat and human studies in which myoepithelial cells not only clone readily but also grow more rapidly than their luminal counterparts. Possible explanations for their absence from mouse cultures include either an intrinsic inability to form clones, a lability such that they can no longer be distinguished from luminal clones using cytostructural markers, or a failure to provide appropriate culture conditions. Nevertheless, the retrieval of purified, sorted luminal cells does provide a valuable resource for future lineage analysis using a combination of transgenic mice and the cleared fat-pad system.

ACKNOWLEDGEMENTS

The authors would like to thank Dr Hugh Paterson for his assistance with the confocal microscopy. This work was funded by the Cancer Research Campaign.

REFERENCES

Dundas, S.R, Ormerod, M.G., Gusterson, B.A. and O'Hare, M.J., 1991, Characterization of luminal and basal cells flow-sorted from the adult rat mammary parenchyma, *J. Cell Sci.* 100:459.

O'Hare, M.J., Ormerod, M.G., Monaghan, P.M., Lane, E.B. and Gusterson, B.A., 1991, Characterization *in vitro* of luminal and myoepithelial cells isolated from the human mammary gland by cell sorting, *Differentiation* 46:209.

Sonnenberg, A., Daams, H., Van der Valk, M.A., Hilkens, J. and Hilgers, H., 1986, Development of mouse mammary gland: identification of stages in differentiation of luminal and myoepithelial cells using monoclonal antibodies and polyvalent antiserum against keratin, *J. Histochem. Cytochem.* 34:1037.

CYTOKINE EXPRESSION IN NORMAL AND NEOPLASTIC BREAST TISSUE

Valerie Speirs, Andrew R. Green, Victoria L. Todd
and Michael C. White

Department of Medicine,
Medical Research Laboratory, Wolfson Building,
University of Hull, Hull HU6 7RX, England, U.K.

INTRODUCTION

In recent years, much information has demonstrated that growth factors, including various cytokines, are involved, either in paracrine or autocrine fashions, in the initiation, maintenance and progression of breast cancer. This has been demonstrated *in vitro*, where cultured breast cell lines have been shown to respond to these factors, both in terms of growth and stimulation of the enzyme 17ß-hydroxysteroid dehydrogenase, which is responsible for the interconversion of estrone to the biologically more potent 17ß-estradiol. Populations of fibroblasts derived from both normal, malignant and benign primary breast lesions have been shown to express various cytokines *in vitro*, particularly interleukin (IL)-6. We have therefore investigated expression of cytokine transcripts in normal and malignant breast tissue using a reverse transcriptase linked PCR. Immediately after surgery, breast tissue was transported to the laboratory and snap frozen in liquid nitrogen. Total RNA was extracted using standard procedures and reverse transcribed to yield cDNA. PCR was performed using cytokine oligonucleotide primers designed to detect IL-1α/ß to IL-8 inclusive, transforming growth factor (TGF)ß$_{1-3}$ inclusive, tumour necrosis factor (TNF)-α/ß and Interferon (IFN)-γ with the following thermal cycle: a denaturation step of 94°C for 2 minutes, followed by 35 cycles of 94°C for 30 seconds, 55°C for 30 seconds, 72°C for 30 seconds, and a primer extension step of 72°C for 5 minutes. PCR products were analysed by agarose gel electrophoresis and visualised by ethidium bromide staining under UV light. The results are shown in Table 1.

With the exception of IL-2 and IFN-γ, all cytokine transcripts were expressed by at least one of the three categories, although the percentage expression varied. This may be explained by age differences or possibly steroid receptor status of the sample. A notable feature was the complete absence of IL-6 mRNA in mastectomy samples. IL-6 has previously been demonstrated to have an important role in estrogen accumulation in breast,

Table 1. Percentage expression of cytokine mRNA in human breast tissues

Cytokine transcript	RM (n=9)	Carcinoma (n=10)	Mastectomy (n=5)
IL-1α	33	20	60
IL-lß	0	20	0
IL-2	0	0	0
IL-3	88	100	100
IL-4	11	60	40
IL-5	0	10	0
IL-6	55	40	0
IL-7	0	10	0
IL-8	33	70	40
TNF-α	44	30	20
TNF-ß	55	60	60
IFN-γ	0	0	0
TGF-ß$_1$	44	100	60
TGF-ß$_2$	66	60	80
TGF-ß$_3$	77	80	40
ß-actin	100	100	100

RM = reduction mammoplasty

a feature associated with breast cancer. The differences in levels of expression of cytokine mRNA may indicate the presence of a cytokine network, with different cytokines being involved in maintenance of normal breast function, compared with breast neoplasms. Breast tissue is comprised of a number of different cell types including epithelial, fibroblastic and endothelial cells, as well as cells of the immune system, and the cell type(s) responsible for the cytokine transcripts detected in the present report have not yet been identified. Confirmation of the specific cell(s) awaits *in situ* hybridisation studies.

PRODUCTION OF AN INSULIN-LIKE GROWTH FACTOR BINDING PROTEIN BY THE INVOLUTING RAT MAMMARY GLAND

Elizabeth Tonner, James Beattie and David J. Flint

Hannah Research Institute, Ayr KA6 5HL, Scotland, U.K.

INTRODUCTION

The insulin-like growth factors (IGFs) regulate proliferation and differentiation of a wide range of cell types and are believed to be important mediators of mammary gland growth and development. IGF receptors increase in the lactating bovine and rat mammary gland and IGF-1 is a potent mitogen for ruminant mammary tissue *in vitro*. The presence of IGFs in milk has been described in several species and a role in the development of the neonatal gastrointestinal tract has been postulated, possibly mediated via IGFs in milk.

Most IGFs in physiological fluids are bound to specific binding proteins (IGFBPs) which modulate IGF bioavailability (Baxter, 1991). IGFBPs have been detected in human, rodent and ruminant milk, and are secreted by mammary tissue in culture. During involution, the basement membrane is degraded, inducing apoptotic cell death of epithelial cells and extensive remodelling occurs as the gland regresses to a resting state. IGF-1 has been proposed as a survival signal or anti-apoptotic factor in cytokine dependent cells and if such a role was evident in the mammary gland then its effects would need to be abrogated during involution. IGFBPs have been suggested to play such a role and we therefore investigated IGFBP production in milk during involution.

Involution was initiated at peak lactation, either by litter removal or by ablation of lactogenic hormones using CB154 to decrease prolactin secretion and an antiserum to GH (anti-rGH). IGFBP activity was measured by: solution phase binding of radiolabelled IGFs, using unlabelled peptides for competition studies; binding of [^{125}I]IGF-1 followed by size exclusion chromatography; and western ligand blotting followed by immunostaining with an antibody to IGFBP-2.

The IGFBP level in milk increased after litter removal and was substantially greater than after hormone ablation thus correlating well with the varying degrees of involution induced by the different treatments (Figure 1A). In competition studies IGFBP in milk from litter-removed rats exhibited preferential binding for IGF-2. Since insulin did not compete even at high concentrations, IGF binding was not to an IGF-1 receptor. Likewise the analogue Long-R$_3$-IGF-1, which has low affinity for IGFBPs but reasonable affinity for IGF-1 receptors, failed to compete for radioligand binding even at concentrations which would bind to the IGF-1 receptor (Figure 1B). Size exclusion chromatography showed that IGFBPs in milk from litter-removed rats formed principally a small molecular weight complex with IGF-1, and this complex was essentially absent in control milk.

Western ligand blotting confirmed the presence of an IGFBP of about 30 kDa in litter-removed milk but there was no evidence of increased amounts of this IGFBP in maternal serum, suggesting local production within the mammary gland. The milk IGFBP did not,

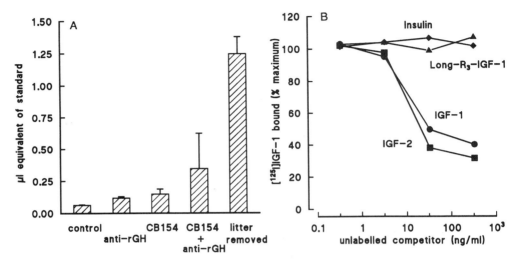

Figure 1. IGFBP activity in milk from treated lactating rats was measured by incubating milk with [^{125}I]IGF-1 and separating free and bound tracer using charcoal. (A) Comparison of milk from various treatments. (B) Competition of [^{125}I]IGF-1 binding to litter-removed milk by unlabelled peptides.

however, react with an antibody to IGFBP-2, which has previously been detected in milk (Donovan *et al.*, 1991).

The function of the IGFBPs is incompletely understood; they enhance or inhibit IGF actions in different *in vitro* systems, and they can also have direct effects on cells (Cohick and Clemmons, 1993). The IGFBP in the involuting gland may have been produced locally in response to involution, though, as yet, a role for IGF/IGFBPs has not been demonstrated. Serum IGF-1 concentrations are low in lactation and rise on litter removal. In the involuting gland this IGFBP may sequester IGF, thus blocking a potential stimulatory action of IGF-1 on cell survival. The purification and characterisation of this IGFBP is required to elucidate such a role in involution.

REFERENCES

Baxter, R.C., 1991, Insulin-like growth factor (IGF) binding proteins: the role of serum IGFBPs in regulating IGF availability, *Acta Paediatr. Scand.* 372(Suppl):107.

Cohick, W.S., and Clemmons, D.R., 1993, The insulin-like growth factors, *Ann. Rev. Physiol.* 55:131.

Donovan, S.M., Hintz, R.L., Wilson, D.M., and Rosenfeld, R.G., 1991, Insulin-like growth factors I and II and their binding proteins in rat milk, *Paediatr. Res.* 29:50.

DEVELOPMENTAL AND HORMONAL REGULATION OF *WNT* GENE EXPRESSION IN THE MOUSE MAMMARY GLAND

Stephen Weber-Hall, Deborah Phippard,
Christina Niemeyer and Trevor Dale

Institute of Cancer Research
Haddow Laboratories
Sutton, Surrey SM2 5NG, U.K.

INTRODUCTION

Ectopic expression of *Wnt-1* in the mammary epithelium causes hyperplasias and increases the frequency of tumour formation. Other members of the *Wnt* gene family (which encode for secreted glycoproteins) are naturally expressed in the breast and are thought to be involved in controlling the major developmental stages of the mammary gland.

RESULTS AND DISCUSSION

Using northern and *in-situ* hybridisation, spatial and temporal expression of different *Wnt* family members was found to be complex (Table 1). *Wnt-4*, *Wnt-5b* and *Wnt-6* are associated with pregnancy development but have different expression profiles. *Wnt-7b*, *Wnt-2* and *Wnt-5b* are expressed during virgin developmental stages and down-regulated during pregnancy. All of the *Wnt*s are down-regulated during lactation and some (*Wnt-2*, *Wnt-5a*, *Wnt-5b* and *Wnt-7b*) but not all (*Wnt-4* and *Wnt-6*) are up-regulated during involution. *In-Situ* hybridisation localised *Wnt-5b* expression to the epithelial ducts and lobulo/alveolar structures. Additional localisation of *Wnt* expression was obtained from separated primary epithelial and fibroblast cells and from northern analysis of epithelium-free fat pads (Table 1). Finally, ovariectomy of 6 and 12 week old mice showed that *Wnt-5b* and *Wnt-4* expression was down-regulated suggesting that they require the presence of ovarian hormones for their expression. *Wnt-5a* and *Wnt-7b* were not regulated by ovarian hormones and *Wnt-2* may be regulated by ovarian hormones at 6 weeks but not at 12 weeks of age.

This work extends previous studies by Gavin and McMahon (1992) and Bühler *et al.*

Table. 1 Summary of *Wnt* gene expression in the mouse mammary gland.

	Wnt-2	Wnt-4[*]	Wnt-5a	Wnt-5b	Wnt-6	Wnt-7b
Early virgin (Hormonal regulation)	+ + (yes)	+/- (yes)	+ + (no)	+/- (yes)	+ (ND)	+ + (no)
Late virgin (Hormonal regulation)	+ + (no)	+ (yes)	+ + + (no)	+ + (yes)	+ (ND)	+ + (no)
Early pregnant (2.5 days)	+ + +	+ + +	+ + +	+	+ +	+ +
Mid pregnant (12.5 days)	+	+	+	+ + +	+ + +	-
Late pregnant (17.5 days)	-	-	-	+ +	+	-
Lactation	-	-	-	-	+	-
Involution	+ +	+/-	+ + +	+ + +	+	+ +
mRNA localisation	Stro[1#]	Epi[1]	Stro[1,]	Ep i[1,2,3]	epi[1,3]/ Stro[1,3]	Epi[1,3]

Combined data from northern and *in-situ* analysis of *Wnt* expression in the mammary gland. Signal intensity is relative to maximal levels for each transcript. - no detectable signal, +/- very weak signal, + weak signal, + + medium signal, + + + strong signal, ND Not determined. RNA localisation: Epi - Epithelium expression, Stro - Stromal expression.
[1] Data from epithelial free 5 week virgin fat pad. Does not preclude possibility of epithelial expression or of stromal expression in the presence of epithelium.
[2] Data from *in-situ* hybridisation.
[3] Data from differential *in vitro* culture of epithelial and fibroblast cells.
[*] *Wnt-4* transcripts are differentially regulated. The summary is for the 4.4 Kb transcript. Expression of the 1.8 Kb transcript is similar to that for *Wnt-5b*.
[#] Expression of *Wnt-2* has previously been described in end buds (Bühler *et al.*, 1993).

al. (1993) and strongly suggests a non-redundant role for the *Wnt* genes in controlling mammary developmental processes. The *Wnt* genes encode secreted glycoproteins which are likely to be involved in cell-cell signalling. Our data shows expression of *Wnt* genes in both the epithelial and mesenchymal mammary compartments implicating individual *Wnt*s in specific communication processes between multiple cell types. A correlation between *Wnt*-5b expression and lobule development suggests that *Wnt*-5b plays a role in this developmental process. Finally, different *Wnt* family members are differentially regulated by ovarian hormones suggesting the possibility that some of the *Wnt*s modulate the mammary glands response to ovarian hormones.

REFERENCES

Bühler, T.A., Dale, T.C., Kieback, C., Humphreys, R.C. and Rosen, J.M., 1993, Localisation and quantification of *Wnt-2* Gene expression in mouse mammary development, *Dev. Biol.* 155:87.
Gavin, B.J. and McMahon, A.P., 1992, Differential regulation of the *Wnt* gene family during pregnancy and lactation suggests a role in postnatal development of the mammary gland, *Mol. Cell. Biol.* 12:2418.

EXTRACELLULAR MATRIX DEPENDENT GENE REGULATION IN MAMMARY EPITHELIAL CELLS

Christian Schmidhauser[1], Connie A. Myers[2],
Romina Mossi[1], Gerald F. Casperson[3] and Mina J. Bissell[2]

[1]Biochemistry I, ETH-Zürich, Switzerland
[2]Cell and Molecular Biology Department, Life Science Division, Lawrence Berkeley Laboratories, Berkeley, California, USA
[3]Molecular and Cell Biology, Searle, St.Louis, Missouri, USA

INTRODUCTION

The dynamic and reciprocal interactions of cells with their extracellular tissue microenvironment has provided the basis for the hypothesis that external cues such as extracellular matrix (ECM) proteins, and in particular the basement membrane underlying the epithelial tissue, acts as a key regulator of many aspects of cell behaviour (Bissell *et al.*, 1982). Ample evidence shows that controlled cellular interactions with the basement membrane are essential for the development and maintenance of the differentiated phenotype in many cell types (Stocker *et al.*, 1990; Wicha *et al.*, 1982; Lee *et al.*, 1984; Durban *et al.*, 1985; Li *et al.*, 1987; Blum *et al.*, 1987; Schmidhauser *et al.*, 1990; Taub *et al.*, 1990; Streuli and Bissell, 1990; Wang *et al.*, 1990; Takahashi and Nogawa, 1991; Montesano *et al.*, 1991; Peterson *et al.*, 1992). Such signals control at least in part growth, morphology, and biochemical differentiation. A number of systems provide evidence for the requirement of ECM-derived signals to regulate their differentiation programmes. One such system is the mammary epithelial cell in which the expression of milk proteins is tightly controlled by the interactions of the alveolar cells with their underlying basement membrane (Streuli, 1993). In particular the regulation of the milk proteins β-casein (Barcellos-Hoff *et al.*, 1989; Schmidhauser *et al.*, 1990) and whey acidic protein (WAP; Chen and Bissell, 1989) have been well-characterized.

Hepatocytes provide another versatile model for studying ECM-cell interactions. A number of studies have shown that the control of albumin gene expression is regulated through interactions with laminin (Bissell *et al.*, 1987, 1990; Caron, 1990). Part of the regulatory mechanisms were found to lie within a transcriptional enhancer of the albumin promoter where transcription factors bind in a matrix-dependent manner to specific sequences (DiPeriso *et al.*, 1991; Liu *et al.*, 1991).

A third well analyzed system is represented by keratinocyte-ECM interactions. Here, terminal differentiation is initiated by loss of contact with the basement membrane when cells stratify into the suprabasal layers (Watt *et al.*, 1988; Adams and Watt, 1989, 1990, 1993; Kubler *et al.*, 1991). The molecular basis for these events involves the loss of adhesiveness to fibronectin, laminin, and type I and IV collagens preceded by the loss of the integrins $\alpha2\beta1$, $\alpha3\beta1$, and $\alpha5\beta1$ on the cell surface (Adams and Watt, 1990). Culture studies demonstrated that addition of fibronectin to basement membrane-deprived keratinocytes inhibited the onset of terminal differentiation measured by the expression of involucrin.

Many other systems are known to direct gene expression and control differentiation programmes through ECM-cell interaction (Guyette *et al.*, 1979; Sorokin *et al.*, 1990; Damsky *et al.*, 1992; Behrendtsen *et al.*, 1992; Bobel *et al.*, 1992). A common theme derived from these systems is that ECM may be thought of as a ligand that interacts with cellular receptors, and signals through an as yet unknown pathway in order to induce tissue- and development-specific responses. In all these systems analysis of the molecular events relies strongly on sophisticated cell culture systems that permit physiologically relevant, *in vivo*-like situations. We will summarize some of the work on the influence of ECM on cell function and gene expression in mouse mammary epithelial cells and will include recent evidence that ECM directs transcription through both a β-casein enhancer and various viral enhancers.

Differentiation of the mammary gland

The mammary gland is one of the few tissues that undergoes repeated cycles of development and regression in the adult animal. Each stage of this differentiation process is tightly controlled by both soluble factors such as hormones and growth factors (Guyette *et al.*, 1979; Topper and Freeman, 1980; Hobbs *et al.*, 1982; Robinson *et al.*, 1993) and by the interactions of cells with the microenvironment such as ECM. In the virgin animal, the ductal tubes are compactly surrounded by a myoepithelial layer that prevents the ductal epithelium from directly interacting with the basement membrane underlying the myoepithelial sheet. During pregnancy, with increased levels of progestins, estrogen, and placental lactogen, the alveolar epithelium begins to bud out from the branching duct system and replaces the fatty stroma. The resulting alveolar structures are enclosed by their own basement membrane and are therefore separated from the surrounding stroma. Most of the epithelial cells comprising the alveoli are in close contact with this basement membrane. Only a few myoepithelial cells are located basally, building a loose network that wraps the alveolar structures. The protecting sheet of myoepithelial cells around the ductal tubes, and therefore the lack of interaction of the ductal epithelium with the basement membrane may be a reason for the incapacity of the ductal epithelial cells to synthesize milk proteins.

During pregnancy the appearance of distinct milk proteins represents different stages of development. β-casein, for example, appears at approximately day 8 of pregnancy, whey acidic protein (WAP) at approximately day 14, and α-lactalbumin is only synthesized immediately after parturition. The fully-lactating gland marks a temporary endpoint of differentiation when the highest amount of milk proteins are produced, secreted, and drained from the lumen of the alveoli through the duct system into the nipple. During involution the alveolar structure starts to disintegrate and the basement membrane is removed by ECM-degrading metalloproteinases. A delicate balance between the expression of these proteinases and their inhibitors, such as tissue inhibitor of metalloproteinases (TIMP), controls the correct breakdown of the gland structure (Talhouk *et al.*, 1991, 1992). The loss of the suckling stimulus results in inhibition of the production of lactogenic hormones and activates the expression of metalloproteinases. Initially, the proteinases are inhibited by a temporary increase in TIMP, which prevents the immediate breakdown of the basal membrane. After one to two days, TIMP expression decreases while metalloproteinases stay high and the disruption of the basement membrane takes place. At this point milk protein synthesis is switched off and is accompanied by a regression of the alveoli and an overall reduction in tissue mass. This tissue remodelling process leads to a phenotype essentially identical to that seen in the virgin animal.

In summary, the appearance of the alveolar structure that allows direct contact of the epithelial cells with the extracellular matrix, and the expression of milk proteins, changes according to the stage of differentiation. Maintenance of the differentiated phenotype appears to be dependent on epithelial cell contact with an intact basement membrane.

Differentiation of mammary cells in culture

The contribution of ECM to the correct functional differentiation of mammary epithelial cells has been compared in a variety of cell culture systems. A convincing argument that factors other than hormones are essential for the differentiation process comes from data derived from conventional tissue culture studies. When mammary epithelial cells from midpregnant mice are cultured on tissue culture plastic in the presence of the lactogenic hormones insulin, hydrocortisone, and prolactin, they rapidly lose the ability to produce milk and acquire a flat and elongated morphology with little or no polarization (Emerman et al., 1977; Lee et al., 1984). Although cells on plastic express and secrete ECM proteins such as laminin and collagen type IV they are not capable of organizing a functional basement membrane (Streuli and Bissell, 1990). However, if these cells are plated on a thick gel of collagen type I (isolated from rat tails) and subsequently floated, the cells regain their polarized epithelial morphology, basally deposit a basement membrane, and produce large amounts of caseins. Surprisingly, WAP, an abundant whey protein in rodent milk, is not synthesized in cells either on tissue culture plastic or floating collagen gels. WAP expression requires an additional level of structural complexity that can only be achieved in EHS cultures (Chen and Bissell, 1989) using EHS matrix or Matrigel (Kleinman et al., 1986). EHS is a predominantly laminin-rich matrix derived from a purified extract of the Engelbreth-Holm-Swarm tumour. Mammary cells plated on top of EHS gels undergo additional morphological changes which ultimately result in a spherical alveolar-like structure that, to a large extent, resembles the alveoli of the gland (Figures 1 and 2). Each of the cultured alveolar structures consists of highly-polarized cells with basal nuclei and apical microvilli separating a sealed central lumen into which the caseins and WAP are secreted unidirectionally. Transferrin, on the other hand, is secreted bidirectionally into the lumen and the surrounding medium (Chen and Bissell, 1989; Barcellos-Hoff et al., 1989). The level of β-casein mRNA in such cultures is comparable to the level in the gland.

While WAP expression requires a polarized cell within a multicellular three-dimensional structure, this complex organization is not a prerequisite for expression of β-casein. Recent studies have made it possible to distinguish between a requirement for cell polarity and the need for cell-cell interaction in the expression of ß-casein. Single cells embedded inside a reconstituted basement membrane (EHS matrix) re-expressed β-casein even though they were separated from each other and not polarized (Streuli et al., 1991).

It is known that the EHS matrix contains growth and differentiation factors which may contribute to the induction of differentiation in these cells (Bradley and Brown, 1990; Klagsbrun, 1990). Although such factors are clearly involved in the morphogenesis of the gland (Robinson et al., 1991; Taverna et al., 1991), accumulating data argues against their direct involvement in the regulation of β-casein gene expression. First, factor-free EHS matrix does not affect either the level of casein expression or its regulation in mammary cells (Streuli *et al.*, 1991). Second, antibodies against the β1 subunit of integrins block induction of the β-casein gene, indicating that the signal transduction is through integrins and not through receptors for growth or other factors (Streuli *et al.*, 1991). Third, purified laminin is sufficient to induce β-casein gene expression, whereas other components of the EHS matrix do not show an inducing effect (Streuli *et al.*, submitted).

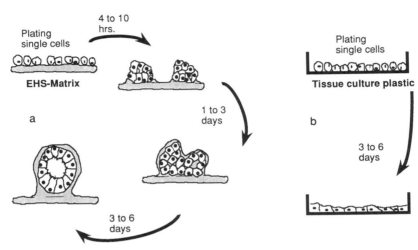

Figure 1. Mouse mammary epithelial cells plated on EHS matrix (a) or tissue culture plastic (b). Cells on EHS matrix start to aggregate within hours after plating. These aggregates may fuse if they are close to each other. Single or fused aggregates cover themselves with EHS matrix and go through an extensive remodelling process. This reorganization results in alveolar-like structures with a central lumen in which the milk proteins are secreted. Cells plated on tissue culture plastic remain flat and elongated without much polarisation and are not capable of producing milk proteins.

The EHS culture system is highly advanced in its potential to induce morphological and biochemical differentiation of mouse mammary epithelial cells. Other systems are not sufficient to induce abundant expression of WAP in mammary cell lines. However, α-lactalbumin brings even this system to its limits. We have never been able to activate transfected α-lactalbumin promoters in mammary cell lines (C. Schmidhauser, unpublished results). Thus, additional fine tuning must be done in order to establish the culture requirements of late milk protein expression such as α-lactalbumin.

The models described above indicate that cells removed from their natural microenvironmental signals are perturbed and lose their differentiated phenotype. However, exogenously-supplied ECM allows the re-establishment of function. Several fundamental questions arise from those observations. What is the signal that links a solid phase regulator with the expression of specific genes? What is the receptor for such signals? What is the signal transduction pathway and what are the final nuclear targets for these signals? Is the nature of this regulation transcriptional or post-transcriptional? In the remainder of this chapter we will address some of these questions and present some of our recent data.

RESULTS AND DISCUSSION

The β-casein promoter contains an ECM-responsive enhancer

Now that several functional mammary epithelial cell lines are available, it is possible to characterize the DNA elements involved in ECM-dependent gene regulation. CID 9 cells (Schmidhauser *et al.*, 1990), an epithelium enriched subpopulation of COMMA 1D cells (Danielson *et al.*, 1984), behave functionally and morphologically in a very similar fashion to primary cell cultures of midpregnant mice. In the presence of lactogenic hormones and ECM, more than 50% of these cells express β-casein, and are capable of secreting milk proteins into a central lumen of an alveolar structure. Little or no β-casein is produced by cells on tissue culture plastic even in the presence of lactogenic hormones (Figure 3). Nuclear run-on analysis with CID 9 and primary cells indicate that the ECM and prolactin effects are controlled primarily at the transcriptional level (Schmidhauser *et al.*, 1990; Goodman and Rosen, 1990). Additionally, CID 9 cells retain their ability to differentiate and function in a ECM-dependent manner even after a proliferation and selection period of more than four weeks after transfection. CID 9 cells are therefore an excellent tool to study ECM-responsive gene elements.

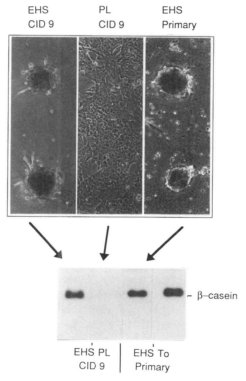

Figure 2. Morphological and biochemical comparison of cells differentiated on EHS-matrix (EHS) or tissue culture plastic (PL). CID 9 or primary mammary epithelial cells were differentiated for six days in the presence of lactogenic hormones without serum. Total RNA was isolated from these cultures and from freshly isolated primary epithelial cells of midpregnant mice (To), separated by gel electrophoresis and probed by northern analysis for the presence of β-casein mRNA. (Reproduced from Schmidhauser *et al.*, 1990).

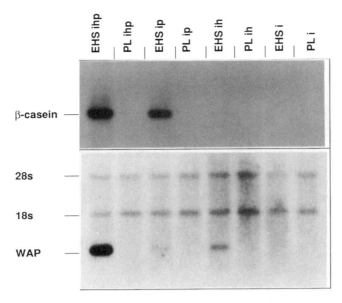

Figure 3. Northern blot showing matrix- and hormone- dependent regulation of β-casein and WAP in CID 9 cells differentiated on EHS-matrix (EHS) or tissue culture plastic (PL). The hormones are indicated as **i** for insulin, **h** for hydrocortisone, and **p** for prolactin. Background hybridization with 18s and 28s RNA confirms equal gel loading.

We constructed a series of bovine β-casein-CAT fusion genes containing various lengths of β-casein 5' flanking regions (Schmidhauser *et al.*, 1990). These constructs were stably transfected into CID 9 cells and CAT activity was measured after culturing the cells on either EHS-matrix or tissue culture plastic. Both matrix and hormonal control of transcription are conferred by the first 1790 base pairs upstream of the transcriptional start site. The regulation of CAT by both ECM and hormones is identical to that of the endogenous β-casein (including transcriptional activity in the absence of glucocorticoids; Figure 3). 5'deletions within the active promoter resulted in the localization of a transcriptional enhancer of 160 bp which contains the regulatory elements for both ECM and prolactin dependent regulation (Schmidhauser *et al.*, 1992). This novel enhancer, termed BCE1, reconstitutes the complete transcriptional activity and regulation in a cell type specific manner even when fused to a truncated and inactive β-casein promoter of 121 or 89 bp, designated ER-1 and ER-2 respectively (Figure 4). Indeed, BCE1 confers this control of transcriptional regulation on a heterologous promoter, a truncated and weakly active MMTV LTR (Figure 5). Interestingly, as is the case for the endogenous β-casein, BCE1, whether fused to ER-1 or the MMTV promoter, mediates ECM responsiveness in a glucocorticoid-independent manner. In fact, a prolactin independent ECM effect can be observed with the BCE1/MMTV constructs, suggesting that prolactin and ECM effects are functionally separable and may be mediated by distinct sequence elements within BCE1.

To identify precisely the sequence requirements for ECM-dependent regulation, site-directed mutagenesis was performed within BCE1 by replacing consecutive ten base pairs through the entire 160 bp enhancer. We found two regions that, when mutated, essentially abolished the activity of the enhancer. One comprised 20 bp in the distal region and the other contained 30 bp in the proximal region of BCE1 (C. Schmidhauser, in preparation). This data correlates with earlier experiments in which neither 100bp of the distal side nor

61bp of the proximal side of BCE1 was transcriptionally active when fused to the truncated promoter (C. Schmidhauser, unpublished). Thus, both the distal and the proximal part of

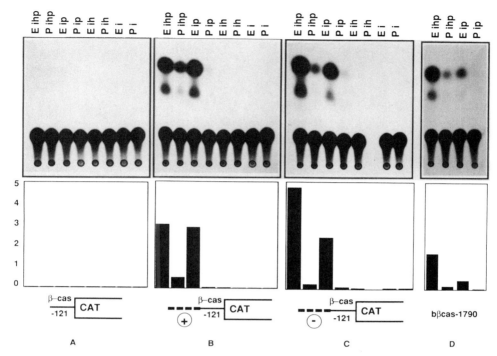

Figure 4. Effect of BCE1, substratum, and hormones on CAT expression by the 121bp promoter (ER-1). CID 9 cells were stably transfected with ER-1 (A), ER-1 plus the BCE1 enhancer in the positive orientation (B), or ER-1 plus the BCE1 enhancer in the negative orientation (C). Stably transfected CID 9 cells were differentiated on EHS-matrix (E) or tissue culture plastic (P) with the hormones insulin (i), hydrocortisone (h), prolactin (p). βcas-1790 is the construct that contains BCE1 in its natural context at around 1600bp upstream of the transcriptional start site (D). (Reproduced from Schmidhauser et al., 1992).

BCE1 are essential for its activity. Interestingly, part of the proximal region is a 6 out of 9 bp consensus for the mammary gland factor (MGF) binding site (Schmitt-Ney et al., 1991). However, other known transcription factor binding sites were not identified by sequence comparison analysis. Our results indicate that both regions are necessary but neither is sufficient for the transcriptional activation of BCE1. Characterization of the proteins that bind to these functionally important regions is now in progress.

To date the matrix and hormonal regulation conferred by BCE1 has only been shown in CID 9 cells. CID 9 cells are a heterogeneous cell strain enriched for the epithelial subpopulation of the COMMA 1D cells. When BCE1/ER-1 constructs are transfected into cloned mammary epithelial cells such as SCP2 (Desperez et al., 1993) or HC11 (Ball et al., 1988) cells, the enhancer is not active. Additionally, the fusion of BCE1 to the distal end of a rat β-casein promoter which had been inactivated by mutations in its MGF binding site (construct 7.1 and 7.2, a 330 bp promoter with a mutation at positions between -75 to -95bp; Schmitt-Ney et al., 1991), can rescue the activity and the regulation of the otherwise inactive promoter only in CID 9 cells. The same constructs fail to be active in cloned mammary cells such as SCP2 and HC11 cell lines (C. Schmidhauser, unpublished

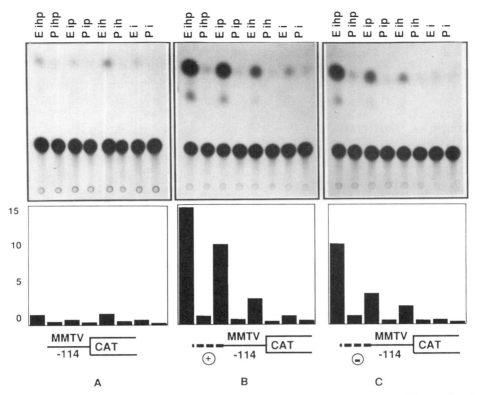

Figure 5. Effect of BCE1 on CAT expression with a viral promoter. CID 9 cells were stably transfected with a truncated MMTV promoter (-114bp) linked to CAT (A) or MMTV with BCE1 fused in either positive orientation (B) or negative orientation (C) to the 5' end of the viral promoter. Conditions are as in Figure 4. Note the matrix-dependent regulation in the absence of prolactin for the E ih conditions. (Reproduced from Schmidhauser *et al.*, 1992).

observations). However, SCP2 cells stably transfected with BCE1/ER-1, and placed in co-culture with SCG6 cells (another cell type clonally derived from the CID 9 cells; Desperez *et al.*, 1993) are capable of activating BCE1 with the same regulation pattern as in CID 9 cells. Conditioned media from SCG6 cells is also capable of restoring the activity of BCE1 in SCP2 cells plated on top of EHS matrix (C.Myers, unpublished observation). These results indicate that the activity of BCE1 requires another cell type not present in cloned cell lines and is dependent on a soluble factor secreted by these cells.

Activation of viral enhancers is modulated by ECM-cell interaction in mammary cells

Many genes have been shown to be positively or negatively regulated by ECM. We found, for example, that the expression of transferrin and WAP are dependent on the presence of an ECM-cell contact (Chen and Bissell, 1989), whereas the expression of TGFβ1 is inhibited when cells interact with a reconstituted basement membrane (Streuli *et al.*, 1993). Since the casein and the albumin genes have the control of matrix-dependent regulation localized within trancriptional enhancers, we investigated the possibility that other enhancers may display ECM responsiveness. Additionally, we were interested in enhancers which might act in a prolactin- and glucocorticoid- independent manner. This could possibly allow for the functional separation of the ECM and the hormone effects. We therefore turned our attention to three widely used viral enhancers (from mouse mammary tumor virus [MMTV], cytomegalovirus [CMV], and simian virus 40 [SV40]) and found that each is modulated in the presence or absence of ECM (Schmidhauser *et al.*, 1994). The MMTV enhancer either linked to its own promoter or to the truncated inactive β-casein promoter (ER-1) drives transcription efficiently only when the cells are in contact with ECM (Figure 6A). Similarly, the CMV enhancer has higher activity when cells are in contact with ECM (Figure 6B). In contrast, the SV40 enhancer is more active in cells on tissue culture plastic and its activity is strongly reduced when cells interact with ECM (Figure 6C). While the MMTV enhancer (and to a certain extent the CMV enhancer as well) demonstrate a hormone dependency, the SV40 enhancer is responsive to neither prolactin or glucocorticoid.

These results also indicate that the truncated β-casein promoter (ER-1) is a "blank slate" that imparts little or no additional transcriptional regulation to enhancers placed upstream. First, ER-1 is not independently active in CID 9 cells. Second, a number of hybrid enhancer/ER-1 constructs closely resemble, in their pattern of regulation, the promoters from which the enhancers are derived. For example the BCE1/ER-1 fusion precisely mimics the regulation pattern of the full length promoter where BCE1 is located in its natural context. The MMTV/ER-1 fusion closely matches the regulation of the intact promoter. Likewise, the CMV/ER-1 and SV40/ER-1 fusions resemble the CMV and the SV40 promoters in their regulation pattern (Figure 6). Finally, the various viral enhancers have completely different patterns of regulation, both quantitatively and qualitatively, when coupled to ER-1.

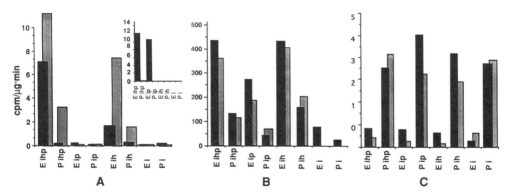

Figure 6. Substratum and hormonal dependency of CAT activity in CID 9 cells transfected with a construct containing either the MMTV enhancer (A), the CMV enhancer (B), or the SV40 enhancer (C) fused to the 5' of 121bp promoter ER-1 (black columns). The gray columns set behind illustrate the activity of the intact viral promoters with their respective enhancers located in its original location. The matrix and hormone conditions are as in Figure 4. The inserted graph in A represents the regulation of the ER-1 promoter linked to the BCE1 enhancer.

It is reasonable to expect to find ECM responsive enhancers in tissue specific genes that are expressed in epithelial cells attached to a well defined basement membrane. The discovery that such enhancers also exist within commonly used viral promoters with little cell-type specificity in culture is both surprising and instructive. It implies that caution must be exercised when evaluating a given promoter / enhancer combination for expression systems. Transcriptional regulation in tissue culture is often aberrant unless the microenvironment is properly defined and controlled. A good example for such a situation has been described for the β-lactoglobulin promoter driving a human serum albumin gene: this promoter was shown to be highly active in the mammary gland of transgenic mice (Shani *et al.*, 1992). However, the same construct is not active in conventional cell cultures of mammary epithelial cells, perhaps due to lack of an appropriate ECM. Placing the SV40 enhancer into the β-lactoglobulin promoter activated the expression in the cell culture system but resulted in loss of activity in the transgenic animals (I. Barash, ARO, Bet Dagen, Israel, personal communication). These data now make sense in light of our results demonstrating the ECM dependent suppression of the SV40 enhancer. The SV40 enhancer represses tissue-specific transcription in the animal, when the cells are in proper contact with their basement membrane. These data may help in understanding why the expression of transfected genes in cell culture is often different than that in transgenic animals (Palmiter *et al.*, 1985; Kollias *et al.*, 1987; Pinker *et al.*, 1987; Brinster *et al.*, 1988; Stewart *et al.*, 1988).

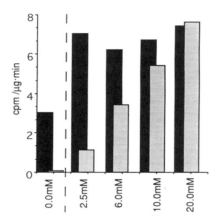

Figure 7. Effect of blocking histone deacetylase on transcriptional activity of BCE1/ER-1. CID 9 cells were stably transfected with BCE1/ER-1 enhancer-promoter fusion linked to the CAT reporter. Cells were differentiated for 6 days on EHS matrix (black columns) or tissue culture plastic (gray columns) in the presence of insulin, hydrocortisone, and prolactin. 24 hours prior to cell harvesting, deacetylase was inhibited by adding sodium butyrate to the culture medium at the concentration indicated.

It is still difficult to interpret how the ECM regulates such a wide variety of different genes. One possibility is that ECM-cell contact activates an entire genetic program, altering the activation states of many *trans*-activating proteins. Alternatively or additionally, nuclear or chromatin organization may be affected. Both ideas are consistent with the fact that ECM-cell contact initiates profound structural, morphological, and biochemical changes in mammary epithelial cells. Preliminary evidence for both concepts exists:
A) Bandshift analysis of nuclear proteins from CID 9 cells with a series of consensus sequences for known nuclear binding proteins results in i) ECM independent mobility shifts for the consensus of Oct-1, NF-κB, SP1, GRE, and TFIID (complex formation is not affected) ii) ECM dependent or preferential shifts for the consensus of AP2, CREB, and MGF (complex formation positively affected by ECM), and iii) bandshifts preferentially with nuclear proteins from extracts of cells without ECM interaction for the consensus of AP1, AP3, and CTF/NF1, with complex formation negatively affected by ECM (C. Schmidhauser, unpublished observations).
B) Altering the acetylation state of histones is known to derepress HML and HMR in yeast

(Johnsen *et al.*, 1990; Megee *et al.*, 1990; Braunstein *et al.*, 1993). Blocking the deacetylase in CID 9 cells with sodium butyrate induces activity of otherwise inactive BCE1/ER-1 constructs in CID 9 cells differentiated on tissue culture plastic. The activity in cells on plastic correlates with the extent of acetylation of histone H4 (Figure 7; R.Mossi and C.Schmidhauser, unpublished results). Further evidence for the concept of nuclear organization comes from the observation that the ECM dependent regulation of the transfected β-casein gene constructs holds only for stable transfections but not for transient transfections. A mechanism involving nuclear and/or chromatin organization may explain the need for integration of the transfected DNA into the host genome. Nevertheless experimental data for such models are quite scarce and confirmation will depend on gathering further data.

REFERENCES

Adams, J.C. and Watt, F.A., 1993, Regulation of development and differentiation by extracellular matrix, *Development* 117:1137.

Adams, J.C. and Watt, F.A., 1990, Changes in keratinocyte adhesion during terminal differentiation: reduction in fibronectin binding precedes α5β1 integrin loss from the cell surface, *Cell* 63:425.

Adams, J.C. and Watt, F.M., 1989, Fibronectin inhibits the terminal differentiation of human keratinocytes, *Nature* 340:307.

Ball, R.K., Friis, R.R, Schoenenberger, C.A., Doppler, W. and Groner, B., 1988, Prolactin regulation of ß casein gene expression and of a cytosolic 120-kd protein in a cloned mouse mammary epithelial cell line, *EMBO J.* 7:2089.

Barcellos-Hoff, M.H., Ram, T.G. and Bissell, M.J., 1989, Functional differentiation and alveolar morphogenesis of primary mammary epithelial cells cultured on a basement membrane, *Development* 105:223.

Behrendtsen, O., Alexander, C.M. and Werb, Z., 1992, Metalloproteinases mediated extracellular matrix degradation by cells form mouse blastocyst outgrowths, *Development* 114:447.

Bissell, M.D., Caron, J.M., Babiss, L.E. and Friedmann, J.M., 1990, Transcriptional regulation of the albumin gene in cultured rat hepatocytes, role of basement-membrane matrix, *Mol. Biol. Med.* 7:178.

Bissell, M.D., Arenson, M.D. and Roll, J.F., 1987, Support of cultured hepatocytes by a laminin rich gel, *J. Clin. Invest.* 79:801.

Bissell, M.J., Hall, H.G. and Parry, G., 1982, How does extracellular matrix regulate gene expression? *J. Theor. Biol.* 99:31.

Blum, L.J., Zeigler, E.M. and Wicha, S.M., 1987, Regulation of rat mammary gene expression by extracellular components, *Exp. Cell Res.* 173:322.

Bobel, C.P., Wolsberg, T.G., Turck, C.W., Myles, D.G., Primakoff, P. and White, J.M., 1992, A potential fusion peptide and an integrin ligand domain in a protein active in sperm-egg fusion, *Nature* 356:248.

Bradly, R.S. and Brown, A.M., 1990, The proto-oncogene *int-1* encodes a secreted protein associated with the extracellular matrix, *EMBO J.* 9:1569.

Braunstein, M., Rose, A.B., Holmes, S.G., Allis, C.D. and Broach, J.R., 1993, Transcriptional silencing in yeast is associated with reduced nucleosome acetylation, *Genes Dev.* 7:592.

Brinster, R.L., Allen, J., Behringer, R.R., Gelinas, R.E. and Palmiter, R., 1988, Introns increase transcriptional efficiency in transgenic mice, *Proc. Natl. Acad. Sci. USA* 85:836.

Caron, J.M., 1990, Induction of albumin gene transcription in hepatocytes by extracellular matrix proteins, *Mol. Cell. Biol.* 10:1239.

Chen, L.H. and Bissell, M.J., 1989, A novel regulatory mechanism for the whey acidic protein gene expression, *Cell Regulation* 1:45.

Damsky, C.H., Fitzgerald, M.L. and Fisher, S.J., Distribution patterns of extracellular matrix components and adhesion receptors are intricately modulated during first trimester cytotrophoblast differentiation along the invasive pathway, *in vivo. J. Clin. Invest.* 89:210.

Danielson, K.G., Oborn, C.J., Durban, E.M., Buetel, J.S. and, Medina, D., 1984, Epithelial mouse mammary cell line exhibiting normal morphogenesis *in vivo* and functional differentiation *in vitro*. *Proc. Natl. Acad. Sci. USA* 81:3756.

Desperez, P.Y., Roskelly, C., Campisi, J. and Bissell, M.J., 1993, Isolation of functional cell lines from a mouse mammary epithelial cell strain: importance of basement membrane and cell-cell interaction, *Mol. Cell. Different.* 1:99.

DiPeriso, C.M., Jackson, A.D. and Zarret, S.K., 1991, The extracellular matrix co-ordinately modulates liver transcription factors and hepatocyte morphology, *Mol. Cell. Biol.* 11:4405.

Durban, E.M., Medina, D. and Butel, J.S., 1985, Comparative analysis of casein synthesis during mammary cell differentiation in collagen and mammary gland development *in vivo*, *Dev. Biol.* 109:288.

Emerman, J.T., Enami, J., Pitelka, R.D. and Nandi, S., 1977, Hormonal effects on intracellular and secreted casein in cultures of mouse mammary epithelial cells on floating collagen membranes. *Proc. Natl. Acad. Sci. USA* 74:4466.

Goodman, H.S. and Rosen, J.M., 1990, Transcriptional analysis of the mouse ß-casein gene, *Mol. Endocrinolology* 4:1661.

Guyette, W.A., Matusik, R.J. and Rosen, J.M., 1979, Prolactin mediated transcriptional control of casein gene expression, *Cell* 17:1013.

Hobbs, A.A., Richards, D.A., Kessler, D.J. and Rosen, J.M., 1982, Complex hormonal regulation of rat casein gene expression, *J. Cell Biol.* 257:3598.

Johnsen, L.M., Kayne, P.S., Kahn, E.S. and Grunstein, M., 1990, Genetic evidence for an interaction between SIR3 and histone H4 in the repression of the silent mating loci in *Saccharomyces cerevisiae*, *Proc. Natl. Acad. Sci. USA* 87:6286.

Klagsbrun, M., 1990, The affinity of fibroblast growth factor for heparin, FGF-heparan sulfate interactions in cells and extracellular matrix, *Curr Opin. Cell Biol.* 2:875.

Kleinman, H.K., McGarvey, M.L., Hassell, J.L., Star, V.L., Cannon, F.B., Laurie, G.W. and Martin, G.R., 1986, Basement membrane complexes with biological activity, *Biochemistry* 25:312.

Kollias, G., Hurst, J., deBoer, E. and Grosfeld, F., 1987, The human ß- globin gene contains a downstream developmental specific enhancer, *Nucleic Acids Res.* 15:5739.

Kubler, D.M., Jordan, P.W., O'Neill, C.H. and Watt, F.M., 1991, Changes in the abundance and distribution of actin and associated proteins during terminal differentiation of human keratinocytes, *J. Cell Sci.* 100:153.

Lee, EY-H, Parry, G. and Bissell, M.J., 1984, Modulation of secreted proteins of mouse mammary epithelial cells by the collagenous substrata, *J. Cell Biol.* 98:146.

Li, M.L., Aggeler, J., Farson, D.A., Hatier, C., Hassell, J. and Bissell, M.J., 1987, Influence of reconstituted basement membrane and its components on casein gene expression and secretion in mouse mammary epithelial cells, *Proc. Natl. Acad. Sci. USA* 84:136.

Liu, J-K, DiPeriso, M.C. and Zarret, K.S., 1991, Extracellular signals that regulate liver transcription factors during hepatic differentiation *in vitro*, *Mol. Cell. Biol.* 11:773.

Megee, P.C., Morgan, B.A., Mittman, B.A. and Mitchell Smith, M., 1990, Genetic analysis of histone H4: Essential role of lysines subject to reversible acetylation, *Science* 247:841.

Montesano, R., Schaller, G. and Orci, L., 1991, Induction of epithelial tubular morphogenesis *in vitro* by fibroblast-derived soluble factor, *Cell* 66:697.

Palmiter, R.D., Chen, H.J., Messing, A. and Brinster, R.L., 1985, SV40 enhancer and large-T antigen are instrumental in development of choroid plexus tumors in transgenic mice, *Nature* 316:457.

Petersen, O.W., Ronnov-Jessen, L., Howlett, A.R. and Bissell, M.J., 1992, Interaction with basement membrane serves to rapidly distinguish growth and differentiation pattern of normal and malignant breast epithelial cells, *Proc. Natl. Acad. Sci. USA* 89:9064.

Pinkert, C.A., Ornitz, D.M., Brinster, R.L. and Palmiter, R.D., 1987, An albumin enhancer located 10kb upstream functions along with its promoter to direct the efficient, liver-specific expression in transgenic mice, *Genes Dev.* 1:268.

Robinson, S.D., Siberstein, G.B., Roberts, A.B. Flanders, K.C. and Daniel, C.W., 1991, Regulated expression and growth inhibitory effects of transforming growth factor ß isoforms in mouse mammary gland development, *Development* 113:867.

Robinson, S.D., Roberts, A.B. and Daniel, C.W., 1993, TGFβ suppresses casein synthesis in mouse mammary explants and may play a role in controlling milk levels during pregnancy, *J. Cell Biol.* 120:245.

Schmidhauser, C., Bissell, M.J., Myers, C.A. and Casperson, G.F., 1990, Extracellular matrix and hormones transcriptionally regulate bovine ß-casein 5' sequences in stably transfected mouse mammary epithelial cells, *Proc. Natl. Acad. Sci. USA* 87:9118.

Schmidhauser, C., Casperson, G.F., Myers, C.A., Sanzo, K.T., Bolten, S. and Bissell, M.J., 1992, A novel transcriptional enhancer is involved in the prolactin- and extracellular matrix-dependent regulation of ß-casein gene expression, *Mol. Biol. Cell.* 3:699.

Schmidhauser, C., Casperson, G.F. and Bissell, M.J., 1994, Transcriptional activation by viral enhancers: Critical dependence on extracellular matrix-cell interactions in mammary epithelial cells, *Mol. Carcinogenesis*, in press.

Schmitt-Ney, M., Doppler, W., Ball, R.K. and Groner, B., 1991, ß-casein promoter activity is regulated by hormone-mediated relief of transcriptional repression and a mammary specific nuclear factor, *Mol. Cell. Biol.* 11:3745.

Shani, M., Barash, I., Nathan, M., Ricca, G., Searfoss, G.H., Dekel, I., Faerman, A., Givol, D. and Hurwitz, D.R., 1992, Expression of human serum albumin in the milk of transgenic mice, *Transgenic Res.* 1:195.

Sorokin, L., Sonnenberg, A., Aumailley, M., Timpl, R. and Ekblom, P., 1990, Recognition of the laminin E8 cell binding site by an integrin possessing the $\alpha6$ subunit is essential for epithelial polarization in developing kidney tubules, *J. Cell Biol.* 111:1265.

Stewart, T.A., Hollingshead, P.G. and Pitts, S.L., 1988, Multiple regulatory domains in the mouse mammary tumor virus long terminal repeat revealed by analysis of fusion genes in transgenic mice, *Mol. Cell. Biol.* 8:473.

Stocker, A.W., Streuli, C.H., Martins-Green, M. and Bissell, M.J., 1990, Designer environment or the analysis of cell and tissue function, *Curr Opinion Cell Biol.* 2:864.

Streuli, C.H. and Bissell, M.J., 1990, Expression of extracellular matrix components is regulated by substratum, *J. Cell Biol.* 110:1405.

Streuli, C.H., Schmidhauser, C., Kobrin, M. and Bissell, M.J., 1993, Derynck R. Extracellular matrix regulates expression of the TGF ß1 gene, *J. Cell Biol.* 120:253.

Streuli C.H., 1993, Extracellular matrix and gene expression in mammary epithelial cells, *Sem. Cell Biol.* 4:203.

Streuli, C.H., Bailey, N. and Bissell, M.J., 1991, Control of mammary epithelial differentiation: basement membrane induces tissue specific gene expression in the absence of cell-cell interaction and morphological polarity, *J. Cell Biol.* 115:1383.

Takahashi, Y. and Nogawa, H., 1991, Branching morphogenesis of mouse salivary epithelium in basement membrane like substratum separated from mesenchyme by the membrane filter, *Development* 111:327.

Talhouk, R.S., Bissell, M.J. and Werb, Z., 1992, Co-ordinated expression of ECM degrading proteinases and their inhibitors regulate mammary epithelial function during involution, *J. Cell Biol.* 118:1271.

Talhouk, R.S., Chin, J.R., Unemori, E.N., Werb, Z. and Bissell, M.J., 1991, Proteinases of the mammary gland: developmental regulation *in vivo* and vectorial secretion in culture, *Development* 112:493.

Taub, M., Wang, Y., Szczesny, T.M. and Kleinman, H.K., 1990, Epidermal growth factor or transforming growth factor α is required for kidney tubulogenesis in Matrigel cultures in serum free medium, *Proc. Natl. Acad. Sci. USA* 87:4002.

Taverna, D., Groner, B. and Hynes, N.E., 1991, Epidermal growth factor receptor, platelet derived growth factor receptor, and *c-erbB-2* receptor activation all promote growth but have distinct effects upon mouse mammary epithelial cell differentiation, *Cell Growth Different.* 2:145.

Topper, Y.J. and Freeman, C.S., 1980, Multiple hormonal interactions in the developmental biology of the mammary gland, *Physiol. Rev.* 60:1049.

Wang, A.Z., Ojakin, G.K. and Nelson, W.J., 1990, Steps in the morphogenesis of a polarized epithelium.I. Uncoupling the roles of cell-cell and cell-substratum contact in establishing plasma membrane polarity in multicellular epithelial (MDCK) cysts, *J. Cell Sci.* 95:137.

Watt, F.M., Jordan, P.W. and O'Neill, C.H., 1988, Cell shape controls terminal differentiation of human epidermal keratinocytes. *Proc. Natl. Acad. Sci. USA* 85:5576.

Wicha, M.S., Lowrie, G., Kohn, E., Bagavandoss, P. and Mahn, T., 1982, Extracellular matrix promotes mammary epithelial growth and differentiation *in vitro*, *Proc. Natl. Acad. Sci. USA* 79:3213.

α-LACTALBUMIN REGULATION AND ITS ROLE IN LACTATION

Robert D. Bremel

Department of Dairy Science
Endocrinology-Reproductive Physiology Program
University of Wisconsin-Madison
Madison, Wisconsin 53706 USA

INTRODUCTION

In most, if not all milks, α-lactalbumin is a protein of relatively low abundance. In cow's milk it represents only about 3% of the total protein. In contrast to the other high abundance proteins whose function is to provide the amino acid and mineral sources for the young growing animal, α-lactalbumin plays a critical role in the process of milk secretion *per se*. It is a subunit of the enzyme lactose synthase. It forms a complex with galactosyl transferase in the Golgi and changes the affinity of this enzyme for glucose. The result is the synthesis of lactose. In this paper, I will review mainly the work from my laboratory. For further background the reader is referred to reviews by Tucker (1987, 1988).

The expression of α-lactalbumin coincides with the onset of copious milk secretion at the time of parturition. This time period and the physiological changes taking place have sometimes been called Phase II lactogenesis. Early work showed that the expression of α-lactalbumin is inhibited by progesterone. Thus, the concept is that the elevated serum concentration of progesterone during pregnancy keeps the level of α-lactalbumin expression low and that the drop in serum progesterone just prior to parturition relieves this inhibition and is critical to the onset of lactation and copious secretion of milk. There is a second requirement. In addition to the drop in progesterone there is also a concomitant surge in prolactin (PRL) which is maintained for a period of about a day in cattle. If this PRL surge is blocked by the ergot alkaloid CB 154 a normal lactation does not occur. The water transport across the mammary epithelium that is required for the onset of lactation and the copious secretion of milk does not occur and cannot be initiated with PRL given at a later time. The requirement for PRL in a very narrow time window is absolute. The work of Akers *et al.* (1981) showed in a classical endocrine replacement protocol that a normal lactation occurred if cows that had been treated with CB 154 were infused with PRL during the period when PRL was normally high, to maintain a normal level of the hormone. An interesting characteristic of this response is that once the lactation is established, PRL is not required to maintain lactation in cattle. This is not the case for all species, some require PRL throughout their lactation to maintain milk secretion and administration of CB 154 during lactation causes lactation to cease. In cattle, administration of CB 154 has no impact on lactation performance.

Recent work by the laboratory of Goodman and Schanbacher (1991) has shown the unique pattern of expression of α-lactalbumin mRNA in cows as compared to other milk proteins. They showed that α-lactalbumin mRNA could not be detected prior to parturition. This was in contrast to the other milk proteins whose mRNAs were readily detected in mid pregnancy. Likewise, when lactation ceased the other milk protein mRNAs were maintained at high levels for several weeks while the α-lactalbumin mRNA dropped precipitously on the day lactation ceased. The pattern they observed was similar to that which we found by measuring the level of ß-lactoglobulin and α-lactalbumin in the blood of animals throughout pregnancy, lactation and the subsequent involution period (Mao and Bremel, 1990, 1991). Because the tight junctions are not established between the mammary epithelial cells until the onset of lactation milk proteins are present in the blood of pregnant animals. We found α-lactalbumin only at very low levels, but, it increased very dramatically just prior to parturition whereas ß-lactoglobulin was detected very early in pregnancy and its concentration seemed to increase in a fashion typical of the geometric growth curve of the mammary gland itself.

Synthesis of lactose and lactation performance

The secretion of lactose into the alveolar lumen is a fundamental requirement for the onset of copious milk secretion (reviewed by Peaker, 1977). The ability to synthesize and partition lactose across the lumenal epithelial boundary is critical to the osmotic flow of water. Once in the alveolar lumen, lactose, being a disaccharide, is incapable of re-entering the cell. Thus, the high concentration of this impermient solute provides a large osmotic gradient and water moves passively to dilute out the lactose and maintain milk in an isosmotic state with blood. The timing of the onset of lactose synthesis is intimately regulated by the expression of α-lactalbumin and with its resultant alteration in the synthetic specificity of galactosyltransferase.

Movement of water across the mammary epithelium is critical to milk production. The coordinated regulation of the α-lactalbumin gene and the appearance and disappearance of the mRNA of α-lactalbumin concomitant with the onset and cessation of lactation is consistent with its pivotal role in water movement and in lactation in general.

Genetic selection for milk production

Since artificial insemination (AI) began to be used in the 1950s milk production of cattle in countries using AI has increased dramatically. Statistical procedures established by population geneticists enabled the detection of unique animals in the population and made it possible to track their progeny. The genetic merit of a bull for milk production is inferred from calculations relating the milk production of his daughters to their contemporaries in the same herd. Superior milk-producing daughters in national herds are identified and then used in special mating schemes to produce the next generation of males. Since 1950, the average milk production per cow has more than doubled in the US. Some of the increase is attributable to improvements in management and feeding techniques but a major factor is the genetic improvement that has occurred over time.

This genetic improvement in component yield has largely been due to the increased ability of the animals to transport water across the mammary epithelium while maintaining a relatively constant gross composition of fat and protein. Genetic selection pressure by dairy producers has been encouraged by economic incentives. This has led to very large increases in fat and protein component yields (fractional composition volume). In Holland, there has been a further economic incentive to increase the fractional composition of protein and fat in milk. The result is that cattle from Holland have the highest fat and protein test in the European Union. However, the increase in fractional composition of milk has been very modest, only about 5% of the starting mean, while yield has more than doubled during the same time period. It is as though the movement of water is much easier to accomplish than the movement of other nutrients. An examination of the metabolic pathways involved in the synthesis gives clues to why this might be. The cellular energy required to synthesize lactose is a fraction of that required for either fat or protein. Also, the synthesis of lactose which is the driving force is but a single enzymatic step. Thus, it could be argued that there is a substantially larger thermodynamic barrier to be overcome to increase the fractional composition of fat and protein in milk as compared to lactose. Clearly some species have acquired the ability to accomplish this. The fat and protein content of mouse milk is 4-6 times that of cows.

RESULTS

α-lactalbumin and lactation performance in cattle

The discussion above suggests that certain characteristics of α-lactalbumin regulation may have played an important part in the selection for milk yield. Work with explant cultures of mammary tissue showed that in response to PRL the mammary tissue of beef cattle produced smaller amounts of α-lactalbumin than that of dairy cattle (McFadden *et al.*, 1989). Subsequently, work in my laboratory indicated that there were substantial differences between the PRL-responsiveness of mammary tissue of dairy cattle. In our case the cattle had been selected from two different breeding lines. This led us to further experiments using animals of known genetic merit for milk production[1]. The experiments are much more difficult in dairy cattle than in laboratory animals but we have been able to obtain statistically valid results by random selection of animals from a herd of known genetic merit in such a way as to obtain a wide dispersion in genetic capabilities. From population genetics experiments it is known that the first lactation performance of animals is a strong indicator of the genetic merit of the individual and accurate genetic predictions can be made with as little as 90 days of lactation data. We therefore collected biopsies of mammary tissue from heifers in the 6th month of their first pregnancy. Multiple core samples were taken and the cores then diced into small pieces, and randomly assigned to different explant culture treatments.

Typical results obtained from this type of experiment is seen in Figure 1. Clearly, the tissue responds to what would be considered physiological levels of PRL. Further experiments showed that PRL was required continually during the culture period. Explant cultures can only be maintained for a few days and their response increases to a maximal level and then declines over time. The maximal secretion of α-lactalbumin (and ß-lactoglobulin, not shown) of the tissue is reached between three and four days of culture. Thereafter, fibroblast outgrowths begin to predominate and the secretion of α-lactalbumin

---

[1]Genetic merit for milk production calculation based on the milk production on the male and female merits in the animal's pedigree as well as their own ability.

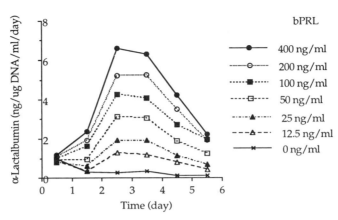

Figure 1. Least squares means of prolactin dependent α-lactalbumin secretion in a six day culture period from 12 heifers. The least squares means standard error is 0.2

(and ß-lactoglobulin) declines. We have compared the PRL-responsiveness of the tissue to the observed milk production phenotype of the animals during their subsequent lactation. Tissue from all animals was subjected to the same treatments: PRL dose response over 6 days. Statistically significant correlations between the response of the mammary tissue the animal's genetic merit and milk production phenotype were detected. The significance was not, however, in the maximal response of the tissue, but rather in what we interpret to be the ability of the tissue to sustain a response over time. Highest correlations between milk production phenotype and PRL-response were with the day 5 and day 6 α-lactalbumin (and ß-lactoglobulin) secretions. These results suggest either that there may be a different responsiveness to feedback inhibition or that some other changes have taken place in the tissue. These changes are somehow related to the milk production capacity of the mammary tissue.

α-lactalbumin gene control

Some of the properties of α-lactalbumin described above suggested to us that the regulatory regions of the gene would be likely to have properties making it useful for mammary-specific transgene expression. We therefore set out to isolate a clone from a bovine library that had a large amount of 5' regulatory sequence. In the course of this work we discovered a point mutation in the 5' untranslated region (UTR) of α-lactalbumin (Bleck and Bremel, 1993b, 1993c). Further analysis of animals of unrelated pedigrees in other breeds of cattle showed that this G > A mutation at +15 (relative to the transcription start site) was only present in Holsteins. Given the importance of this region in both transcription and translation we sought to determine whether this point mutation had an impact on the lactation performance of the cattle. We attempted to determine the effect in two ways. First, we examined milk production in relation to genotype in two different research herds of females. The genetic merit of the females was used as a method of ranking the animals. Animals were then selected in a stratified random sampling scheme to give equal numbers of animals spanning the genetic capabilities within the herds. In a second study, 398 bulls in young sire programs of US bull studs were genotyped for the α-lactalbumin allele and subsequently the production records of their daughters were used to estimate the effect of the point mutation. Bulls in the young sire program are used in a random manner within the dairy population at large. In this case, the lactation performance records of over 20,000 animals were used in the evaluation. Both methods of estimation of the effect showed positive affects on milk production in animals with an A at a +15 position of the 5' UTR (Table 1; Bleck and Bremel, 1993a). In analysis of the herds of females, it appeared that the A-allele exhibited partial dominance. Animals with a single copy of the AB[2]-genotype produced 80% of the amount of milk as their AA counterparts. The magnitude of the effect was smaller in the male population, however it should be pointed out that these males represent a very elite sub-sample of the population (~ 2 percentile) and so the effects are expected to be biased for that reason. In the males, the females and the half-sib males the trends were similar. A further interesting property was that milk composition also differed by genotype. Our working hypothesis is that the A-allele leads to an elevated level of α-lactalbumin secretion which, in turn, results in a higher lactose synthesis and a greater movement of water. If this hypothesis is correct then we might see a more dilute milk in the A-allele animals as compared to the B-allele. That is precisely what is observed.

Table 1. Association of α-lactalbumin genotype with genetic merit for milk production (Adapted from Bleck and Bremel, 1993a).

Genotype	All Males		Half Sib Males		Females	
	n	PDM (kg)	n	PDM (kg)	n	PDM (kg)
AA	27	628	8	638	7	453
AB	144	557	34	577	27	317
BB	116	501	21	479	36	81
AA - BB		127		160		372
AB - BB		56		98		236

Unfortunately, there is limited access to germplasm for retrospective analysis to determine the original animal that had the mutation. Analysis of a group of bulls that were widely used in the 1960s indicated that two of the popular bulls were heterozygotes. Moreover, from our work we infer that the gene frequency of the A-allele appears to be increasing. In the two herds of females, the frequency of the A-allele was approximately 0.3 with indication that it was in Hardy-Weinberg equilibrium. For the young sire population, the elite of the Holstein breed, the frequency was higher. The female population would be expected to lag behind the male population because selection is being carried out in the male population. The inference we draw from this is that the statistical selection schemes used in the dairy cattle industry have been capable of detecting the presence of this point mutation, and because of its strong effect on milk production phenotype the frequency in the population at large is increasing. Given the predominance of the Holstein breed that has developed over the period of use of artificial insemination it

---

[2]The nomenclature adopted for this genotype conforms to conventions. A is an adenosine at the +15 location and is NOT A. The wild type in cattle is a G, however the PCR-based genotyping we developed is only capable of detecting whether there is an A at that location or not.

seems plausible that this mutation may have partly been responsible for the emergence of the Holstein as the dominant breed in North America.

Molecular control of the α-lactalbumin gene

What is unique about the α-lactalbumin gene and how is it different from other milk protein genes which show substantially different patterns of expression? How does a point mutation in the 5' UTR cause such a large effect on milk production phenotype? Examination of the α-lactalbumin sequence suggested, however, that the point mutation might lie within a binding domain of the (relatively common) nuclear factor, NF1. Appropriate oligonucleotides were therefore constructed to test this hypothesis (Gorodetsky, Ahrens and Bremel, unpublished). Further, a comparison of this region in a number of species (Table 2) shows that two ends of the NF1 domain are quite highly conserved. It is also clear that nature has provided many other sequences to test. Interestingly, human α-lactalbumin, which is found at a concentration in human milk 3-4 times that of cows milk has a deletion in the region, resulting in the proximal and distal NF1 binding domains being closer together. Whether this is the reason for the difference remains to be determined. Experiments to this end are in progress.

Table 2. Variation in the sequences of α-lactalbumin from different animals in the region around the +15 mutation in bovine α-lactalbumin (from GenBank).

..**TTG.........CAA....**	NFI consensus
..**TTG**aggggt**AACCAA**aatg	A Variant
..**TTG**ggggt**AACCAA**aatg	B Variant
atttcagaatc**TTG**ggggt**AACCAA**aatg	bovine mRNA-1 France
t**AACCAA**aatg	bovine mRNA-2, WI, IL
..**TTG**gggagt**AACCAA**aatg	Brahman variant
gatgcttccatttcaggttc**TTG**ggggtagc**CAA**aatg	Human
cccatttcaggatct**TTG**ggggt**AACCAA**aatg	*Capra Hircus* two start codons
ttccaggatc**TTG**ggggt**AACCAA**aatg	*Ovis Aries*
ttcaggttc**ACA**gcagcagc**CAA**aatg	Guinea Pig two start codons
cagttccatgttccagaccc**CTG**aagcagc**CAG**gatg	*Macropus eugenii*
gggatccacattcaaggt**CTG**ggagcagg**CAA**aatg	*Rattus norvegicus*

Use of transgenic animals to study α-lactalbumin regulation

As noted above we have developed various DNA constructs containing the α-lactalbumin regulatory sequences. In the course of these studies we noted wide variations in bovine α-lactalbumin expression between full sisters within transgenic line. Variation in expression among transgenic lines is common and is generally attributed to insertion site. However, since these variations were occurring within line they could not be attributed to insertion site differences. We have, therefore, undertaken a breeding experiment to evaluate this observation further. We have used classical animal breeding methodologies in an attempt to determine whether it is possible to establish different lines with different mean levels of the bovine transgene expression. Methodology has been developed which makes it possible to measure simultaneously both the expression of the bovine transgene and the mouse's endogenous α-lactalbumin. Further work makes it possible to examine the levels of the different milk protein gene mRNAs from biopsy material as well.

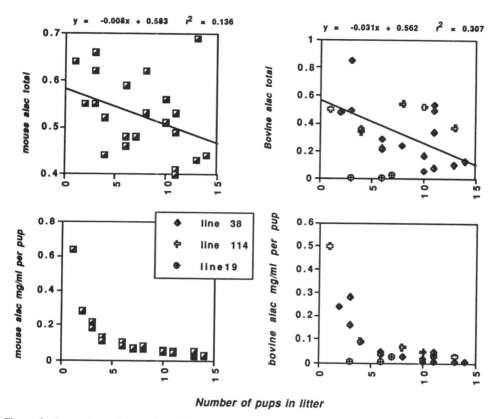

Figure 2. Comparison of expression of the endogenous mouse α-lactalbumin with the bovine α-lactalbumin transgene in relation to litter size in 3 different founder lines.

Preliminary results have been obtained from twenty litters of three different transgenic founder lines. The first thing that was noted was that levels of both the mouse α-lactalbumin and the bovine transgene were varying considerably. A simple comparison of the two levels of expression while positive and statistically significant showed only a very weak association. Earlier observations with several individual mice in which transgene expression had been followed through multiple lactations suggested that when the mouse had bigger litters the level of transgene expression was greater. Therefore, the transgene and endogenous gene were normalized to litter size. The results are shown in Figures 2 and 3. The effect of normalizing the data in this way is striking and shows that both the transgene and the mouse's endogenous gene are responding to the same physiological signals which are in some way associated with litter size. Whether this effect is due to nursing intensity

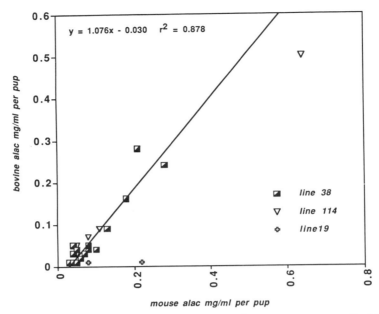

Figure 3. Relative expression of the endogenous gene with the bovine transgene normalized for litter size.

or endocrine effects during pregnancy is currently under investigation. A second feature of this data is worth noting. The shape of the curve relating transgene to endogenous gene expression is effectively an inverse saturation curve. At low litter sizes the incremental effect of one more pup on expression is greater than at high litter sizes. In fact, at high litter sizes the expression remains relatively constant.

CONCLUSION

I have attempted to demonstrate some of the varied mechanisms of control of the α-lactalbumin gene. Some of the data suggest that this gene is tightly regulated and responding to a variety of endocrine and other intra- and inter-cellular signalling. Although α-lactalbumin has been studied for some time, some of the surprising results with the transgene expression suggest that further understanding of the regulation of this gene will provide important insights in the process of lactation in mammals.

ACKNOWLEDGMENTS

I must acknowledge the work of a number of individuals who have worked in my laboratory on different phases of this project. John Byatt who introduced α-lactalbumin measurements to my laboratory while attempting to develop bioassays for bovine placental lactogen, Juan Vega who noted the first differences in α-lactalbumin expression while doing his Ph.D. project, Frank Mao whose Ph.D. project developed biopsy techniques for measuring α-lactalbumin expression in explant culture, Greg Bleck whose Ph.D. project was to develop the α-lactalbumin transgenic model and who discovered the mutation and its association with production, Peta Madgwick whose Ph.D. project is to utilize animal breeding methodology to study transgene expression, Helga Ahrens and Stanislav Gorodetsky who are collaborating on the nuclear factor project. Dr. Gorodetsky is a Visiting Professor from the Russian Academy of Science and has brought to the laboratory a seemingly endless bag of molecular tricks, and introduced nuclear factor analysis to the laboratory. The work reported here has been supported by grants from the Wisconsin Milk Marketing Board, USDA-Hatch #5124, and USDA-NCGRI #92-02445.

REFERENCES

Akers, R.M., Bauman, D.E., Capuco, A.V., Goodman, G.T. and Tucker, H.A., 1981, Prolactin regulation of milk secretion and biochemical differentiation of mammary epithelial cells in periparturient cows, *Endocrinology* 109:23.

Bleck, G.T. and Bremel, R.D., 1993a, Correlation of the α-lactalbumin ($+15$) polymorphism to milk production and milk composition in Holstein cattle, *J. Dairy Sci.* 76:2292.

Bleck, G.T. and Bremel, R.D., 1993b, Nucleotide sequence of the *bos indicus* α-lactalbumin 5' flanking region: comparison with the *bos taurus* sequence, *Animal Biotech.* 4(1):109.

Bleck, G.T. and Bremel, R.D., 1993c, Sequence and single-base polymorphisms of the bovine α-lactalbumin 5' flanking region, *Gene* 126:213.

Goodman, R.E. and Schanbacher, F.L., 1991, Bovine lactoferrin mRNA: sequence, analysis and expression in the mammary gland, *Biochem. Biophys. Res. Comm.* 180:175.

Johke, T. and Hodate, K., 1978, Effects of CB154 on serum hormone level and lactogenesis in dairy cows, *Endocrinol. Japon.* 25:67.

Mao, F.C. and Bremel, R.D., 1990, Development of sensitive α-lactalbumin, ß-lactoglobulin, and α_{S1} casein ELISAs for mammary explant culture study, *J. Dairy Sci.* 73(Suppl. 1):241.

Mao, F.C., Bremel, R.D. and Dentine, M.R., 1991, Serum concentrations of the milk proteins α-lactalbumin and ß-lactoglobulin in pregnancy and lactation: correlations with milk and fat yields in dairy cattle, *J. Dairy Sci.* 74: 2952.

McFadden, T.B., Akers, R.M. and Beal, E.E., 1989, Influence of breed and hormones on production of milk proteins by mammary explants from prepubertal heifers, *J. Dairy Sci.* 72:1754.

Peaker, M., 1977, Acqueous phase of milk: ion and water transport. *in* "Comparative Aspects of Lactation", Peaker, M., ed., Symposia of the Zoological Society of London, number 41, Academic Press, London.

Tucker, H.A., 1987, Quantitative estimates of mammary growth during various physiological states: a review, *J Dairy Sci.* 70:1958.

Tucker, H.A., 1988, Lactation and its hormonal control, *in*: "The Physiology of Reproduction", Knovil, E. and Neill, J., eds., Raven Press, New York, 2235.

REGULATION OF MILK SECRETION AND COMPOSITION BY GROWTH HORMONE AND PROLACTIN

David J. Flint
Hannah Research Institute
Ayr, KA6 5HL, U.K.

INTRODUCTION

Prolactin, as its name suggests, has long been recognized as the major hormone concerned with both lactogenesis (the onset of copious milk secretion) and galactopoiesis (the maintenance of milk secretion). Such roles have been clearly demonstrated from mouse to man. In ruminants, however, growth hormone (GH) was considered to be much more important in the regulation of milk secretion since it could maintain milk secretion in the absence of prolactin in hypophysectomized lactating goats (Cowie et al., 1964) and stimulate milk production in intact goats (Knight et al., 1990) and cows (Bauman et al., 1985). More recently, however, we have demonstrated an important role for GH in maintaining milk production in the rat when circulating prolactin concentrations are low (Madon et al., 1986; Barber et al., 1992; Flint et al., 1992) and Knight and co-workers have shown that prolactin does play a significant role in maintaining milk production in the goat (Knight, 1993). It has also been shown that in women, where prolactin is thought to be the major lactogenic hormone, hGH can also increase milk production (Milsom et al., 1992). The initial aim of this review is to explore these overlapping roles of GH and prolactin in order to define more precisely their modes of action. In addition, it was hoped that the identification of individual roles for each hormone might offer insights into potential ways of manipulating milk composition.

Even though an important role for GH in maintaining milk yield in ruminants is well established, the precise manner in which GH achieves this is unclear. Initial proposals included homeorhetic mechanisms such as changes in energy partitioning, away from storage in adipose tissue, directed instead towards mammary gland utilization (Bauman and Currie, 1980) although such a mechanism could only contribute about 5-10% of total milk energy in rodents (Sadurskis et al., 1991). An alternative proposal has suggested that GH increases milk production indirectly via stimulation of insulin-like growth factor-1 (IGF-1) production principally from the liver, which in turn increases milk production (Prosser et al., 1990). This latter proposal stemmed from the observations that there are IGF-I receptors but not GH receptors on mammary epithelial cells (Dehoff *et al.*, 1988; Disenhaus *et al.*, 1988) and GH increases circulating concentrations of IGF-I. The second aim of this review was therefore to examine the evidence for direct versus indirect effects of GH and in particular to examine the role, if any, of IGFs in mediating the effect of GH on galactopoiesis.

GH AND PROLACTIN ARE IMPORTANT FOR MAINTENANCE OF LACTATION IN THE RAT

We have previously shown that although prolactin is the major hormone driving milk synthesis in the rat, GH also plays a role, the importance of which increases as lactation progresses (Flint *et al.*, 1992). Thus in the first 2 days of lactation bromocriptine, which suppresses prolactin secretion, inhibited lactation, whilst a specific antiserum to rat GH (anti-rGH) failed to influence milk synthesis either alone or in combination with bromocriptine. By contrast, on day 6 of lactation, both anti-rGH and bromocriptine were required to completely suppress milk synthesis, although once again anti-rGH was ineffective alone. By day 14 of lactation anti-rGH treatment alone led to a 20% reduction in milk yield and again it required both anti-rGH and bromocriptine to suppress milk production (Figure 1). If animals given anti-rGH and bromocriptine were given prolactin concurrently these effects were completely suppressed, whilst even high doses of recombinant bGH (free of prolactin contamination) only led to a partial recovery of milk secretion. Thus about half of the milk yield was maintained by GH alone whilst prolactin appeared to have an additional, specific role which elicited the additional milk synthesis required for full milk production. In order to address these differences we examined various parameters of mammary gland (and adipose tissue) function. By reducing milk yield using bromocriptine we were able to examine mammary gland biochemistry solely under the influence of GH.

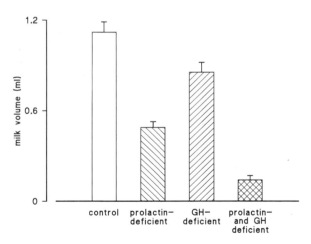

Figure 1. Milk yields in single mammary glands of lactating rats treated for 48 hours with bromocriptine (prolactin-deficient), anti-rGH serum (GH-deficient) or a combination of both.

Since lactose is the major osmolite in milk it usually determines milk water content and thus milk yield. Its synthesis is thought to be limited by glucose uptake into the cell which is mediated by a specific glucose transporter of the GLUT-1 isoform in rats (Madon *et al.*, 1990) and cows (Zhao *et al.*, 1993). Indeed changes in GLUT-1 concentration in the mammary gland supported this proposal since milk yield changes were broadly paralleled by changes in GLUT-1 in the plasma membrane (Madon *et al.*, 1990; Fawcett *et al.*, 1991). Similar correlations with milk yield were also found for amino acid transport and for concentrations of acetyl CoA carboxylase (thought to be rate-limiting for lipogenesis) in the mammary gland (Barber *et al.*, 1992). Thus regulatory steps for all of the major synthetic pathways of milk synthesis (lactose, protein and fat) appeared to be affected to approximately the same extent as milk yield. This suggested that, whilst GH regulates these 3 parameters, prolactin appears to be regulating each of them in a specific fashion which GH cannot. Are there specifically prolactin-responsive glucose and amino acid transporters and prolactin-specific responses in terms of acetyl CoA carboxylase synthesis?

GH PREFERENTIALLY MAINTAINS MILK FAT SYNTHESIS

Despite these similarities of action, two factors provided pointers towards a possible specific role for GH in mammary gland metabolism. The first was that in prolactin-deficient rats, despite the large decrease in milk yield, mammary gland lipoprotein lipase activity (LPL) was unaffected and secondly that although total acetyl CoA carboxylase activity was decreased the amount in the active state (since it exists in both active and inactive forms) was not (Barber *et al.*, 1992; Figure 2). This suggested that milk lipid

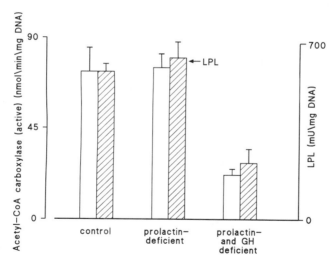

Figure 2. Changes in acetyl CoA carboxylase in the active state and lipoprotein lipase in the mammary glands of lactating rats treated for 48 hours with bromocriptine (prolactin-deficient) or anti-rGH serum (GH-deficient) alone or in combination.

secretion, which is derived approximately equally from *de novo* fatty acid synthesis and uptake of preformed dietary triacylglycerol (via the action of LPL), would be unaffected. For this reason we examined milk composition in prolactin-deficient rats, where milk secretion was maintained by GH, with milk from control rats. The results were consistent with the enzyme data, with milk volume decreased by 55% in the absence of prolactin but without any reduction in milk fat output, so that milk fat concentration was doubled from around 12% to 25% (Flint and Gardner, 1994; Figure 3). Interestingly, milk lactose output also decreased dramatically (around 80%) so that milk lactose concentration, which is normally relatively stable, also decreased significantly. This suggested that other factors must make a major contribution to maintenance of milk osmolarity under these conditions.

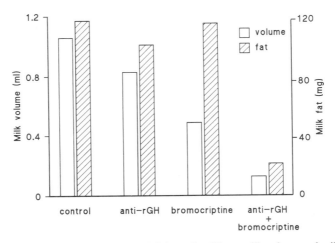

Figure 3. Effects of GH- and prolactin-deficiency for 48h on milk volume and milk fat yield.

Thus GH appeared to be responsible for maintaining a milk with a high energy and low water content and as such it might be an important regulator of milk secretion in periods of water shortage. But how does GH maintain these aspects of milk triacylglycerol metabolism? One possibility is that it is an indirect effect mediated via insulin. Prolactin-deficient lactating rats do have increased serum insulin concentrations (Agius et al., 1979; Flint, 1982; Barber et al., 1992) and insulin is known to increase both the uptake of preformed triacylglycerol (Da Costa and Williamson, 1993) and de novo fatty acid synthesis in the mammary gland (Robinson et al., 1978; Munday and Hardie, 1986). Thus in the face of decreased glucose entry into the cell (due to decreased GLUT-1 transporters) insulin may maximise glucose utilization for lipogenesis at the expense of glucose utilization for lactose synthesis. Despite the distinct possibility that insulin maintains milk fat output during prolactin-deficient states, it is not responsible for maintenance of glucose transporters or the residual lactose and protein secretion, since if prolactin-deficient rats are also rendered GH deficient using anti-rGH, GLUT-1 transporters on the plasma membrane of the mammary gland fall to 10% of control values and milk yield, lactose, protein and fat outputs are similarly decreased, despite the fact that serum insulin concentrations are significantly elevated (Flint and Gardner, 1994). The most likely explanation of these results is that GH maintains GLUT-1 concentrations and thereby glucose transport at high levels in order to permit sufficient lactose synthesis and thereby water secretion to maintain continued secretion of triacylglycerol synthesized under the control of insulin. In support of this proposal, mice which have the α-lactalbumin gene deleted and thus cannot synthesize significant amounts of lactose, fail to lactate adequately and possess a highly viscous milk (Stacey and Schnieke, 1995).

This role of GH in maintaining milk triacylglycerol secretion may also be present in ruminants since, if goats are made prolactin-deficient with bromocriptine, milk yield also falls, whilst milk fat percentage increases, so that total milk fat output is maintained (Knight, unpublished observations). Thus the differences in different species regarding the importance of prolactin and GH may be relative, rather than absolute. Further support for this proposal comes from studies in both lactating beef cows and pigs. Armstrong *et al.* (1994) have utilized the approach of active immunization against growth-hormone release factor (GRF), which leads to inhibition of GRF action and thus to large decreases in circulating concentrations of both GH and IGF-I (which is GH-dependent). Despite the fact that GH is able to increase milk yield in cattle, reducing serum GH in this manner had a remarkably small (though significant) effect on milk production. It is tempting to speculate that under these circumstances of GH deficiency, prolactin may assume a more important role, in analogous fashion to GH in prolactin-deficient rats. Support for the fact that GRF immunoneutralization was effective came from similar studies in growing animals in which puberty was significantly delayed in cattle (GH is thought to play a role in sensitizing the ovary to gonadotrophins) (Armstrong *et al.*, 1994).

PROLACTIN MAINTAINS CELLULAR INTEGRITY AND SURVIVAL

Prolactin is clearly a more effective lactogenic hormone than GH in the rat, so what is its precise mechanism of action? It appears that prolactin plays a major role in maintaining cellular functions by effects on cell survival. For example in prolactin-deficient rats, tight junctions open, total DNA content decreases (Flint and Gardner, 1994; Figure 4) and apoptosis (as judged by the occurrence of DNA "ladders") increases (Quarrie, Flint and Wilde, unpublished observations). By contrast GH-deficiency had no effect on these parameters. Prolactin has also been demonstrated to inhibit apoptotic cell death in mouse mammary gland (Sheffield and Kotolski, 1992) and prevented the occurrence of DNA ladders in our studies, whereas IGF-I was ineffective. In apparent contrast, IGF-I has been shown to inhibit secretion of plasminogen activator which plays a role in the tissue remodelling which occurs during involution of the mammary gland (Turner and Huynh, 1991).

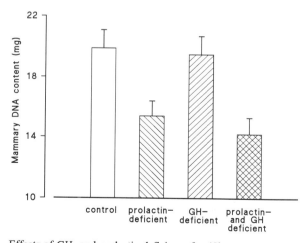

Figure 4. Effects of GH- and prolactin-deficiency for 48h upon mammary DNA content.

The fact that GH appears to play a qualitatively different role from prolactin argues against GH acting as a partial agonist at the prolactin receptor. Two other factors also mitigate against such a proposal; firstly, the affinity of GH for the prolactin receptor is several orders of magnitude lower than prolactin so that extremely high circulating concentrations of GH would have to be attained to achieve effects through the prolactin receptor and secondly, the more direct observation that high doses of recombinant bGH are unable to increase milk yield in bromocriptine-treated, prolactin-deficient rats. The question then becomes, how does GH increase milk secretion in the apparent absence of GH receptors on the mammary secretory epithelium?

DO THE INSULIN-LIKE GROWTH FACTORS (IGFS) MEDIATE THE ACTIONS OF GH ON MAMMARY GLAND MILK SECRETION?

The fact that GH increases serum IGF-I concentrations in dairy cows at the same time as increasing milk production has led to the proposal that IGF-I is responsible for mediating GH action (Breier et al., 1991). Indeed there are IGF-I receptors on the mammary gland (Dehoff et al., 1988; Disenhaus et al., 1988; Lavandero et al., 1990) and IGF-I is mitogenic in vitro for mammary cells (Furlanetto and Di Carlo, 1983; McGrath et al., 1991; Peri et al., 1992). The only evidence for an effect of IGF-I on milk production, however, is in goats given close-arterial infusions into the mammary gland (Prosser et al., 1990) and even these authors have subsequently concluded that IGF-I does not mediate the effects of GH, since if goats are milked at frequent intervals the response to IGF-I is abolished (Prosser and Davis, 1992) whereas that to GH is not (Davis et al., 1991).

It has not proved possible to study the IGF system in detail in large animals because many of the reagents are not available in the quantities required. We therefore examined the effects of IGF-I and IGF-II in the lactating rat model described earlier in this article. By treating lactating rats with anti-rGH and bromocriptine for 48h milk production was reduced by 90%. If these animals were given recombinant bGH concurrently, milk yield was maintained at almost 50% of control values (Madon et al., 1986). If animals were first treated with anti-rGH and bromocriptine for 48h to suppress milk production and then subsequently given prolactin or GH to reinitiate it, prolactin acted rapidly (4-6h) whereas GH required 8-12h suggesting an indirect effect (Flint et al., 1992; Figure 5). Indeed the time course of GH action on milk secretion matched the increase in serum IGF-I concentration which occurred between 6 and 12h after injection. However, IGF-I treatment in this same model failed to influence milk secretion (Flint et al., 1992). Further studies revealed that in lactating rats which were both GH and prolactin-deficient not only serum IGF-I, but also IGF-II and the major serum IGF binding protein-3 (IGFBP-3) were all decreased but were maintained if GH replacement therapy was given. We therefore examined the effects of IGF-I and IGF-II administered precomplexed to IGFBP-3 on milk production but once again the complex was ineffective (Flint et al., 1994). Recent studies revealed that, during involution, the mammary gland contains large amounts of an IGFBP which has potentially inhibitory actions on IGF-I action (Tonner et al., 1995). It was thus conceivable that exogenous IGF action was blocked by this IGFBP. We therefore took advantage of the availability of three IGF analogues, R_3 IGF-I, Long IGF-I and Long R_3 IGF-I which bind to IGF-I receptors but not to IGFBPs and should thus circumvent any inhibition by IGFBP (Francis et al., 1992). None, however, were able to stimulate milk production (Flint et al., 1994). GH was also effective in animals given antiserum to IGF-I providing further evidence that IGF-I was not mediating the effects of GH (D.J. Flint, unpublished observations), and even high doses of IGF-I administered directly into the mammary gland were ineffective in stimulating milk production (D.J. Flint, unpublished observations).

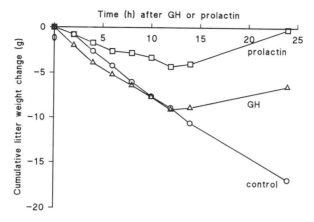

Figure 5. Time-course of milk production response to treatment with prolactin, GH or saline (control) in lactating rats whose milk secretion had been inhibited by bromocriptine and anti-rGH treatment for 2d previously.

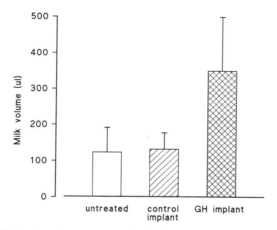

Figure 6. Effect of GH implants into mammary glands of lactating rats pre-treated for 48h with bromocriptine and anti-rGH to inhibit milk synthesis. Contralateral glands received control implants whilst some rats received no additional treatment (untreated).

GH ACTS DIRECTLY WITHIN THE MAMMARY GLAND

We finally decided to administer GH directly into individual mammary glands allowing the contralateral gland to serve as a control. Under such circumstances GH stimulated milk production in the treated gland (but not the control gland) to the same extent as systemic GH (Flint and Gardner, 1994; Figure 6). Thus GH action appeared to be occurring within the mammary gland. Recent studies have demonstrated that GH also stimulates mammary growth after being implanted into individual mammary glands and that there is a coincident increase in IGF-I mRNA concentrations in the gland (Kleinberg *et al.*, 1990).

Although no evidence exists for GH receptors on the secretory epithelial cells, numerous studies have described GH receptor mRNA within the gland (Hauser *et al.*, 1990; Lincoln *et al.*, 1990; Jammes *et al.*, 1991). In fact the gland contains numerous other cell types such as adipocytes and fibroblasts which possess GH receptors and the possibility that GH mediates its effects indirectly through cell types other than the secretory epithelial cells is currently under investigation.

CONCLUSIONS

Prolactin and GH appear to play complementary roles in maintaining synthesis and secretion of all the major components of milk in the rat. GH appears to maintain a high fat, energy-rich milk in the absence of prolactin, whereas prolactin appears to play an important role in maintaining cell integrity and survival and may be an important anti-apoptotic factor in this respect. The fact that GH can only maintain glucose transporters, amino acid transport and lipogenic enzymes at half the maximal rate (whereas prolactin can maintain them fully) implies mechanisms of action which involve separate pools of transporters, one pool which responds to GH and prolactin whilst the other responds only to prolactin. Alternatively the quantitative difference in prolactin and GH action could be explained in terms of prolactin's ability to maintain cellular integrity and cell survival. In the absence of prolactin a reduced level of cellular differentiation may limit the maximal metabolic responses attainable (see Streuli, 1993; Howlett and Bissell, 1993) and thereby limit the effectiveness of GH. Finally GH appears to stimulate milk production directly within the mammary gland but the precise target cell remains to be determined.

ACKNOWLEDGEMENTS

I am deeply grateful to numerous colleagues at the Hannah Research Institute for their expert skills and for the use of unpublished observations, in particular Dick Vernon, Mike Barber, Jim Beattie, Roger Clegg, Hilary Fawcett, Margaret Gardner, Chris Knight, Rus Madon, Lynda Quarrie, Dave Shennan, Elizabeth Tonner, Maureen Travers, Colin Wilde, and to Steve Baldwin (Leeds), John Ballard (Adelaide), Bob Collier (Monsanto), Steve Hodgkinson (Ruakura, NZ), and Chris Maack (Celltrix) for invaluable reagents. My thanks to Mrs M Knight for preparation of the manuscript. Much of this work was funded by the Scottish Office Agriculture and Fisheries Department.

REFERENCES

Agius, L., Robinson, A.M, Girard, J.R., and Williamson, D.H., 1979, Alterations in the rate of lipogenesis *in vivo* in maternal liver and adipose tissue on premature weaning of lactating rats, *Biochem. J.* 180:689.

Armstrong, J.D., Harvey, R.W., Stanko, R.L., Cohick, W.S., Simpson, R.B., Moore, K.L, Schoppee, P.D., Clemmons, D.R., Whitacre, M.D., Britt, H.J., Lucy, M.C., Heimer, E.P., and Campbell, R.M., 1994, Active immunisation against growth hormone releasing factor: Effects on growth, metabolism and reproduction in cattle and swine, *in:* "Vaccines in Agriculture: Immunological applications to Animal Health and Production," Wood, P., Willadsen, P., Vercoe, J., Hoskinson, R. and Demeyer, D. eds., CSIRO Press (in press).

Barber, M.C., Clegg, R.A., Finley, E., Vernon, R.G., and Flint, D.J., 1992, The role of growth hormone, prolactin and insulin-like growth factors in the regulation of rat mammary gland and adipose tissue metabolism during lactation, *J. Endocrinol.* 135:195.

Bauman, D.E., and Currie, W.B., 1980, Partition of nutrients during pregnancy and lactation: a review of mechanisms involving homeostatis and homeorhesis, *J. Dairy Sci.* 63:1514.

Bauman, D.E., Eppard, P.J., DeGeeter, M.J., and Lanza, G.M., 1985, Responses of high-producing dairy cows to long-term treatment with pituitary somatotropin and recombinant somatotropin, *J. Dairy Sci.* 68:1352.

Breier, B.H., McCutcheon, S.N., Davis, S.R., and Gluckman, P.D., 1991, Physiological responses to somatotropin in the ruminant, *J. Dairy Sci.* 74(Suppl 2):20.

Cowie, A.T., Knaggs, G.S., and Tindal, J.S., 1964, Complete restoration of lactation in the goat after hypophysectomy, *J. Endocrinol.* 28:267.

DaCosta, T.H.M., and Williamson, D.H., 1993, Effects of exogenous insulin or vanadate on disposal of dietary triacylglycerols between mammary gland and adipose tissue in the lactating rat: insulin resistance in white adipose tissue, *Biochem. J.* 290:557.

Davis, S.R., Hodgkinson, S.C., and Farr, V.C., 1991, The time-course of milk yield and hormonal responses following growth hormone injections in hourly-milked goats, *Proc. Soc. Anim. Prod.* 51:239.

Dehoff, M.H., Elgin, R.G., Collier, R.J., and Clemmons, D.R., 1988, Both type I and type II insulin-like growth factor receptors increase during lactogenesis in bovine mammary tissue, *Endocrinology* 122:33.

Disenhaus, C., Belair, L., and Djiane, J., 1988, Characterization et évolution physiologique des récepteurs pour les "insulin-like growth factors" I et II dans la gland mammaire de la brebis, *Reprod. Nutr. Develop.* 28:241.

Fawcett, H.A.C., Baldwin, S.A., and Flint, D.J., 1991, Hormonal regulation of the glucose transporter in the lactating rat mammary gland, *Biochem. Soc. Trans.* 20:17S.

Flint, D.J., 1982, Regulation of insulin receptors by prolactin in lactating rat mammary gland, *J. Endocrinol.* 93:279.

Flint, D.J., Tonner, E., Beattie, J., and Panton, D., 1992, Investigation of the mechanism of action of growth hormone in stimulating lactation in the rat, *J. Endocrinol.* 134:377.

Flint, D.J., Tonner, E., Beattie, J., and Gardner, M., 1994, Several insulin-like growth factor-I analogues and complexes of insulin-like growth factors-I and -II with insulin-like growth factor-binding protein-3 fail to mimic the effect of growth hormone upon lactation in the rat, *J. Endocrinol.* 140:211.

Flint, D.J., and Gardner, M.J., 1994, Evidence that growth hormone stimulates milk synthesis by direct action on the mammary gland and that prolactin exerts effects on milk secretion by maintenance of mammary DNA content and tight junction status, *Endocrinology* (in press).

Francis, G.L., Ross, M., Ballard, F.J., Milner, S.J., Senn, C., McNeil, K.A., Wallace, J.C., King, R., and Wells, J.R.E., 1992, Novel recombinant fusion protein analogues of insulin-like growth factor (IGF-I) indicate the relative importance of IGF-binding protein and receptor binding for enhanced biological potency, *J. Mol. Endocrinol.* 8:213.

Furlanetto, R.W., and Di Carlo, J.N., 1984, Somatomedin-C receptors and growth effects in human breast cells maintained in long-term culture, *Cancer Res.* 44:2122.

Hauser, S.D., McGrath, M.F., Collier, R.J., and Krivi, G.G., 1990, Cloning and *in vivo* expression of bovine growth hormone receptor mRNA, *Mol. Cell. Endocrinol.* 72:187.

Howlett, A.R., and Bissell, M.J., 1993, The influence of tissue microenvironment (stroma and extracellular matrix) on the development and function of mammary epithelium, *Epith. Cell Biol.* 2:79.

Jammes, H., Gaye, P., Belair, L., and Djiane, J., 1991, Identification and characterization of growth hormone receptor mRNA in the mammary gland, *Mol. Cell. Endocrinol.* 75:27.

Kleinberg, D.L., Ruan, W., Catanese, V., Newman, C.B., and Feldman, M., 1990, Non-lactogenic effects of growth hormone on growth and insulin-like growth factor-I messenger ribonucleic acid of rat mammary gland, *Endocrinology* 126: 3274.

Knight, C.H., Fowler, P.A., and Wilde, C.J., 1990, Galactopoietic and mammogenic effects of long-term treatment with bovine growth hormone and thrice-daily milking in goats, *J. Endocrinol.* 127:129.

Knight, C.H., 1993, Prolactin revisited, *in:* Hannah Research Institute Yearbook 1993, pp 72-80. E. Taylor, ed., Thomson, Glasgow.

Lavandero, S., Santibanez, J.F., Ocaranza, M.P., and Sapag-Hagar, M., 1990, Insulin-like growth factor-

I receptor levels during the lactogenic cycle in rat mammary gland, *Biochem. Soc. Trans.* 18:576.

Lincoln, D.T., Waters, M.J., Breipol, W., Sinowatz, F., and Lobie, P.E., 1990, Growth hormone receptor expression in the proliferating rat mammary gland, *Acta Histochemica* S40:47.

Madon, R.J., Ensor, D.M., Knight, C.H., and Flint, D.J., 1986, Effects of an antiserum to rat growth hormone on lactation in the rat, *J. Endocrinol.* 111:117.

Madon, R.J., Martin, S., Davies, A., Fawcett, H.A.C., Flint, D.J., and Baldwin, S.A., 1990, Identification and characterization of glucose transport proteins in plasma membrane- and Golgi vesicle-enriched fractions prepared from lactating rat mammary gland, *Biochem. J.*, 272:99.

McGrath, M.F., Collier, R.J., Clemmons, D.R., Busby, W.H., Sweeny, C.A., and Krivi, G.G., 1991, The direct *in vitro* effect of insulin-like growth factor (IGFs) on normal bovine mammary cell proliferation and production of IGF binding proteins, *Endocrinology* 129:671.

Milsom, S.R., Breier, B.H., Gallaher, B.W., Cox, V.A., Gunn, A.J., and Gluckman, P.D., 1992, Growth hormone stimulates galactopoiesis in healthy lactating women, *Acta Endocrinologica* 127:337.

Munday, M.R., and Hardie, D.G., 1986, The role of acetyl-CoA carboxylase phosphorylation in the control of mammary gland fatty acid synthesis during the starvation and re-feeding of lactating rats, *Biochem. J.* 237:85.

Peri, I., Shamay, A., McGrath, M.F., Collier, R.J., and Gertler, A., 1992, Comparative mitogenic and galactopoietic effects of IGF-I, IGF-II and Des-3-IGF-I in bovine mammary gland *in vitro*, *Cell Biol. Internat. Rep.* 16:359.

Prosser, C.G., and Davis, S.R., 1992, Milking frequency alters the milk yield and mammary blood flow response to intramammary infusion of insulin-like growth factor-I in the goat, *J. Endocrinol.* 135:311.

Prosser, C.G., Fleet, I.R., Corps, A.N., Froesch, E.R., and Heap, R.B., 1990, Increase in milk secretion and mammary blood flow by intra-arterial infusion of insulin-like growth factor-I into the mammary gland of the goat, *J. Endocrinol.* 126:437.

Robinson, A.M., Girard, J.R., and Williamson, D.H., 1978, Evidence for a role of insulin in the regulation of lipogenesis in lactating rat mammary gland, *Biochem. J.* 176:343.

Sadurskis, A., Sohlström, A., Kabir, N., and Forsum, E., 1991, Energy restriction and the partitioning of energy among the costs of reproduction in rats in relation to growth of the progeny, *J. Nutr.* 121:1798.

Sheffield, I.G., and Kotolski, L-C., 1992, Prolactin inhibits programmed cell death during mammary gland involution, *FASEB J.* 6:A1184.

Stacey, A., and Schnieke, A., 1995, Murine α-lactalbumin gene inactivation and replacement, *in*: "Intercellular Signalling in the Mammary Gland," Wilde, C.J., Peaker, M. and Knight, C.H. eds., Plenum Publishing Company, New York.

Streuli, C.H., 1993, Extracellular matrix and gene expression in mammary epithelium, *Sem. Cell. Biol.* 4:307.

Tonner, E., Beattie, J., and Flint, D.J., 1995, Production of an insulin-like growth factor binding protein by the involuting rat mammary gland, *in*: "Intercellular Signalling in the Mammary Gland," Wilde, C.J., Peaker, M. and Knight, C.H. eds., Plenum Publishing Company, New York.

Turner, J.D., and Huynh, H.T., 1991, Role of tissue remodelling in mammary epithelial cell proliferation and morphogenesis, *J. Dairy Sci.* 74:2801.

Zhao, F.-Q., Glimm, D.R., and Kennelly, J.J., 1993, Distribution of mammalian facilitative glucose transporter messenger RNA in bovine tissues, *Int. J. Biochem.* 25:1897.

BASEMENT MEMBRANE IN THE CONTROL OF MAMMARY GLAND FUNCTION

Charles H. Streuli

School of Biological Sciences
University of Manchester
3.239 Stopford Building
Oxford Road, Manchester, M13 9PT, UK

INTRODUCTION

Milk is produced in the lactating mammary gland by the alveolar epithelium. Alveoli are composed of a single layer of secretory polarized cells that surround a lumen and interact both with one another and with their basement membrane. Caseins and other milk proteins are synthesized and secreted in response to lactogenic hormones. However, milk production also requires cell-cell and cell-extracellular matrix associations. In this article, I discuss how the use of culture models has unravelled different facets of the molecular control of mammary gland differentiation. I show that milk production is dependent on a specific component of the basement membrane and is under coordinate regulation by the matrix as well as hormones.

RESULTS AND DISCUSSION

Requirement of basement membrane and hormones for mammary gland differentiation

One way to understand how the local environment of a cell regulates its function is to remove cells from that environment, ask how their behaviour is altered, and then use the principles learnt to reconstruct the tissue in culture. Studies dating back many years illustrate that such manipulations dramatically affect the phenotype of mouse mammary epithelial cells. Cultured on plastic dishes, functional epithelial cells from pregnant or lactating mammary glands no longer synthesize milk proteins such as caseins, even under hormonal conditions optimal for lactation. Although such cells interact closely with each

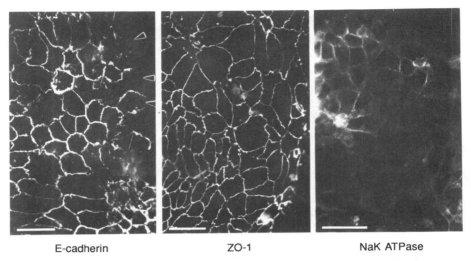

E-cadherin   ZO-1   NaK ATPase

Figure 1. Cell-cell and cell-ECM interactions in monolayer culture of primary epithelial cells. Cell-cell interactions are shown by immunofluorescence localization of molecules that segregate to intercellular junctions (E-cadherin) and basal-lateral surfaces (Na$^+$,K$^+$-ATPase). The tight junction antigen, ZO-1, is exclusively localized to the apical surface of these monolayers, indicating that the cells are polarized. Bars = 30 μm.

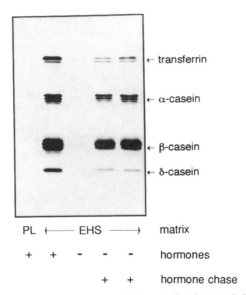

	← transferrin
	← α-casein
	← β-casein
	← δ-casein

PL ⟵——— EHS ———⟶   matrix

+   +   -   -   -   hormones

+   +   hormone chase

Figure 2. Milk protein synthesis is dependent on ECM as well as lactogenic hormones. First passage mammary cells were plated on plastic dishes or on EHS matrix and cultured for 4.5 days with hydrocortisone and insulin, and in some cases prolactin, before being pulse-labelled for 1 hour with [^{35}S]-methionine. Newly-synthesized milk proteins were immunoprecipitated with a polyclonal anti-mouse milk antibody, separated by PAGE, and the gels fluorographed. Note that both the basement membrane matrix and prolactin are required for milk production. In the EHS cultures lacking prolactin, extensive milk synthesis is induced within 12 hours of adding back this hormone.

other and are to some extent polarized (Figure 1), they do not have a columnar morphology and are not in contact with extracellular matrix (ECM) molecules such as laminin (Streuli and Bissell, 1990; Streuli, 1993).

In contrast with this phenotype on tissue culture plastic, cells cultured on laminin-rich basement membrane extracts, such as those isolated from the EHS (Engelbreth-Holm-Swarm) tumour, recreate their alveolar structure and synthetic phenotype (Li *et al.*, 1987; Barcellos-Hoff *et al.*, 1989; Aggeler *et al.*, 1991). Even isolated single cells can aggregate, undergo multicellular morphogenesis, and form milk-producing alveoli, thus reconstructing essential elements of functional mammary tissue in the culture dish (Streuli *et al.*, 1991). Differentiation is dependent on the presence of both prolactin and basement membrane (Figure 2), illustrating that the establishment of a lactational phenotype is complex and depends on coordinate signals both from soluble factors and from solid phase cues in the local microenvironment (Stoker *et al.*, 1990).

Studies such as these indicate that basement membrane has a prominent role in maintaining tissue-specific gene expression in mammary epithelium, certainly for those cells that are already committed to become lactational (Hay, 1993). This correlates with the situation *in vivo* since *in situ* hybridization of tissue sections reveals that only alveolar epithelium, and not ductal epithelium, is competent to transcribe β-casein, and it is only the alveolar epithelial cells, not luminal cells of the ducts, that interact directly with basement membrane.

Further evidence for the role of basement membrane comes from culturing cells on thick gels of collagen I (Emerman *et al.*, 1977; Lee *et al.*, 1984; Streuli and Bissell, 1990). On anchored gels the cells behave just as they do on plastic dishes, forming squamous monolayers and loosing the capacity for milk production. However, after floatation into the culture medium, the cells contract the gel, assume a columnar morphology, and deposit an endogenously-synthesised basement membrane. Under these conditions, the cells initiate milk synthesis (Streuli and Bissell, 1990). A number of mouse mammary cell strains can also express milk proteins in culture, but again this occurs only in conjunction with a laminin-containing basement membrane (Reichmann *et al.*, 1989; Schmidhauser *et al.*, 1990; Chammas *et al.*, 1994).

Direct signalling of casein expression by basement membrane

Three features of these cells are thus important for defining a lactational phenotype, in addition to the requirement for hormones such as prolactin. Cell-cell contact, cell-basement interactions, and the consequent establishment of columnar polarity, all correlate with the full expression of differentiation (Figure 3). But are all these microenvironmental regulators needed simultaneously? Or do different extracellular signals from adjacent cells and from basement membrane contribute individually to unique aspects of the differentiation programme?

To answer this question, we separated cells from one another and asked whether signals from basement membrane could induce differentiation in the absence of cell-cell contact and morphological polarity. Single cells and small cell clusters were isolated from non-functional casein-negative monolayer cultures, embedded within EHS matrix or collagen I gels, and cultured with lactogenic hormones (Streuli *et al.*, 1991). Many of the single cells remained viable for several days and did not divide. The cells were not morphologically polarized, and distributed cell surface markers evenly around their surfaces. Immunostained sections of these cultures revealed that single cells which were unable to contact each other could be induced to express β-casein, but only by the basement membrane matrix (Figure 4). Single cells embedded in collagen I were not triggered to synthesize β-casein, unless the gels were supplemented with small amounts of EHS matrix, when a differentiation threshold could be reached at 3% EHS (Table 1).

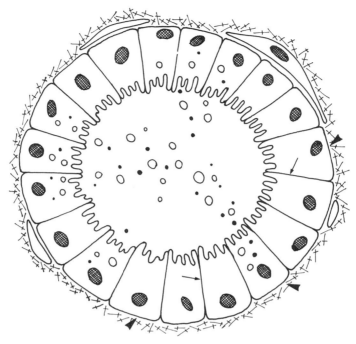

Figure 3. The structure of lactating alveolar epithelium. The spherical mammary gland alveolus, consisting of a single layer of polarized secretory cells, is shown in schematic cross section. Casein micelles and fat droplets are formed intracellularly and secreted apically into the lumen. Critical features of this structure required for full differentiation are interactions between epithelial cells and the basement membrane (arrowheads), and cell-cell contact (arrows).

Table 1. Effect of supplementing collagen I gels with basement membrane.

		percentage of ß-casein staining		
% EHS	% collagen	single cells	2 cell clusters	> 4 cell clusters
0	100	7	51	74
1	99	7	33	85
3	97	28	41	87
10	90	39	63	87
100	0	71	94	95

Mammary cells were cultured inside collagen I - EHS matrix mixtures for 6 days, then sectioned and stained as in Figure 4. The total number of single cells and small cell clusters were counted, and the percentage of β–casein positives were scored. In each case, 150 - 300 clusters were counted.

Thus, as long as basement membrane components are present, intercellular contact is not required for inducing tissue-specific gene expression, at least for the β-casein gene.

Support for this came from culturing EHS-suspended cell clusters in medium containing low calcium concentrations (Streuli *et al.*, 1991). Under these conditions, cell-cell contact was significantly impaired. E–cadherin was absent from intercellular junctions, and in cell clusters individual cells appeared morphologically rounded rather than compacted together. Yet these cells expressed abundant β-casein, providing further evidence that basement membrane alone can confer unique differentiation signals.

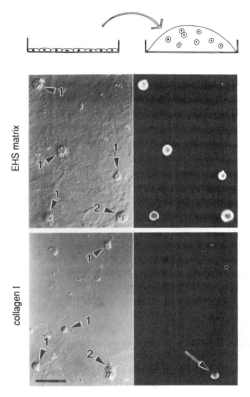

Figure 4. Differentiation of single mammary cells is triggered by basement membrane. Single mammary cells, and small cell clusters containing between 2 and 10 cells, were isolated from non-functional monolayers on plastic dishes and cultured inside EHS matrix or collagen I gels for 6 days with lactogenic hormones, as illustrated schematically in the upper panel. The gels were fixed with paraformaldehyde and then 25 μm sections were cut and stained by immuno-fluorescence for the presence of β-casein (lower panels). The number of cells in each cluster are indicated on the Nomarski images. Single cells in the basement membrane gel were induced to synthesize casein. When identical cells were seeded into collagen I gels, only cell clusters, but no single cells, expressed β-casein. Bar = 50 μm.

Cell-cell adhesion does, however, contribute significantly to differentiation in another context. Intercellular adhesion is likely to provide the necessary signals for establishing polarity, which itself is necessary for apical secretion of milk. For example, examination of our embedded cultures suggests that casein is retained intracellularly by single cells, but is distributed to discrete areas within cell clusters. In addition, it had already been shown that on top of collagen I gels, mammary cells only deposit endogenous basement membrane when they interact sufficiently with one another to produce cytostructural changes that lead to columnar polarization (Streuli and Bissell, 1990). From the evidence presented above, we now believe that it is this new basement membrane that provides signals for production of milk proteins. Indeed, the small cell clusters embedded within pure collagen I gels that stained positive for β-casein, were usually associated with some laminin at the cell periphery (Streuli *et al.*, 1991).

Basement membrane signals casein production through integrins

If basement membrane genuinely provides local environmental cues that control differentiation, one might expect ECM receptors on mammary cell surfaces to mediate the transduction of such signals. Integrins are transmembrane heterodimeric ECM receptors, and we have shown that $\alpha_2\beta_1$, $\alpha_3\beta_1$, $\alpha_5\beta_1$, $\alpha_6\beta_1$ and $\alpha_6\beta_4$ integrins are synthesized by alveolar epithelial cells (Delcommenne and Streuli, submitted). To ask whether any of these integrins were involved with basement membrane signalling, we interfered directly with cell-ECM interactions using a broad-spectrum anti-integrin antibody that blocks the function of β_1 integrins (Tomaselli *et al.*, 1988). When single cells were cultured inside EHS supplemented with this antibody, β-casein synthesis was reduced dramatically (Streuli *et al.*, 1991). Even the differentiation of large cell clusters was blocked (Figure 5). Thus, β_1 integrins transduce the basement membrane-derived signals that control casein expression in mammary epithelial cells.

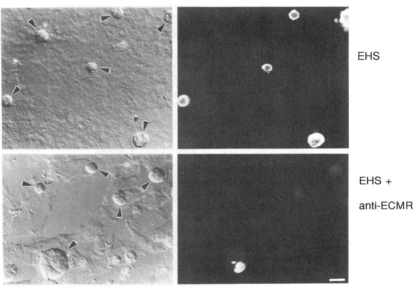

EHS

EHS +

anti-ECMR

Figure 5. Anti-integrin antibodies block β–casein expression. Mammary cells were cultured inside EHS matrix as in Figure 4. In the lower panel, a function-blocking anti-integrin antibody (anti-ECMR) was included in the gel throughout the culture period. Arrowheads in the Nomarski images denote single cells or cell clusters. β-casein expression occurred in nearly all the cells/clusters in the EHS gel (upper panel), but was blocked by the anti-integrin antibody (lower panel). In one large cluster, a single cell escaped the antibody block. Bar = 20 μm.

Laminin directs transcription of the β-casein gene

Basement membranes consist of supramolecular networks of laminin and collagen IV, and also contain glycoproteins such as entactin and perlecan (Yurchenco and Schittny, 1990). All of these molecules can independently interact with epithelial cells through a variety of cell surface receptors including integrins. Since it is only possible to obtain small

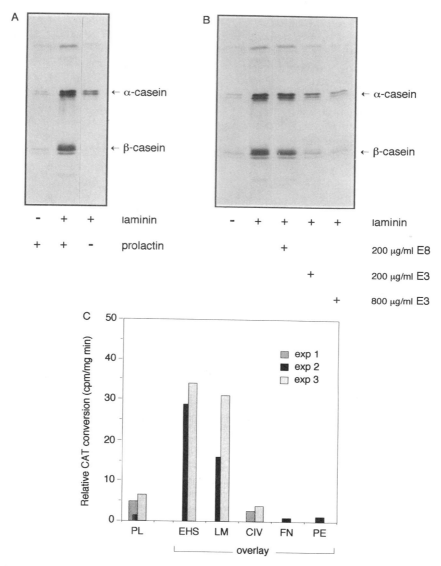

Figure 6. Laminin triggers mammary-specific gene expression. A, B. Mammary cells were cultured on glass coverslips for 5 days in differentiation medium containing lactogenic hormones. Purified laminin was diluted into the culture medium to a concentration of 100 μg/ml, and included for the last 4 days of culture. At the end of the culture period, casein synthesis was examined as in Figure 2. A. Induction of casein expression by laminin. In the absence of prolactin, β-casein was not synthesized. B. The purified elastase fragments of laminin, the E3 and E8 domains, were included as competitive inhibitors in this differentiation assay. Only the E3 fragment, not the E8 fragment was able to block casein expression. C. The β-casein promoter contains a transcriptional enhancer normally located 1517 bp upstream from the transcription initiation site. This enhancer was linked to a non-functional minimal casein promoter and used to drive the chloramphenicol reporter gene (Schmidhauser *et al.*, 1992). CID-9 cells stably expressing this gene were cultured on plastic dishes (PL) with lactogenic hormones, and treated for 3.5 days with 200 μg/ml EHS matrix (EHS), purified laminin (LM), collagen IV (CIV), or fibronectin (FN), or 20 μg/ml perlecan (PE), diluted into the culture medium. The relative levels of CAT indicate a dramatic laminin dependence of promoter activity.

quantities of purified basement membrane protein, we developed a novel assay to identify the precise components that trigger the signalling pathway between basement membrane and milk protein expression (Streuli *et al.*, submitted). Mammary epithelial cells were seeded on glass or plastic and then treated with soluble ECM proteins. We used this assay to show that purified laminin, but not other ECM proteins, could induce casein expression in mammary cells (Figure 6A). Low levels of matrix proteins, around 100 μg/ml in the culture medium, triggered differentiation. But this could only be induced in cooperation with lactogenic hormones, confirming the requirement of coordinate controls from matrix and hormones. We also demonstrated that a specific region within the globular domain of the long arm of the laminin α_1 chain (Burgeson *et al.*, 1994), the E3 region, was required for this activity (Figure 6B).

To determine whether the laminin-induced increase in casein expression occurred at the transcriptional level, we measured the activity of the β-casein promoter using this assay. A reporter gene containing elements of the bovine β-casein promoter, linked to the chloramphenicol acetyl-transferase reporter gene (Schmidhauser *et al.*, 1992), was transfected into a functional mammary epithelial cell strain, CID-9, and stably transfected cells were treated with soluble ECM proteins (Streuli *et al.*, submitted; Schmidhauser *et al.*, 1995). We found that both the complete basement membrane matrix and purified laminin could activate transcription from this promoter, but that other ECM components such as collagen IV, fibronectin, and perlecan were ineffective, corroborating our previous results with the β-casein protein (Figure 6C).

Mechanisms for ECM-directed control on gene expression

Several conclusions can be drawn from this work. Our results provide strong evidence for the direct role of basement membrane in maintaining differentiation. Specifically, we have shown that laminin, through a site in its globular domain, can interact functionally with mammary epithelial cells to activate gene transcription. The ECM signal is transferred to the cells by integrins, and does not depend on cell-cell interactions. The mammary gland model is therefore an elegant example of how epithelial differentiation can be regulated cooperatively, by microenvironmental solid phase cues together with systemic hormonal signals (Figure 7).

Since ectopic expression of ECM-degrading matrix metalloproteinases lowers casein synthesis in transgenic mice (Sympson *et al.*, 1994), and since the E3 domain of laminin is present in basement membranes surrounding alveoli in the mammary gland (C.H. Streuli unpublished work), our results indicate that the cell-laminin interaction is also likely to be a control point for tissue-specific gene expression *in vivo*.

What is the nature of the signal transduction mechanism from ECM? Two ideas are worth considering. One possibility is that basement membrane to cell signalling is required to set up the correct intracellular cytostructure for lactogenic hormones to activate transcription of the casein gene. Elements of the cytoskeleton constitute a major class of structures that interact with integrin-ECM complexes to regulate shape and locomotion (Hynes, 1992). It is now apparent the correct cell shape may additionally be required for milk protein production (Roskelley *et al.*, 1994) so laminin may induce the formation of cytoskeletal structures that signal directly to the nucleus through mechano-chemical transmission (Bissell *et al.*, 1982; Ingber, 1993).

An alternative possibility is that true second messengers are involved. For example, tyrosine kinases such as pp125[FAK] associate with clustered integrin receptors, and may trigger downstream signalling events (Kornberg *et al.*, 1991; Hildebrand *et al.*, 1993). Part of the regulation of casein transcription occurs through a STAT-related transcription factor,

148

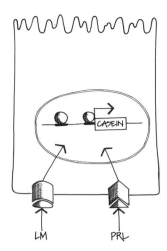

Figure 7. Model for coordinate regulation of mammary gland function by laminin and prolactin. Our model for the control of mammary gland function is that activation of transcription of milk protein genes is dependent on separate pathways from laminin through integrin receptors, and from lactogenic hormones such as prolactin. The control on transcription may operate entirely through independent routes, or alternatively the components of these signalling pathways may interact.

mammary gland factor (Wakao *et al.*, 1994). Prolactin interacts with its receptor to activate the cytoplasmic tyrosine kinase, Jak–2 (Rui *et al.*, 1994), and as these kinases phosphorylate STAT transcription factors (Darnell *et al.*, 1994), the prolactin signal now appears to be mediated through Jak kinases and mammary gland factor. An intriguing possibility for the control of mammary differentiation by laminin is that occupation of integrin receptors is required for the phosphorylation and therefore the full transcriptional activity of mammary gland factor.

ACKNOWLEDGEMENTS

Charles Streuli is a Wellcome Senior Research Fellow in Basic Biomedical Science.

REFERENCES

Aggeler, J., Ward, J., Blackie, L.M., Barcellos-Hoff, M.H., Streuli, C.H. and Bissell, M.J., 1991, Cytodifferentiation of mouse mammary epithelial cells cultured on a reconstituted basement membrane reveals striking similarities to development *in vivo*, *J. Cell Sci.* 99:407.

Barcellos-Hoff, M.H., Aggeler, J., Ram, T.G. and Bissell, M.J., 1989, Functional differentiation and alveolar morphogenesis of primary mammary cultures on reconstituted basement membrane, *Development* 105:223.

Bissell, M.J., Hall, H.G. and Parry, G., 1982, How does extracellular matrix direct gene expression? *J. Theor. Biol.* 99:31.

Burgeson, R.E., Chiquet, M., Deutzmann, R., Ekblom, P., Engel, J., Kleinman, H., Martin, G.R., Meneguzzi, G., Paulsson, M., Sanes, J., Timpl, R., Tryggvason, K., Yamada, Y. and Yurchenco, P.D., 1994, A new nomenclature for the laminins, *Matrix Biology* 14:209.

Chammas, R., Taverna, D., Cella, N., Santos, C. and Hynes, N.E., 1994, Laminin and tenascin assembly and expression regulate HC11 mouse mammary cell differentiation, *J. Cell Sci.* 107:1031.

Darnell, J.E., Kerr, I.M. and Stark, G.R., 1994, Jak-STAT pathways and transcriptional activation in response to IFNs and other extracellular signalling proteins, *Science* 264:1415.

Delcommenne, M. and Streuli, C.H., Expression of β_1 integrins is regulated by the extracellular matrix, *Submitted for publication.*

Emerman, J.T., Enami, J., Pitelka, D.R. and Nandi, S., 1977, Hormonal effects on intracellular and secreted casein in cultures of mouse mammary epithelial cells on floating collagen membranes, *Proc. Natl. Acad. Sci. USA* 74:4466.

Hay, E.D., 1993, Extracellular matrix alters epithelial differentiation, *Curr. Opin. Cell Biol.* 5:1029.

Hildebrand, J.D., Schaller, M.D. and Parsons, J.T., 1993, Identification of sequences required for the efficient localization of the focal adhesion kinase, pp125[FAK], to cellular focal contacts, *J. Cell Biol.* 123:993.

Hynes, R.O., 1992, Integrins: versatility, modulation, and signalling in cell adhesion, *Cell* 69:11.

Ingber, D.E., 1993, The riddle of morphogenesis: A question of solution chemistry or molecular cell engineering? *Cell* 75:1249.

Kornberg, L.J., Earp, H.S., Turner, C.E., Prockcop, C. and Juliano, R.L., 1991, Signal transduction by integrins: increased protein tyrosine phosphorylation caused by clustering of β_1 integrins, *Proc. Natl. Acad. Sci.* 88:8392.

Lee, E.Y.-H., Parry, G. and Bissell, M.J., 1984, Modulation of secreted proteins of mouse mammary epithelial cells by the collagenous substrata, *J. Cell Biol.* 98:146.

Li, M.L., Aggeler, J., Farson, D.A., Hatier, C., Hassell, J. and Bissell, M.J., 1987, Influence of a reconstituted basement membrane and its components on casein gene expression and secretion in mouse mammary epithelial cells, *Proc. Natl. Acad. Sci. USA* 84:136.

Reichmann, E., Ball, R., Groner, B. and Friis, R.R., 1989, New mammary epithelial and fibroblastic clones in culture form structures competent to differentiate functionally, *J. Cell Biol.* 108:1127.

Roskelley, C.D., Desprez, P.Y. and Bissell, M.J., 1994, Extracellular matrix-dependent tissue specific gene expression in mammary epithelial cells requires both physical and biochemical signal transduction, *Proc. Natl. Acad. Sci. USA, in press.*

Rui, H., Kirken, R.A. and Farrar, W.L., 1994, Activation of receptor-associated tyrosine kinase JAK2 by prolactin, *J. Biol. Chem.* 269:5364.

Schmidhauser, C., Myers, C., Mossi, R., Casperon, G.F. and Bissell, M.J., 1995, Extracellular matrix dependent gene regulation in mammary epithelial cells, *in* "Intercellular Signalling in the Mammary Gland", Plenum Publishing Company, London.

Schmidhauser, C., Casperson, G.F., Myers, C.A., Sanzo, K.T., Bolten, S. and Bissell, M.J., 1992, A novel transcriptional enhancer is involved in the prolactin- and extracellular matrix-dependent regulation of beta-casein gene expression, *Mol. Biol. Cell.* 3:699.

Schmidhauser, C., Bissell, M.J., Myers, C.A. and Casperson, G.F., 1990, Extracellular matrix and hormones transcriptionally regulate bovine beta-casein 5' sequences in stably transfected mouse mammary cells, *Proc. Natl. Acad. Sci. USA* 87: 9118.

Stoker, A., Streuli, C.H., Martins-Green, M. and Bissell, M.J., 1990, Designer microenvironments for the analysis of cell and tissue function, *Curr. Opin. Cell Biol.* 2:864.

Streuli, C.H. and Bissell, M.J., 1990, Expression of extracellular matrix components is regulated by substratum, *J. Cell Biol.* 110: 1405.

Streuli, C.H., 1993, Extracellular matrix and gene expression in mammary epithelium, *Sem. Cell Biol.* 4:203.

Streuli, C.H., Schmidhauser, C., Roskelley, C.D., Bailey, N., Yurchenco, P., Skubitz, A.P.N. and Bissell, M.J., A domain within laminin that mediates tissue-specific gene expression in mammary epithelia, *Submitted for publication.*

Streuli, C.H., Bailey, N. and Bissell, M.J., 1991, Control of mammary epithelial differentiation: the separate roles of cell-substratum and cell-cell interaction, *J. Cell Biol.*, 115:1383.

Sympson, C.J., Talhouk, R.S., Alexander, C.M., Chin, S.K., Clift, S.M., Bissell, M.J. and Werb, Z., 1994, Targeted expression of stromelysin-1 in mammary gland provides evidence for a role of proteinases in branching morphogenesis and the requirement for an intact basement membrane for tissue-specific gene expression, *J. Cell Biol.* 125:681.

Tomaselli, K.J., Damsky, C.H. and Reichardt, L.F., 1988, Purification and characterization of mammalian integrins expressed by a rat neuronal cell line (PC12): evidence that they function as alpha/beta heterodimeric receptors for laminin and type IV collagen, *J. Cell Biol.* 107:1241.

Wakao, H., Gouilleux, F. and Groner, B., 1994, Mammary gland factor (MGF) is a novel member of the

cytokine regulated transcription factor gene family and confers the prolactin response. *EMBO J.*
13:2182.

Yurchenco, P.D. and Schittny, J.C., 1990, Molecular architecture of basement membranes. *FASEB J.*
4:1577.

ASYNCHRONOUS CONCURRENT SECRETION OF MILK PROTEINS IN THE TAMMAR WALLABY (*MACROPUS EUGENII*)

Kevin R. Nicholas[1], Colin J. Wilde[2], Peter H. Bird[1,2],
Kay A.K. Hendry[2], Karen Tregenza[1,3] and Beverley Warner[1,3]

[1]CSIRO, Division of Wildlife and Ecology,
PO Box 84, Lyneham, ACT 2602 Australia
[2]Hannah Research Institute, Ayr, KA6 5HL, UK
[3]Division of Biochemistry and Molecular Biology,
Australian National University, Canberra, ACT 2600 Australia

INTRODUCTION

Lactation plays a central role in the reproductive cycle of all mammals. Eutherian species have evolved a lactational strategy that results in the birth of a relatively large young and whereas the milk provided to the young does not change significantly in composition, its synthesis and secretion is controlled by a complex interplay of endocrine, autocrine and paracrine influences (Topper and Freeman, 1980; Wilde and Peaker, 1990; Wilde *et al.*, 1990; Streuli, 1993). In contrast to eutherian mammals, marsupials have adopted a reproductive strategy that includes a short gestation with the birth of an immature young. This is followed by extensive growth and physiological development of the dependent young during a comparatively long period of lactation (see Tyndale-Biscoe and Janssens, 1988). The need to provide the appropriate nutrition for this development requires that the lactating mother progressively alters the volume and composition of the milk made available to the sucking young (Green, 1984; Dove and Cork, 1989).

The lactation cycle in the tammar wallaby (*Macropus eugenii*), a macropodid marsupial, can be divided into three distinct phases (Figure 1). The first phase comprises a 28-day pregnancy followed by parturition and lactogenesis. Phase 2 of lactation in the tammar (parturition to 200 days) is characterised by the secretion of small volumes of dilute milk which is high in carbohydrate and low in fat. The transition from phase 2 to phase 3 is associated with an increased rate of milk production (Dove and Cork, 1989) and an accelerated rate of growth of the young as it begins to leave the pouch. The milk secreted in phase 3 of lactation (200-330 days) is concentrated, with elevated levels of fat and decreased carbohydrate (Green and Merchant, 1988). In addition to these changes, the protein content of the milk increases and the composition of the whey proteins change as lactation progresses (Nicholas, 1988b).

In common with eutherian mammals, it has been established that lactation in marsupials is controlled by endocrine stimuli, particularly prolactin (Nicholas, 1988a, b; Hinds, 1988). However, macropod marsupials such as the tammar wallaby (*Macropus eugenii*) are capable of asynchronous concurrent lactation whereby the lactating mother provides a concentrated milk for the older young at heel which has vacated the pouch, while providing a more dilute milk from an adjacent mammary gland for a newborn young (Nicholas, 1988b). This phenomenon suggests that, in addition to endocrine influences, it is likely that the mammary gland is controlled by factors which are local to the gland. Therefore, the marsupial offers the opportunity to study the mechanisms which specifically influence changes in milk composition during established lactation.

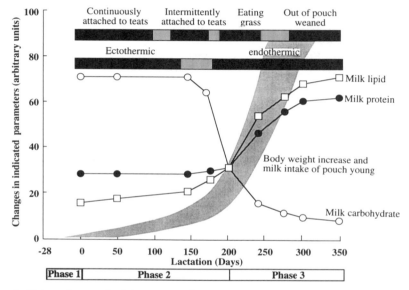

Figure 1. Changes in growth of the young, milk secretion, growth of the mammary gland and milk production during the three phases of the lactation cycle in the tammar wallaby.

The focus of this chapter is the temporal changes in secretory pattern of the milk proteins and the potential mechanisms for control of the genes which code for these proteins.

MAMMARY GLAND DEVELOPMENT, GENE EXPRESSION AND MILK SECRETION

Mammary gland development

During the transition from phase 1 to phase 2 of lactation the weight of the mammary gland increased from 228 mg at day 20 to approximately 567 mg at parturition, and then increased further to a maximum of 810 mg in the suckled mammary glands (Figure 2). Additional mammary development occurs during the transition from phase 2 to phase 3 of lactation. The weight of the mammary gland increased 7-fold from 99 days of lactation to a maximum at 239 days (Bird *et al.*, 1994). Earlier studies (Nicholas, 1988a) showed the rate of DNA synthesis was elevated in mammary explants from phase 2 tissue and low in phase 3 tissue, confirming the findings of Findlay (1982) which show an increased appearance of mitotic indices associated with mammary cell growth in sections from phase 2 mammary tissue.

Mammary alveoli and extensive connective tissue stroma are present in sections of tissue from tammars at day 24 of gestation (phase 1), and with the onset of phase 2 of lactation the alveoli become distended and are more typical of secretory tissue (Nicholas *et*

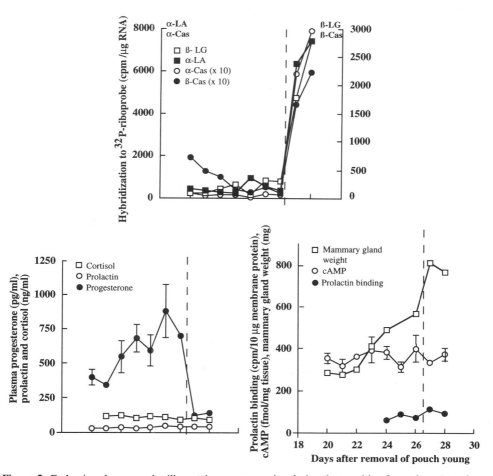

Figure 2. Endocrine changes and milk protein gene expression during the transition from phase 1 to phase 2 of lactation. All analysis was done in the same mammary tissue and in matched plasma samples. Pregnancy was initiated and synchronised in tammars by removing the pouch young (RPY) from lactating animals. Parturition occurs about day 27 and is indicated by the dashed line. Values shown are the mean ± SEM for 3-4 tammars.

al., 1994). In sections of phase 3 mammary tissue the secretory epithelia are flattened, alveoli are very distended and contain secretory material, and the stromal content is reduced.

Milk protein gene expression and milk protein secretion

The transition from phase 1 to phase 2 of lactation coincides with lactogenesis. Figure 2 shows the induction of α-lactalbumin (α-LA), ß-lactoglobulin (ßLG), α-casein and ß-casein gene expression during this period. The accumulation of mRNA for each of these genes was almost undetectable prior to parturition but thereafter each gene was induced in a coordinate fashion by day one *post partum* (Figure 2).

The transition from phase 2 to phase 3 of lactation in the tammar at 200 days *post partum* is characterized by several significant changes in milk protein secretion (Nicholas, 1988b). In contrast to the coordinated induction of milk protein gene expression at

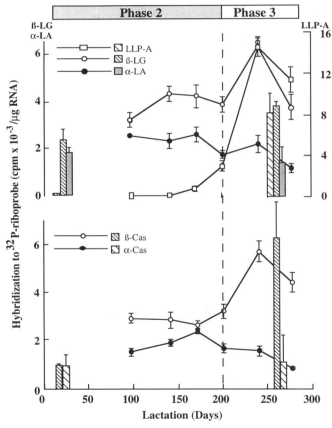

Figure 3. Progressive changes in milk protein gene expression during phase 2 and phase 3 of lactation. (A) The accumulation of the respective mRNA for Late Lactation Protein-A (LLP-A), β-lactoglobulin (β-LG), α-lactalbumin (α-LA), β-casein (β-Cas) and α-casein (α-Cas) genes in mammary tissue was quantitated by slot blot analysis (Collet *et al.*, 1992). The approximate transition from phase 2 to phase 3 is indicated. Values represent mean ± SEM (n=3). The data presented as bars represent milk protein gene expression in asynchronous concurrent lactation. Phase 2 and phase 3 mammary glands were at day 25 and day 255 of lactation respectively. Values represent mean ± range (n=2).

parturition, the pattern of expression of the same genes during the transition to phase 3 at 200 days of lactation is asynchronous (Figure 3; Bird *et al.*, 1994). For example, the accumulation of ß-LG mRNA increases, whereas the level of α-LA mRNA remains unchanged. The pattern of α-LA gene expression is entirely consistent with the relatively unchanged level of α-LA in the milk (Messer and Elliott, 1987), but in contrast to the marked change in milk carbohydrate content during the lactation cycle. Lactose is secreted for the first 4-6 days of lactation (Messer *et al.*, 1984) after which a ß-1,3 galactosyltransferase (ß-1,3 GAT) is induced in the mammary gland (Messer and Nicholas, 1991), resulting in a change to complex long chain galactosyl-lactose molecules for the remainder of phase 2 of lactation. Phase 3 of lactation is characterized by decreased carbohydrate content and the presence of only the monosacharides glucose and galactose in the milk (Messer and Green, 1979). This latter change results from a decline in ß-1,3 GAT (Messer and Nicholas, 1991) and an increase in the activity of ß-galactosidase at this time (Nicholas, unpublished).

Following the transition to phase 3 lactation, the accumulation of mRNA for ß-casein increased while that for α-casein declined slightly. To determine whether the changes in milk protein gene expression in the mammary gland were reflected in the concentration of the individual milk proteins secreted in the milk, samples were examined by SDS polyacrylamide gel electrophoresis (SDS-PAGE). The identity of the proteins was confirmed by end-terminal sequence analysis. The changes in casein mRNA were reflected in changes in the level of the individual casein proteins (Figure 4; Bird *et al.*, 1994). The implications of these changes for the structure of the casein miscelle is not yet apparent, although it is clear that the structure of the miscelle is retained sufficiently to permit centrifugation from the milk. SDS-PAGE analysis of the caseins indicated that they exist in multiple forms which are present in milk at day 143 and 239 (Figure 4a). The mobility of the two ß-caseins indicated an apparent molecular weight of 41.1 ± 0.4 kD and 39.1 ± 0.2 kD (mean $\pm$ SEM, n=4). These molecular weights were greater than calculated from the amino acid sequence (Collet *et al.*, 1992) and indicates it is likely that post-translational processing had occurred. At 143 days of lactation the slow-moving form of ß-casein stained with greater intensity than the faster moving form. However, both forms increased in concentration with the transition to phase 3 of lactation. The α-caseins were present in milk in three forms with apparent molecular weights of 31.3 ± 0.4 kD, 30.2 ± 0.5 kD and 28.6 ± 0.5 kD. All forms of α-casein appeared to remain unchanged, or increase slightly in concentration, with the progression to phase 3 of lactation.

One of the most striking features associated with the transition to phase 3 of lactation is the induction of a novel milk protein gene called late lactation protein-A (LLP-A; Collet *et al.*, 1989). This protein forms a significant component of the whey in phase 3 lactation and the timing of its appearance in milk (Nicholas *et al.*, 1987) correlates well with the induction of the gene in the mammary gland. A recent report (Collet and Joseph, 1993) has suggested that LLP-A may be grouped with the non-core lipocalins, a family of proteins which include ß-LG and which bind small ligands. SDS-PAGE of the whey proteins during the transition from phase 2 to phase 3 of lactation shows additional changes in milk protein secretion (Figure 4). It can be seen that another protein with a mobility slightly greater than LLP-A is induced after 200 days of lactation. This protein has been partially sequenced and shares significant homology with LLP-A and consequently has been labelled as LLP-B. Although there is homology in the amino acid sequence between these two proteins, it is interesting that there is a significant difference in the temporal pattern of their appearance in milk during lactation.

It is known that the total protein content of the milk increases during the transition from phase 2 to phase 3 of lactation (Nicholas, 1988b). However, our recent studies have

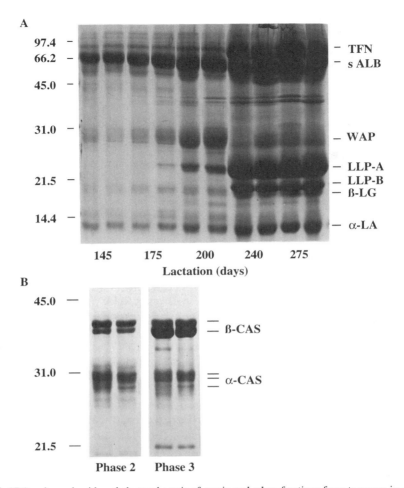

Figure 4. SDS polyacrylamide-gel electrophoresis of casein and whey fractions from tammars in phase 2 and 3 of lactation. (A) Whey proteins from the milk of two animals and the indicated days of lactation were electrophoresed on a 20% gel and visualized with Coomassie stain. The position of α-lactalbumin (α-LA), ß-lactoglobulin (ß-LG), late lactation protein A (LLP-A), late lactation protein-B (LLP-B), serum albumin (S Alb) and transferrin (TFN) is indicated. The molecular weight of the protein standards is shown in kD. (B) Caseins from two animals at 143 days (phase 2) and 239 days (phase 3) of lactation were centrifuged from skim milk, electrophoresed on 15% gels and localised with coomassie stain. The position of α-caseins (α-CAS) and ß-caseins (ß-CAS) is indicated. The amount of protein electrophoresed in each lane corresponded to caseins in 1 µl of milk. The molecular weight of the protein standards is shown in kD.

revealed that some proteins cease to be secreted at this time. For example, tammar whey acidic protein (tWAP) is present in milk during phase 2 of lactation but can no longer be detected in whey by SDS-PAGE analysis of phase 3 whey (Figure 4). This result was confirmed by analysing whey from each phase using reverse-phase HPLC and identifying the tWAP by amino acid sequencing.

Asynchronous concurrent lactation

Asynchronous concurrent lactation (ACL) does not often occur in tammars in the wild but can be experimentally induced by removal of the pouch young at mid-year to alter the normal pattern of reproduction (Tyndale-Biscoe and Janssens, 1988). This results in the mother providing milk for a young animal which is out of the pouch and well advanced in phase 3 of lactation at the beginning of the following year, at the time the mother is seasonally-programmed to give birth to a new pouch young. The retracted pouch of an ACL tammar in Figure 5 shows a 7 day-old young attached to the teat of a mammary gland in phase 2 of lactation and an elongated teat from an adjacent mammary gland in phase 3 of lactation providing milk for an older animal which has vacated the pouch and is at heel. The two remaining teats are from non-secretory quiescent mammary glands.

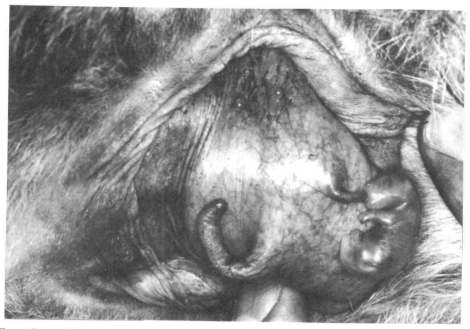

Figure 5. Asynchronous concurrent lactation. The pouch has been retracted to expose the four mammary glands. A 7-day old young is attached to a teat from a mammary gland secreting phase 2 milk. An older animal at approximately 275 days of age has vacated the pouch and sucks from the elongated teat which receives phase 3 milk from the enlarged mammary gland. The remaining two teats are from quiescent mammary glands.

Quantification of milk protein gene expression in ACL tammars showed that the LLP-A gene was expressed only in the phase 3 mammary gland (Figure 3) and that ß-casein and ß-LG gene expression was elevated when compared to phase 2 mammary tissue. However, the level of expression of the α-LA and α-casein genes was similar in both mammary glands. All of these results in ACL tammars are similar to the tissue responses measured at the corresponding time points during the lactation cycle of tammars with a single pouch young.

ENDOCRINE AND LOCAL REGULATION OF MAMMARY DEVELOPMENT AND LACTATION

Endocrine regulation of milk protein gene expression

The coordinate expression of milk protein genes at parturition in the tammar contrasts with results from studies in the rat showing that both α-lactalbumin and casein gene expression appears to occur prior to parturition (Nakhasi and Qasba, 1979; Hobbs *et al.*, 1982). In addition, recent studies in both sheep (Gaye *et al.*, 1986; Harris *et al.*, 1991) and pigs (Shamay *et al.*, 1992) have revealed that ß-lactoglobulin gene expression is detected in mid-pregnancy and levels of ß-LG mRNA continue to increase in a constitutive manner during the second half of pregnancy.

Previous studies on the endocrine control of lactogenesis in a range of species has demonstrated an inhibitory role for progesterone and a prolactational role for various combinations of insulin, cortisol and prolactin (see Topper and Freeman, 1980). An analysis of endocrine changes in plasma samples collected from tammars during the *peri-partum* period (Figure 2) confirmed earlier studies (Hinds and Tyndale-Biscoe, 1982a) showing that the level of progesterone declined from approximately 800 pg/ml just prior to parturition to 100 pg/ml on day 1 and 2 *post partum*. This developmental profile of progesterone secretion is identical to that reported for most mammals which have been studied, and is consistent with withdrawal of the proposed inhibitory influence of progesterone during pregnancy (Kuhn, 1977). The plasma level of prolactin (30-40 ng/ml) and cortisol (100-120 ng/ml) in the tammar remained unchanged during the *peri-partum* period although prolactin binding increased progressively by a total of 50% from day 21 of gestation through to day 1 *post partum* (Figure 2). An assay to measure the level of peripheral insulin has not yet been developed for a marsupial.

A mammary gland explant culture system (Nicholas and Tyndale-Biscoe, 1985) using tissue from tammars at day 24 of gestation (phase 1) was utilised to examine the hormonal control of the individual milk protein genes. The induction of milk protein gene expression was quantified (Collet *et al.*, 1991) by measuring the accumulation of mRNA in mammary explants after culture in the presence of various combinations of hormones. Maximal induction of the whey protein genes ß-LG (Figure 6; Collet *et al.*, 1991) and α-LA (Collet *et al.*, 1990) was evident in tissue after 3 days culture in the presence of prolactin alone. In contrast, the casein genes were more demanding in terms of hormonal requirements for maximal levels of expression. For example, the induction of the α-casein gene required the presence of insulin, cortisol and prolactin (Figure 3; Collet *et al.*, 1992). Expression of the ß-casein gene required both cortisol and prolactin which contrasts an earlier report (Collet *et al.*, 1992) indicating a requirement for insulin and prolactin. The reason for this difference is not clear. However, additional studies (Tregenza and Nicholas, unpublished) have shown that the gene can be induced by cortisol in explants cultured for 2 days in prolactin and that the effect of cortisol can be partially supplanted by replacing cortisol with spermidine in culture media. This latter observation confirms observations in other species which have suggested that polyamines may be involved as intracellular effectors for glucocorticoids for the induction of milk protein synthesis in mammary explants (Oka, 1974).

The inclusion of progesterone in culture media together with insulin, cortisol and prolactin did not inhibit the induction of ß-casein gene expression in mammary explants fom the tammar (Figure 7a), indicating that it is unlikely this hormone plays a negative role in the lactogenic process.

However, cholera toxin was effective in inhibiting ß-casein gene expression. Cholera toxin-sensitive G proteins have been identified in the membranes from the tammar mammary gland (Nicholas and Warner, unpublished) and earlier studies (Maher and Nicholas, 1987) have shown that the inclusion of cAMP in culture media specifically inhibited the prolactin-dependent induction of α-LA synthesis in mammary explants from late pregnant tammars. The results of studies presented in Figure 7b suggest that cAMP may well exert a rapid inhibitory effect, within 6 hours, on ß-LG gene expression. Furthermore,

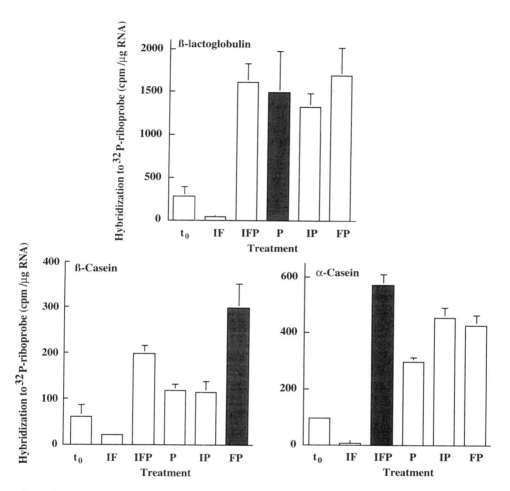

Figure 6. Hormone-dependent induction of milk protein gene expression in mammary gland explants fom tammars at phase 1 of lactation. Mammary explants from tammars at day 24 of gestation were cultured for 3 days in Medium 199 with the indicated combinations of insulin (I; 1 μg/ml), cortisol (F; 50 ng/ml) and prolactin (P; 50 ng/ml). Total RNA was extracted from the tissue and the accumulation of each mRNA quantified (Collet *et al.*, 1992). t_o indicates tissue prior to culture. Values represent mean $\pm$ SEM for 3-6 animals. The treatment which represents the minimal hormonal requirement for maximal induction of the gene is highlighted.

it is likely that that the effect is specific to the milk protein gene since there was no comparable effect on the accumulation of mRNA for cytochrome oxidase (Figure 7c), a gene expressed in a hormone-independent manner (Nicholas *et al.*, 1991). The level of cAMP in mammary tissue from tammars did not change during the *peri-partum* period (Figure 2), indicating that it is unlikely this intracellular messenger is implicated in lactogenesis. However, it may be tempting to speculate that if cAMP has a role in controlling milk protein gene expression in the tammar it may be limited to short term regulation.

Non-coordinate control of the expression of the tammar milk protein genes in established lactation is most likely to occur at the level of transcription and suggests that the regulatory elements of the individual genes differ considerably. This is reflected in the hormonal requirements for expression of these milk protein genes *in vitro*. As discussed

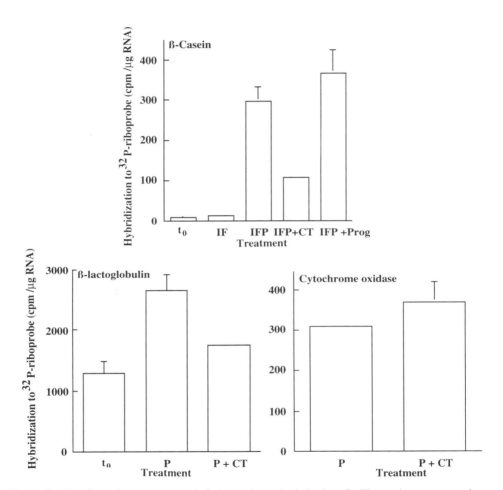

Figure 7. The effect of progesterone and cholera toxin on the induction of milk protein gene expression. Mammary explants from tammars at day 24 of gestation were cultured in the indicated combinations of insulin (I; 1 μg/ml), cortisol (F; 50 ng/ml), prolactin (P; 50 ng/ml), progesterone (prog; 1 μg/ml) and cholera toxin (CT; 1 μg/ml). Values represent mean $\pm$ SEM for 3-4 animals. Top panel, the effect of progesterone and cholera toxin on the induction of ß-casein after 3 days of culture. Bottom left, the effect of cholera toxin on ß-LG gene expression after 6 h of culture. Bottom right, the effect of cholera toxin on cytochrome oxidase gene expression after 6 h of culture.

above, maximal induction of the whey protein genes ß-LG and α-LA in mammary explants from late pregnant tammars required only the presence of prolactin (Nicholas and Tyndale Biscoe, 1985: Collet *et al.*, 1990, 1991), whereas maximal expression of the α-casein gene was dependent on insulin, cortisol and prolactin (Collet *et al.*, 1992) and ß-casein gene expression on cortisol and prolactin. Earlier studies have shown that the peripheral concentration of prolactin increases during the transition to phase 3 lactation (Figure 8; Hinds and Tyndale-Biscoe, 1982; Nicholas, 1988b), whereas the levels of cortisol and thyroid hormone remain unaltered (Figure 3; Nicholas, 1988b). However, culture of phase 2 mammary explants in the presence of insulin, cortisol, thyroid hormone, estrogen and elevated levels of prolactin maintained casein synthesis but failed to induce LLP-A synthesis (Nicholas, 1988b). Clearly, additional endocrine or local intramammary factors must operate to account for the differential regulation of milk protein synthesis and secretion at this time. It may be assumed that unidentified factors must also operate to induce LLP-B gene expression, to inhibit WAP secretion in phase 3 of lactation and to effect the non-coordinate control of the expression of the α-LA, ß-LG and casein genes. Nevertheless, the singular dependence on prolactin for induction of the milk protein genes *in vitro*, and the changes in circulating prolactin and mammary prolactin binding during lactation (see below) have focused attention on this hormone as a likely contributor to the transition between phases of lactation in the tammar.

Endocrine regulation of lactation: the role of prolactin

The primary role of prolactin for lactogenesis in the tammar is consistent with that observed in eutherian mammals (Topper and Freeman, 1980). However, whereas the requirement for prolactin to maintain lactation in the tammar is well established (Hinds, 1988), the role of prolactin in the transition between phase 2 and 3 of lactation remains to be elucidated. As discussed earlier, the circulating concentration of this hormone increased significantly during the transition from phase 2 to phase 3 of lactation (Figure 8; Hinds and Tyndale-Biscoe, 1982b; Nicholas, 1988b). The ontogeny of prolactin binding to membranes from tammar mammary glands during the lactation cycle showed an increment in binding during phase 2 of lactation (Figure 8; Bird *et al.*, 1993), and this increase was attributable to an increase in the number of binding sites for the hormone rather than any change in receptor affinity for its ligand (Bird *et al.*, 1994). Similarly, in ACL tammars the level of prolactin binding was elevated in mammary tissue in phase 3 (day 255) when compared to tissue early in phase 2 (day 25) lactation (Figure 8; Bird *et al.*, 1994). However, in contrast to the earlier report of Stewart (1984), in the present study prolactin binding declined with the onset of phase 3 of lactation. Scatchard analysis of prolactin binding to membranes from mammary tissue at 172 and 239 days of lactation revealed that this decline in prolactin binding resulted from a decrease in the number of binding sites and not from changed affinity of the binding sites for the ligand. Furthermore, this decline was not a consequence of prolactin receptor occupancy associated with the significantly elevated level of prolactin in phase 3 of lactation (Bird *et al.*, 1994).

Studies in several species have shown that the microenvironment of the cell membrane in which the prolactin receptor is located can influence the number of available binding sites (see Vonderhaar *et al.*, 1985). For example, conconavalin A (con A) can perturb membranes from the tissues of several species and result in either an increase in the number of previously cryptic (masked) sites or a decline in the number of exposed sites available for binding (Vonderhaar *et al.*, 1985). In our studies, membranes from the mammary gland of tammars in phase 2 lactation contained a significant number of cryptic prolactin receptors, since the inclusion of con A in binding assays produced a 40% increase in specific prolactin binding to both unfractionated microsomal membranes (Figure 8; Bird *et al.*, 1994) and plasma membrane preparations (data not shown). Con A did not stimulate prolactin binding to receptors from CHAPS-solubilized membranes confirming an earlier report (Vonderhaar *et al.*, 1985) that the effect of con A is elicited by interaction with the membrane and not by direct interaction with the receptor.

In contrast, con A treatment of phase 3 tammar mammary membranes produced a 40% decrease in specific prolactin binding (Figure 8; Bird *et al.*, 1994), similar to that reported for membranes from lactating rat and mouse mammary tissue and liver membranes (Vonderhaar *et al.*, 1985). The differential effects of con A on phase 2 and phase 3 mammary tissue from the tammar indicate that significant developmentally-induced differences exist at the level of the cell membrane.

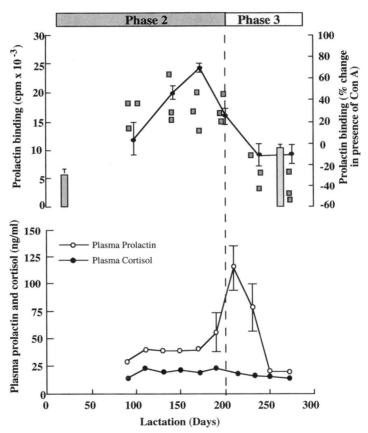

Figure 8. Endocrine changes and prolactin binding to membranes from the mammary glands of tammars during the transition from phase 2 and phase 3 of lactation. (A) Membranes (100 μg) from mammary tissue were incubated with ^{125}I-oPRL (100,000 cpm) and non specific binding determined in the presence of oPRL (1 μg). Values represent mean ± SEM (n=3). The approximate transition from phase 2 to phase 3 is indicated. The bar values show prolactin binding to membranes from phase 2 (25 days) and phase 3 (255 days) mammary glands from ACL tammars. Values represent mean ± SEM (n=3). The effect of concanavalin A (con A) on prolactin binding to membranes from the mammary glands of tammars in phase 2 and 3 of lactation is shown. Membranes (50 μg) were incubated with ^{125}I-oPRL in the presence and absence of con A (1 μg/ml). Non-specific binding was determined in the presence of oPRL (1 μg). Individual values represent prolactin binding in the presence of con A as a percentage of that in the absence of con A. (B) Changes in plasma prolactin and cortisol during phase 2 and phase 3 of lactation.

Endocrine regulation and asynchronous lactation

Asynchronous concurrent lactation in the tammar suggests that the individual mammary glands are functioning independently and that they reflect the expected developmental stage for milk protein gene expression and prolactin binding. One possible explanation for this phenomenon is that the individual mammary glands interpret their immediate environment differently as a result of altered responsiveness from a developmentally regulated change to the mammary epithelium and myoepithelium. For example, Renfree and her colleagues (Findlay and Renfree, 1984) have reported a decline in responsiveness to oxytocin as lactation progresses. Thus, the constant sucking of the newborn does not induce milk "let down" in the adjacent phase 3 mammary gland. Clearly, the different responses to Con A by membranes from phase 2 and phase 3 mammary tissue indicates that there are physical changes in the membrane which permit altered responses to lectins and subsequently altered prolactin binding. However, whether these changes indicate potential altered responsiveness to endogenous external stimuli remains to be investigated.

LOCAL REGULATION OF MILK SECRETION AND COMPOSITION

The ability of adjacent glands to maintain asynchronous concurrent lactations demonstrates that they are functioning independently, despite being within the same hormonal milieu. Secretion of milk of different compositions and at different rates is clearly a reflection of the developmental state of each gland, both in cell number and in pattern of gene expression. Other presentations in this volume emphasise the variety of locally-active growth factors synthesised in mammary tissue, any of which may contribute to development of the tammar mammary gland during the transition from phase 2 to phase 3 of lactation, and could therefore account for differences in number and differentiated status of secretory epithelial cells in asynchronously-lactating glands. The factor(s) involved and the physiological stimulus for gland development and progression to phase 3 lactation have not been identified. However, induction of phase 3 lactation in the tammar is accompanied by a change to less frequent but more vigorous (and presumably more efficient) suckling by the young. Therefore, one possible explanation for the difference between adjacent lactating mammary glands in the tammar is that gland development and stimulation of milk secretion is led by the offspring through changes in the pattern and completeness of milk removal. There is a precedent for this in other species. In goats and dairy cows, the rate of milk secretion is regulated locally within each gland by frequency and efficiency of milking (see Peaker, this volume), and a similar mechanism operates in human lactation, such that the breasts synthesise milk at different rates according to the completeness of gland emptying at the previous feed (Hartmann et al., this volume). In lactating goats, milk secretion rate is regulated by milk removal through autocrine feedback inhibition by a milk protein (Wilde et al., this volume). The same protein regulates prolactin receptor distribution or number in mammary epithelial cell cultures, and when added to milk stored in the mammary gland (C.N. Bennett, C.H. Knight and C.J. Wilde, unpublished work). The transition from phase 2 to phase 3 of tammar lactation was reportedly associated with an increase in mammary prolactin binding with the onset of phase 3 (Stewart, 1984). Therefore, it appeared that an homologous factor might be present in tammar milk and, through depletion with more effective suckling in phase 3, could account for the stimulation of milk secretion and a prolactin-mediated induction of mammary development. Therefore, we screened tammar milk for the presence of factors able to exert feedback inhibition on milk secretion.

Factors in goat's milk (Wilde *et al.*, 1987, 1995a) and human milk (Wilde *et al.*, 1995b) were identified using a rabbit mammary explant bioassay. Initial screening using this system confirmed the presence of an inhibitory factor in whey proteins extracted from phase 3 tammar milk. As with inhibitory activity in goat's milk, this activity was present in a M_r 6,000-30,000 fraction of the whey proteins, based on passage through or retention by ultrafiltration membranes. Immunoblotting showed that this M_r 6,000-30,000 whey fraction contained several molecular variants of ß-LG, α-LA and, as a major component, LLP, the protein characteristic of phase 3 milk. Bioassay of eluted protein pooled arbitrarily in three fractions confirmed that inhibition by the M_r 6,000-30,000 whey fraction was specific to a fraction containing ß-LG and representing approximately 3% of protein in the original whey fraction. Therefore inhibition of milk constituent synthesis in the bioassay was not a non-specific effect of added protein. Further experiments confirmed that inhibitory activity was associated with a minor component of the active fraction. Based on these preliminary observations, it appeared that rabbit explant bioassay had identified a minor component of tammar milk proteins able to inhibit synthesis of milk constituents.

In order to identify the inhibitory factor in phase 3 milk, and to determine if a similar factor was present in phase 2 milk, further screening of milk constituents was performed using suspensions of lactating mouse mammary acini. Milk protein secretion by mouse acini cultures is inhibited by the caprine inhibitor of milk secretion in a concentration-dependent and reversible manner (Rennison *et al.*, 1993), and this system is amenable to testing milk constituents which may be obtainable only in small quantities from phase 2 milk. The acini bioassay confirmed the presence of an inhibitory factor in the milk fraction shown previously to inhibit casein and lactose synthesis in rabbit mammary explants. In mouse mammary acini cultures, inhibition was measured by the milk constituent's effect on protein synthesis and secretion, and it was notable that inhibition of secretion, when observed, was consistently greater than the accompanying inhibition of protein synthesis. Preferential inhibition of protein secretion was also observed with caprine inhibitor of milk secretion in the same bioassay (Rennison *et al.*, 1993; Burgoyne *et al.*, this volume). In that case, the inhibitor was found to act primarily by blocking constitutive secretory protein transport between the endoplasmic reticulum and the Golgi apparatus, an effect which feeds back to inhibit secretory protein synthesis. Inhibition of membrane trafficking may also account for the inhibitor's ability to down-regulate mammary prolactin receptors and epithelial cell differentiation (C.N. Bennett, C.H. Knight and C.J. Wilde, unpublished work). Inasmuch as the tammar factor is also a more effective inhibitor of protein secretion than protein synthesis, it is, therefore, possible that the inhibitory component in phase 3 tammar milk acts in a similar manner to the caprine inhibitor. The active constituent in the inhibitory tammar whey fraction has not been purified to homogeneity, and so structural comparison with proteins identified as feedback inhibitors of lactation (FIL) in other species is not possible. However, a component of this phase 3 milk fraction cross-reacts specifically with antiserum raised against FIL isolated from cow's milk (Addey *et al.*, 1991). The same antiserum also specifically recognises caprine FIL in goat's milk (Wilde *et al.*, 1995a). Therefore, tammar milk appears to contain a protein which bears some structural relation, as well as similar bioactivity, to the FIL proteins identified in other species. This is perhaps not surprising. Regulation of milk secretion by milk removal has been demonstrated in all species where it has been studied, and it may be that this is a mechanism shared by most, if not all, mammals. Structural conservation of the protein responsible might, therefore, be expected, even in species as evolutionarily-distant as the dairy cow, human and marsupial.

The mouse mammary acini bioassay also produced several unexpected results with fractions initially found inactive in the rabbit explant bioassay, and included as putative controls in mouse cultures. One phase 3 tammar whey fraction which contained M_r 6,000-30,000 whey constituents eluting at low salt during anion exchange chromatography also inhibited murine protein synthesis and secretion. This appeared to be due to the presence in the starting material of two activities, one inhibitory, one stimulatory, which had cancelled each other in rabbit explant bioassays. When tested separately in mouse acini cultures, the two activities were then revealed. This second inhibitory activity was also recognised in immunoblotting by antiserum against bovine FIL, suggesting that, as in cow's milk, feedback inhibition is exerted by two structurally-different but immunologically-related proteins (Addey et al., 1991, 1992). The stimulatory factor in this whey fraction has not yet been identified.

If mammary development and the stimulation of milk secretion during the phase 2 to phase 3 transition were due primarily to a change in autocrine inhibition by tammar homologues of caprine and bovine FIL proteins, it would be expected that these proteins' concentrations should be higher in phase 2 milk, and prolactin binding should be lower in phase 2 mammary tissue. Mouse acini bioassay and immunoblotting has detected an inhibitor of secretion in phase 2 milk, but we have as yet obtained no definitive evidence by either method that it is present at higher concentrations than in phase 3 milk. Moreover, the original premise on which these experiments were conducted - that the phase 2 to phase 3 transition is associated with increased mammary prolactin binding (Stewart, 1984) - has been shown by more extensive study to be incorrect. As discussed earlier, mammary prolactin binding is high in late phase 2 mammary tissue and decreases in phase 3 (Bird et al., 1994). Therefore, while autocrine inhibition may operate to regulate the rate of milk secretion in the tammar wallaby, it is possible that other mechanisms may stimulate the extensive gland development required for phase 3 lactation. In any case, unlike the situation in other species where autocrine feedback inhibition regulates milk secretion without affecting milk composition, local control in the tammar must accommodate this species' need to secrete milks of vastly different composition at different stages of lactation.

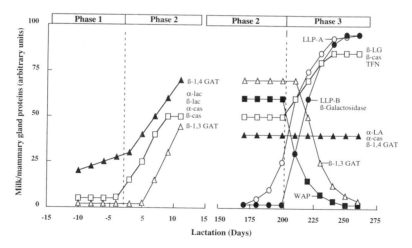

Figure 9. A schematic presentation of the progressive changes in milk protein gene expression, milk protein secretion or enzyme activity in the mammary glands from tammars during the transition from phase 1 to phase 2 of lactation and from phase 2 to phase 3 of lactation. The acronyms are the same as used in previous figures, except GAT (galactosyltransferase).

We have demonstrated that this occurs through an increase in concentration of some milk constituents e.g ß-casein and ß-LG in phase 3, a decrease in other milk constituents, for example, WAP and ß-1,3 galactosyltransferase activity and the induction of a third class, characterised by LLP. Any consideration of local regulation of tammar mammary development and function must take account of these coordinated but differential changes in gene expression. It seems unlikely that this is accommodated by the presence of distinct populations of secretory epithelial cells: phase 3 tammar mammary epithelial cells appear competent to synthesise all milk proteins (Joseph and Collet, 1994). Asynchronous gene expression and the sequential changes observed through phase 2 and phase 3 lactation may instead be the result of concerted changes in a number of regulatory factors elicited by a common physiological stimulus. As, the majority of the changes we have observed coincide with the transitions from phase 1 to phase 2 or phase 2 to phase 3 lactation (Figure 9), when changes in suckling pattern are indeed most pronounced, it remains possible that this is the principal factor cueing this developmental and functional adaptation within a single mammary gland. The presence of a factor or factors which stimulate protein secretion in mouse mammary acini, in addition of one involved in acute regulation of milk secretion, is of considerable interest and suggests that these milk-borne agents warrant further investigation.

CONCLUSION

These data suggest that milk secretion in the Tammar wallaby is regulated through autocrine feedback inhibition by a secreted milk protein (tFIL), in a manner similar to that described in other species. The potential effect of tFIL on secretion of specific tammar milk proteins has yet to be investigated. However, the presence of a factor(s) which stimulate protein secretion in isolated mammary epithelial cells from mice is of considerable interest and indicates that a family of autocrine factors, some inhibitory and others stimulatory, may be present in milk and could conceivably exert specfic effects on milk protein synthesis and secretion. Whereas it appears likely that WAP secretion and ß-1,3 GAT activity are specifically and temporally inhibited with the onset of phase 3 of lactation, it is not yet evident whether the induction of LLP-A gene expression and the presumed induction of genes for LLP-B, ß-1,3 GAT and ß-galactosidase result from a release from inhibition and/or from a stimulatory process. It is clear, however, that the majority of these changes occur around the time of the transition from either phase 1 to phase 2 of lactation, or phase 2 to phase 3 of lactation (Figure 9). Changes in the sucking pattern would be most pronounced at these times and would therefore be a possible candidate for cueing these changes by an autocrine mechanism within the mammary gland.

ACKNOWLEDGEMENTS

This work was funded by a project grant from the Wellcome Trust, and was supported by the CSIRO Division of Wildlife and Ecology and the Scottish Office Agriculture and Fisheries Department.

REFERENCES

Addey, C.V.P., Peaker, M. and Wilde, C.J., 1991, Protein inhibitor which controls secretion of milk, *U.K. Patent Application* GB 2 238 052 A1.

Addey, C.V.P., Peaker, M. and Wilde, C.J., 1992, Control of milk secretion, *UK Patent Application* GB 2 255 562 A1.

Bird, P., Hendry, K.A.K., Shaw, D, Wilde, C. and Nicholas, K.R., 1994, Progressive changes in milk protein gene expression and prolactin binding during lactation in the tammar wallaby (*Macropus eugenii*). *J. Mol. Endocrinol.* 13:117.

Bird, P.H., Nicholas, K.R., Hendry, K.A.K. and Wilde, C.J., 1993, Asynchronous milk protein gene expression in lactating tammars, EAAP/ASAS Workshop on the Biology of Lactation in Farm Animals, Madrid, Elsevier Science Publishers, 35:189.

Collet, C. and Joseph, R., 1993, A novel member of the lipocalin superfamily: tammar wallaby late-lactation protein. *Biochim. Biophys. Acta* 1167:219.

Collet, C., Joseph, R. and, Nicholas, K.R., 1989, Molecular cloning and characterization of a novel marsupial milk protein gene. *Biochem. Biophys. Res. Commun.* 164:1380.

Collet, C., Joseph, R., and Nicholas, K.R., 1990, Cloning, cDNA analysis and prolactin-dependent expression of a marsupial α-lactalbumin. *Rep. Fert. Dev.* 2:693.

Collet, C., Joseph, R. and Nicholas, K.R., 1991, A marsupial β-lactoglobulin gene: characterization and prolactin-dependent expression. *J. Mol. Endocrinol.* 6:9.

Collet, C., Joseph, R., and Nicholas, K.R., 1992, Molecular characterization and *in vitro* hormonal requirements for expression of two casein genes from a marsupial, *J. Mol. Endocrinol.* 8:1.

Dove, H. and Cork, S.J., 1989, Lactation in the tammar wallaby (*Macropus eugenii*). I. Milk consumption and the algebraic description of the lactation curve. *J. Zool. London* 219:385.

Findlay, L.,1982, The mammary glands of the tammar wallaby (*Macropus eugenii*) during pregnancy and lactation. *J. Reprod. Fert.* 65:59.

Findlay, L. and Renfree, M.B.,1984, Growth, development and secretion of the mammary gland of macropodid marsupials, *in* "Physiological Strategies in Lactation", p. 403, Peaker, M., Vernon, R.G. and Knight, C.H., eds., Symposia of the Zoological Society of London, 51, Academic Press, London.

Gaye, P., Hue-Delahaie, D., Mercier, J.C., Soulier, S., Vilotte, J.L., and Furet, J.P., 1986, Ovine ß-lactoglobulin messenger RNA: Nucleotide sequence and mRNA levels during functional differentiation of the mammary gland, *Biochimie* 68:1097.

Green, B., 1984, Composition of milk and energetics of growth in marsupials, *in* "Physiological Strategies in Lactation", p. 369, Peaker, M., Vernon, R.G. and Knight, C.H., eds., Symposia of the Zoological Society of London, 51, Academic Press, London.

Green, B. and Merchant, J., 1988, The composition of marsupial milk, *in* "The Developing Marsupial: Models for Biomedical Research," p. 41, Tyndale-Biscoe, C.H. and Janssens, P.A., eds., Springer-Verlag, Heidelberg.

Harris, S., McClenaghan, M., Simons, J.P., Ali, S., and Clark, A.J., 1991, Developmental regulation of sheep ß-lactoglobulin gene in the mammary gland of transgenic mice, *Devel. Gen.* 12:299.

Hinds, L.A., 1988, Hormonal control of lactation, *in* "Physiological Strategies in Lactation", p. 55, Peaker, M., Vernon, R.G. and Knight, C.H., eds., Symposia of the Zoological Society of London, 51, Academic Press, London.

Hinds, L.A. and Tyndale-Biscoe, C.H., 1982a, Plasma progesterone levels in the pregnant and non-pregnant tammar, *Macropus eugenii*, *J. Endocrinol.* 93:99.

Hinds, L.A. and Tyndale-Biscoe, C.H., 1982b, Prolactin in the marsupial *Macropus eugenii*, during the oestrus cycle, pregnancy and lactation, *Biol. Reprod.* 26:391.

Hobbs, A.A., Richards, D.A., Kessler, D.J., and Rosen, J.F., 1982, Complex hormonal regulation of rat casein gene expression, *J. Biol. Chem.* 257:3598.

Joseph, R. and Collet, C., 1994, Double staining *in situ* study of mRNAs encoding milk proteins in the mammary gland of the tammar wallaby (*Macropus eugenii*). *J. Reprod. Fert.* 101:241.

Kuhn, N.J., 1977, Lactogenesis: the search for trigger mechanisms in different species, *in* "Comparative Aspects of Lactation", p. 165, Peaker, M. ed., Academic Press, New York.

Maher, F. and Nicholas, K.R., 1987, Pituitary-induced lactation in mammary gland explants from the pregnant tammar (*Macropus eugenii*): a negative role for cyclic AMP, *Comp. Biochem. Physiol.* 87A:1107.

Messer, M. and Elliott, C., 1987, Changes in α-lactalbumin, total lactose, UDP-galactose hydrolase and other factors in tammar wallaby (*Macropus eugenii*) milk during lactation, *Aust. J. Biol. Sci.* 40:37.

Messer, M and Green, B, 1979, Milk carbohydrates of marsupials II. Quantitative and qualitative changes in milk carbohydrates during lactation in the tammar wallaby (*Macropus eugenii*), *Aust. J. Biol. Sci.* 32:519.

Messer, M., Griffiths, M. and Green, B., 1984, Changes in milk carbohydrates and electrolytes during early lactation in the tammar wallaby, *Macropus eugenii*, *Aust. J. Biol. Sci.* 137:1.

Messer, M. and Nicholas, K.R., 1991, Biosynthesis of marsupial milk oligosaccharides: characterization and developmental changes of two galactosyltransferases in lactating mammary glands of the tammar wallaby (*Macropus eugenii*), *Biochim. Biophys. Acta* 1077:79.

Nakhasi, H.L. and Qasba, P.K., 1979, Quantitation of milk proteins and their mRNAs in rat mammary gland at various stages of gestation and lactation, *J. Biol. Chem.* 254:6016.

Nicholas, K.R., 1988a, Control of milk protein synthesis in the tammar wallaby: a model system to study prolactin-dependent development, *in* "The Developing Marsupial: Models for Biomedical Research", p. 68, Tyndale-Biscoe, C.H. and Janssens, P.A., eds., Springer-Verlag, Heidelberg.

Nicholas, K.R., 1988b, Asynchronous dual lactation in a marsupial, the tammar wallaby (*M. eugenii*), *Biochem. Biophys. Res. Comms.* 154:529.

Nicholas, K.R., Collet, C., and Joseph, R., 1991, Hormone-responsive survival of mammary gland explants from the pregnant tammar (*Macropus eugenii*) in the absence of exogenous hormones and growth factors, *Comp. Biochem. Physiol.* 100A:163.

Nicholas, K.R., Messer, M., Elliott, C., Maher, F., and Shaw, D.C., 1987, A novel whey protein synthesized only in late lactation by the mammary gland from the tammar (*Macropus eugenii*), *Biochem. J.* 241:891.

Nicholas, K. R. and Tyndale-Biscoe, C.H., 1985, Prolactin-dependent accumulation of α-lactalbumin in mammary gland explants from the pregnant tammar wallaby (*Macropus eugenii*), *J. Endocrinol.* 106:337.

Nicholas, K.R., Wilde, C.J., Bird, P.H., and Hendry, K.A.K., 1994, Asynchronous expression of milk protein genes during lactation in the tammar wallaby (*Macropus eugenii*), *in* "Comparative Biochemistry and Physiology. Research Trends", Trivandrum, India.

Oka, T., 1974, Spermidine in hormone-dependent differentiation of mammary gland *in vitro*, *Science* 184:78.

Rennison M.E., Kerr M., Addey C.V.P., Handel S.E., Turner M.D., Wilde C.J., and Burgoyne R.D., 1993, Inhibition of constitutive protein secretion from lactating mouse mammary epithelial cells by FIL (feedback inhibitor of lactation), a secreted milk protein, *J. Cell Sci.* 106:641.

Stewart, F., 1984, Mammogenesis and changing prolactin receptor concentrations in the mammary glands of the tammar wallaby (*Macropus eugenii*), *J. Reprod. Fert.* 71:131.

Shamay, A., Pursel, V.G., Wall, R.J., and Hennighausen, L., 1992, Induction of lactogenesis in transgenic virgin pigs: evidence for gene and integration site-specific hormonal regulation, *Mol. Endocrinol.* 6:191.

Streuli, C. H., 1993, Extracellular matrix and gene expression in mammary epithelium, *Sem. Cell Biol.* 4:203.

Topper, Y.J. and Freeman, C.S., 1980, Multiple hormone interactions in the developmental biology of the mammary gland, *Physiol. Rev.* 60:1049.

Tyndale-Biscoe, C.H. and Janssens, P.A., 1988, The Developing Marsupial: Models for Biomedical Research, Springer-Verlag, Heidelberg.

Vonderhaar, B.K., Bhattacharya, A., Alhadi, T., Liscia, D.S., Andrew, E.M., Young, J.K., Ginsburg, E., Bhattacharjee, M. and Horn, T.M., 1985, Isolation, characterization, and regulation of the prolactin receptor, *J. Dairy Sci.* 68:466.

Wilde, C.J., Addey, C.V.P., Boddy, L.M. and Peaker, M., 1995a, Autocrine inhibition of milk secretion by a protein in milk, *Biochem. J.*, in press.

Wilde, C.J., Calvert, D.T., Daly, A. and Peaker, M., 1987, The effects of goat milk fractions on synthesis of milk constituents by rabbit mammary explants and on milk yield *in vivo*: evidence for autocrine control of milk secretion, *Biochem. J.* 242:285.

Wilde, C.J., Knight, C.H., Addey, C.V.P., Blatchford, D.R., Travers, M., Bennet, C.N. and Peaker, M., 1990a, Autocrine regulation of mammary cell differentiation, *Protoplasma* 159:112.

Wilde, C.J. and Peaker M., 1990b, Autocrine control in milk secretion. *J. Agric. Sci., Cambridge* 114:235.

Wilde, C.J., Prentice, A. and Peaker, M., 1995b, Breastfeeding: matching supply with demand in human lactation, *Proc. Nutr. Soc.*, in press.

HETEROGENEOUS EXPRESSION AND SYNTHESIS OF HUMAN SERUM ALBUMIN IN THE MAMMARY GLAND OF TRANSGENIC MICE

Itamar Barash[1], Alexander Faerman[1],
Raisa Puzis[1], Margaret Natan[2],
David R. Hurwitz[2] and Moshe Shani[1]

[1]Institute of Animal Science, ARO,
The Volcani Center, Bet-Daga 50250,
Israel
[2]Rhone-Poulenc Rorer Central Research,
500 Arcola Road. P.O. Box 1200
Collegeville, PA, 19426, USA

INTRODUCTION

Limited information on the production and secretion of the individual milk proteins at the cellular level in the mammary gland is available. It is generally assumed that all active epithelial cells within the lactating mammary gland of mice share a comparable ability to produce and secrete all the components of milk. This is based on ultrastructural studies and observations on cell membrane turnover related to fat globule membrane loss and vesicular protein exocytosis.

Transgenic mice provide an *in vivo* approach for studying the regulation of milk protein gene expression in the mammary gland, since the tested gene may acquire the developmentally-regulated chromatin structure assumed by its endogenous counterparts and is required for normal expression. Furthermore, it may reveal cryptic biological interactions that are masked by the expression pattern of the native milk protein genes.

RESULTS AND DISCUSSION

The mode of expression of the human serum albumin (HSA) gene constructs driven by the ovine ß-lactoglobulin promoter (BLG/HSA) in the mammary gland of five transgenic mouse strains, at different stages of development was studied by a combined *in situ* hybridization/immunohistochemical approach. Except for one transgenic strain (#23), which exhibited a uniform pattern of HSA expression throughout the mammary epithelium of virgin, pregnant and lactating mice, the other four strains displayed an heterogeneous pattern of expression. Some expressed the transgene in all mammary lobules (strains #69 and #83), although not in all epithelial cells, while in others only a fraction of the lobules contained HSA expressing cells (strains #91 and #92). However, within expressing lobules of lactating mammary glands, the general pattern of expression was very similar in all four strains. This pattern corresponded to the morphology of the alveolar cells and their physiological state. Thus, in "collapsed" alveoli containing no immunostained material in their lumina, cells expressing HSA messenger RNA and protein were intermingled with non-expressing cells. Similar correspondence between immunostaining and hybridization signals was observed in alveoli containing little immunostained material in their lumina. Alveoli containing heavily stained material within the lumina were lined by immunostained cells, some of which exhibited no detectable hybridization signal.

Taken together, there was a good correlation between the patterns of expression in virgin, pregnant and lactating mice of the strains analyzed. Uniform pattern of expression was observed in the mammary gland of transgenic mice of strain #23 and heterogeneous expression in all other strains. These differences suggest that the modes of expression were already established during the development of the mammary secretory structures.

Of possible biotechnological importance is the fact that there was no direct correlation between the level of HSA in milk and the proportion of the expressing cells in the mammary gland. Thus, lactating transgenic mice of strain #91 produced as much HSA in their milk as animals from the other strains, whereas the relative number of expressing cells in their mammary gland was comparatively low.

In contrast to the heterogeneous pattern of HSA expression in the mammary gland, a uniform pattern of expression was demonstrated for the endogenous α-lactalbumin and casein genes. These proteins were present in most mammary epithelial cells of lactating mice.

CONCLUSION

Our studies demonstrate the complexity and apparently dynamic patterns of gene expression in the mammary gland. It may reflect a cryptic mechanism controlling the production of milk proteins in epithelial cells, possibly affected by local regulatory factors.

POST-TRANSCRIPTIONAL REGULATION OF CASEIN GENE EXPRESSION BY PROLACTIN AND GROWTH HORMONE IN LACTATING RAT MAMMARY GLAND

Michael C. Barber, Amanda J. Vallance,
David J. Flint, Richard G. Vernon and
Maureen T. Travers

Hannah Research Institute, Ayr, KA6 5HL, U.K.

It is well established that prolactin (PRL) is a major lactogenic hormone. Lowering its serum concentration with bromocryptine (Br) results in a marked decrease in milk yield as reflected by a decrease in pup weight gain and by direct measurement. In addition growth hormone (GH) plays an increasingly important role in maintaining milk synthesis as lactation proceeds since combined ablation of PRL and GH (with an antiserum to rat GH (anti-rGH) reduces milk yield further (Barber *et al.*, 1992) to around 20% of control animals. The mechanism by which these hormones regulate milk synthesis and secretion remains unclear. However, a number of experiments conducted *in vitro* with mammary tissue explants and isolated cells has shown that expression of the genes for the major milk proteins, the caseins, are regulated by PRL at the level of transcription and mRNA stability (Guyette *et al.*, 1979). The role of PRL in controlling the expression of casein mRNAs *in vivo* has not been addressed. Indeed casein mRNA abundances are relatively high per mammary cell in the rat during pregnancy at a time when serum prolactin is relatively low (Rosen and Barker, 1976).

In this study we used lactating rats treated for 24h with Br + anti-rGH and the for a further 24h with Br + anti-rGH alone or in conjunction with PRL and/or GH. Lactating animals and animals which had pups removed for 48h were used as controls. Total and polysomal RNA was extracted from the mammary glands of these animals and steady-state levels of casein and acetyl-CoA carboxylase (ACC) transcripts were determined. The distribution of polysomes on sucrose density gradients was also obtained as a means of assessing *in vivo* translational activity.

Combined ablation of PRL and GH resulted in a marked reduction in pup weight gain at 24h and this was accentuated by 48h post-treatment. PRL replacement after 24h combined ablation resulted in a marked increase in pup weight gain compared with both 24h and 48h of combined ablation. GH replacement after 24h combined ablation prevented the fall in pup weight observed after 48h combined ablation although it was not as effective as prolactin.

Intercellular Signalling in the Mammary Gland
Edited by C.J. Wilde *et al.*, Plenum Press, New York, 1995

Expression of the ACC gene, the major enzyme regulating *de novo* lipid synthesis is dramatically reduced in the mammary glands of animals subjected to 24h combined ablation, decreasing to levels ~ 10% of lactating control animals and similar to those observed in mammary tissue after pup removal for 24h. PRL replacement after 24h resulted in a marked increase in the level of mammary ACC gene expression which was similar to that observed in lactating control animals. GH replacement did not significantly increase steady-state levels of ACC mRNA above values observed after 48h combined ablation.

Casein mRNA abundances were relatively unaffected by combined PRL and GH ablation: after 24h combined ablation no reductions in the level of casein mRNA transcripts were observed and by 48h reductions of 30 - 50% were observed. γ-casein mRNA appeared to be more susceptible to combined ablation than the mopre abundant α- and ß-casein mRNAs. PRL and/or GH replacement at 24h post-treatment did not restore casein mRNA abundances to levels seen in lactating control animals. By contrast casein mRNA transcripts were reduced 95% in animals that had their pups removed for 48h and this was associated with a marked reduction in the cellular content of polysomes (cytoplasmic ribonucleoprotein particles including mRNPs, ribosomal subunits and mRNA-multimeric ribosome particles of varying sizes). Polysomes isolated from pup-removed animals resolved on sucrose density gradients in favour of smaller polysomes and monosomes. By contrast combined PRL and GH ablation did not diminish the cellular content of polysomes and did not significantly alter the size distribution from large polysomes to monosomes when compared to lactating control animals. Abundance of casein mRNA in polysomal RNA reflected that observed for total RNA. Interestingly, in all treatments including 48h litter removal, casein mRNA transcripts were associated with medium to larger polysomes.

In conclusion we have demonstrated that PRL regulates ACC gene expression in lactating rat mammary gland by a pre-translational mechanism (transcription rate or mRNA stability) whereas it modulates casein gene expression through a post-transcriptional mechanism, for example at the level of translation, secretion or protein turnover. However, the factors that result in a high level of casein mRNA transcripts when serum PRL concentrations are low, for example, in mid-to-late pregnancy (Rosen and Barker, 1976) and combined PRL and GH ablation in lactating rats remain to be established. Interestingly, pup removal results in a marked reduction in serum prolactin and a dramatic fall in the level of casein mRNA transcripts, suggesting that local control through milk stasis may have contributed to this effect. The mechanism by which GH affects milk synthesis in the absence of PRL is intriguing and as yet remains unclear.

ACKNOWLEDGEMENTS

This work was funded by the Scottish Office Agriculture and Fisheries Department.

REFERENCES

Barber, M.C., Travers, M.T., Finley, E., Flint, D.J. and Vernon, R.G., 1992, Growth hormone-prolactin interactions in the regulation of mammary and adipose-tissue acetyl-CoA carboxylaes activity and gene expression in lactating rats, *Biochemical Journal* 285:469.

Guyette, W.A., Matusik, R.J. and Rosen, J.M., 1979, Prolactin-mediated transcriptional and post-transcriptional control of casein gene expression, *Cell* 17:1013.

Rosen, J.M. and Barker, S.W., 1976, Quantitation of casein messenger ribonucleic acid sequences using a specific complementary DNA hybridisation probe, *Biochemistry* 15:5272.

CHARACTERIZATION OF TRANSCRIPTIONAL CONTROL ELEMENTS IN THE MAMMARY GLAND AND PANCREAS: STUDIES OF THE CARBOXYL ESTER LIPASE GENE

Ulf Lidberg, Jeanette Nilsson and Gunnar Bjursell

Department of Molecular Biology
University of Göteborg
Medicinaregatan 9c
S-413 90 Göteborg, Sweden

INTRODUCTION

The human lactating mammary gland synthesizes and secretes with the milk a bile salt-stimulated lipase (BSSL) that, after specific activation by primary bile salts, contributes to the breast-fed infant's endogenous capacity for intestinal fat digestion. This enzyme, which accounts for approximately 1% of total milk protein, is a very nonspecific lipase. BSSL is not degraded during passage with the milk through the stomach. It is, however, inactivated when the milk is pasteurized. Our group has isolated a 2359 nucleotide (nt) cDNA clone from human lactating mammary gland and deduced from this, the BSSL enzyme consists of 745 amino acid (aa) residues including a 23 aa leader peptide. We have also isolated and analysed the human BSSL gene, which proved to span a region of 9832 nt and contains 11 exons, see Figure 1. The carboxyl ester lipase (CEL) of human pancreatic juice is a product of the same gene. Because of this, the BSSL gene now is called CEL.

During the screening procedure we found a different gene with a striking homology to the CEL gene, hence called the CEL-like gene (CELL). This new gene proved to span 4846 nt. Data revealed that this CELL gene, compared with the CEL gene, is missing a 4.8 kb segment, including the CEL gene exons 2-7, and hence, this gene is functionally a pseudogene. Despite these differences, the CELL gene is transcribed, and we have isolated a transcript which is composed of exons 1 and 8-11. A hypervariable region present in the last exon of the CELL gene has also been characterized. The two genes have been assigned to chromosome 9q34-qter, separated by approximately 10 kb.

The low nucleotide sequence divergence between the CEL gene and corresponding segment in the CELL gene indicates that the duplication and translocation that gave rise to the CELL gene was a relatively recent event in primate evolution. The 2.5% sequence divergence between the two genes in the noncoding regions thus places the duplication event at about 23 million years ago, suggesting that the CELL gene is only present in primates.

RESULT AND DISCUSSION

The CEL enzyme and the CEL gene are quite different compared to the common milk proteins and their genes. The CEL is an enzyme which is functional when it reaches the intestine. Since the CEL gene not only is expressed in lactating mammary gland but also in exocrine pancreas the CEL gene promoter is clearly different from the common milk protein gene promoters. These differences make the CEL gene a quite unique system for studying tissue specific gene regulation and for understanding the mammary gland.

In the sequence analysis of the flanking regions of the CEL gene we found several putative transcription factor binding sites, cis-elements, including one MGF site in intron 1. A very interesting and probably useful finding is that the structure of the 5'-flanking region of the rat CEL gene shows a relatively high level of homology to the 5'-flanking region of the human CEL gene with eight segments of close homology.

As the CELL gene has lost its tissue specificity, we have isolated and sequenced 1.7 kb of its 5'-flanking region and since it, except for a 90 bp insertion, only shows small differences when compared to the same region in the CEL gene, it could provide some information about the fine structure of the regulatory elements of the CEL gene and point at regulatory elements located outside the 5'-flanking region e.g. possible silencer elements within the region of the CEL gene that are missing in the CELL gene. One possible explanation for the conserved structure of the CELL gene is that it provides "extra" enhancer elements (such as the MGF site present in intron 1 which contribute to the total expression level. The putative MGF cis-element in intron 1 is also present in the CELL gene intron 1. Due to these similarities and the different pattern of expression, the CELL gene 5´-flanking region can be used for elucidating the regulation of the CEL gene.

In order to study DNA-protein interaction we have prepared nuclear extract from mammary gland and pancreas of mice in late pregnancy and early lactation, using liver as a negative control. DNA fragments covering different elements of the putative transcription factor binding sites in the CEL gene flanking region have been isolated. Several of these sites have now been analyzed with nuclear extract through DNAse footprint. We have so far found 5 sites of protein binding.

REFERENCES

Lidberg, L., Nilsson, J., Strömberg, K., Stenman, G., Sahlin, P., Enerbäck, S., and Bjursell, G., 1992, Genomic organization, sequence analysis, and chromosomal localization of the human carboxyl ester lipase (CEL) gene and a CEL-like (CELL) gene, *Genomics* 13:630.

Nilsson, J., Bläckberg, L., Carlsson, P., Enerbäck, S., Hernell, O. and Bjursell, G., 1990, cDNA cloning of human-milk bile-salt-stimulated lipase and evidence for its identity to pancreatic carboxylic ester hydrolase, *Eur. J. Biochem.* 192:543.

Nilsson, J., Hellquist, M. and Bjursell, G., 1993, The human carboxyl ester lipase-like (CELL) gene is ubiquitously expressed and contains a hypervariable region, *Genomics* 17:416.

OLIGOSACCHARIDE ANALYSIS OF EPITHELIAL MAMMARY TISSUES

Anthony J.C. Leathem[1], Susan A. Brooks[1],
Miriam V. Dwek[1] and Udo Schumacher[2]

[1]Breast Cancer Research Group
Department of Surgery
University College London Medical School
London W1P 7PN, U.K.
[2] Department of Human Morphology
University of Southampton Medical School

Oligosaccharides, or short chains of sugars, form a major component of the cell surface, linked to proteins (as glycoproteins) and to lipids (as glycolipids). Multiple rôles have been assigned to the oligosaccharide part of these molecules, particularly in cell recognition and signalling through specific sugar receptors, also called lectins, selectins and carbohydrate reactive proteins (CRP). Of over 100 different monosaccharides described, only about 7 have been detected in the breast: glucose, mannose, fucose, galactose, N-acetyl-galactosamine, N-acetyl-glucosamine and sialic acid. However, about 11 different linkages possible between each sugar allows a vaste number of permutations in sugar chains, for example 34,000 possible forms for a trisaccharide. In practice, the linkage and assembly of these chains is determined by the enzymes of glycosylation available and limited oligosaccharide structures may be conserved by a wide range of organisms in closely related forms which show differing functions, for example the mammalian Stage Specific Embryonic Antigen-1 (SSEA1) and carbohydrate antigens Le^x, Le^y and sialyl-Le^x. Most oligosaccharides have no known function, but a major rôle of many membrane oligosaccharides is adhesion to each other, to other cells and to extracellular matrix.

RESULTS AND DISCUSSION

To understand the structure and function of these sugars during development, we have been exploring technology for mapping oligosaccharides of breast epithelial cells.

The two main approaches for cleaving intact oligosaccharides are:

(a) Enzyme cleavage using either PNGase F, to cleave all classes of N-linked carbohydrate chains at the peptide-carbohydrate linkage, or Endo H, to cleave high-mannose N-linked chains between innermost N-acetyl-glucosamine.

(b) Hydrazinolysis for cleavage at glycosylamine linkage, leaving glycosidic linkages untouched. This presents several practical problems, especially for manual methods by being potentially toxic and explosive. However, automated hydrazinolysis systems are available, to release O and N- linked oliogosaccharides from purified glycoproteins for recombinant biotechnology. By varying temperature and times of hydrazinolysis, we can release O- linked sugars specifically and separately at a lower temperature (eg. 80°C) and N-linked at a higher temperature (eg. 90°C).

The oligosaccharide mixture released is labelled with ^{3}H and detected with a highly sensitive ß counter or by scintillation counting.

Oligosaccharides are separated according to hydrodynamic volume (size) using a P4 gel column for neutral sugars and Dionex carbohydrate HPLC with pulsed amperometric detection for charged sugars. The hydrodynamic volume gives a reasonable idea of the number of monosaccharides present if simultaneously run with glucose-polymers (detected by refractive index). The oligosaccharide peaks are then identified relative to a glucose unit, eg. 2 gu. The presence of aminyl sulphate groups, or other reactive groups, increases the hydrodynamic volume, as does the tritium label.

We have initiated a study of carbohydrate recognition systems on the mammary cell surface by mapping the oligosaccharides of the milk fat globule membrane (MFGM). Using the methods outlined above, MFGM oligosaccharides were isolated, and sialic acid was removed by neuraminidase treatment. The asialylglycans were separated by size chromatography on a Biogel P4 column. Human MFGM oligosaccharides were found to range in hydrodynamic volume (measured in glucose units, "gu") from 1.4 gu (presumably free Glc or Gal) through O-glycans (peaking at 6.5 gu) up to large N-glycans in excess of 20 gu. The oligosaccharide of 6.5 gu has been isolated by further P4 chromatography and then subjected to sequential exoglycosidase digestion, being run through the P4 column after each digestion to determine effect of the digesting enzyme. As a result of sequential treatment with ß-galactosidase, ß-hexoseaminidase, then ß-galactosidase again, the 6.5 oligosaccharide was reduced to a monosaccharide, indicating the structure:

Gal ß 1-3 GlcNAc ß 1-2 Gal ß 1-3 GalNAc

SUMMARY

Our initial results suggest that automated methods, although high in capital costs, provide an effective means of mapping oligosaccharides in mammary epithelial cells. They offer considerable advantage over manual methods which are simple but hazardous. Different strategies for enzymic cleavage of oligosaccharide may be required for each tissue, but other procedures e.g. oligosaccharide hydrazinolysis can be standardised for quantitative release of O- and N-linked sugars.

HUMAN BILE SALT-ACTIVATED LIPASE:
STRUCTURAL ORGANISATION AT THE C-TERMINUS

Kerry M. Loomes

Biochemistry and Molecular Biology Group
School of Biological Sciences, University of Auckland
Auckland, New Zealand

Bile salt-activated lipase is an abundant human milk protein (Wang and Hartsuck, 1993). The primary structure is known as deduced from cDNA (Nilsson *et al.*, 1990; Hui and Kissel, 1990; Baba *et al.*, 1991; Reue *et al.*, 1991) and expression of the murine (Kissel *et al.*, 1990) and human (Hansson *et al.*, 1993) isoforms have been reported. Structural comparisons show that bile salt-activated lipase is homologous to the cholinesterase protein family over approximately 500 N-terminal residues (Krejci *et al.*, 1991). However, bile salt-activated lipase possesses a unique repeated consensus sequence Gly-Ala-Pro-Pro-Val -Pro-Pro-Thr-Gly-Asp-Ser with minor residue substitutions at the C-terminus. This region is rich in O-linked carbohydrate and the human isoform has the largest number of these repeats with 16 compared to 4 and 3 for the murine and bovine isoforms respectively (Figure 1).

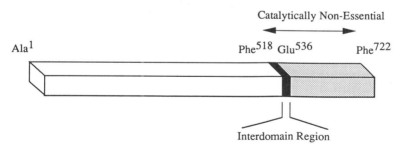

Figure 1. Structural Organisation at the C-terminus

We have isolated and expressed cDNA sequences encoding recombinant human milk bile salt-activated lipase. A full length cDNA sequence and a truncated cDNA sequence encoding a protein which lacked the C-terminus from Leu[519] were constructed by splicing overlapping partial cDNA sequences through a unique *SmaI* restriction site. Both the cDNA constructs encoded a 20 amino acid signal sequence which was correctly cleaved during secretion of the protein. The secreted active recombinant bile salt-activated lipases were conclusively identified from transfected COS-7 and BHK cells by lipase and esterase activity measurements, Western blotting with antisera raised against bile salt-activated lipase purified from human milk, and by N-terminal sequence analysis.

The lipase and esterase activities exhibited by the truncated recombinant bile salt-activated lipase show that the polypeptide region from $Leu^{519} \rightarrow Gln^{535}$ together with the proline-rich repeats starting from Glu^{536} are not catalytically essential (Figure 1). This finding correlates with an expected structural division between the cholinesterase-related part of the tertiary structure and the proline-rich repeats as shown by theoretical comparisons (stippling) (Wang and Hartsuck, 1993). Consequently, we propose that the segment $Leu^{519} \rightarrow Gln^{535}$ (shaded black) comprises part of an interdomain region which links these two parts of the protein tertiary structure.

ACKNOWLEDGMENTS

The early part of this work was carried out under an Auckland University postdoctoral fellowship with Dr. David Christie (Biochemistry and Molecular Biology Group, School of Biological Sciences, University of Auckland) and Prof. Chairman J. O'Connor (Department of Chemistry, University of Auckland) with support from the New Zealand Foundation for the Newborn and the Auckland University Research Committee. The remainder of the work was carried out with an independent research fellowship to Dr. Loomes from the Health Research Council of New Zealand and the Auckland Medical Research Foundation.

REFERENCES

Baba, T., Downs, D., Jackson, K.W., Tang, J. and Wang, C.-S., 1991, Structure of human milk bile salt activated lipase, *Biochemistry* 30:500.

Hansson, L., Blackberg, L., Edlund, M., Lundberg, L., Stromqvist, M. and Hernell, O., 1993, Recombinant human milk bile salt-stimulated lipase. Catalytic activity is retained in the absence of glycosylation and the unique proline-rich repeats, *J. Biol. Chem.* 268:26692.

Hui D. and Kissel, J., 1990, Sequence identity between pancreatic cholesterol esterase and bile salt-stimulated lipase, *FEBS Letts.* 276:131.

Kissel, J.A., Fontaine, R.N., Turck, C.W., Brockman, H.L. and Hui, D.Y., 1989, Molecular cloning and expression of cDNA for rat pancreatic cholesterol esterase, *Biochim. Biophys. Acta* 1006:227.

Krejci, E., Duval, N., Chatonnet, A., Vincens, P. and Massoulie, J., 1991, Cholinesterase-like domains in enzymes and structural proteins: functional and evolutionary relationships and identification of a catalytically essential aspartic acid, *Proc. Natl. Acad. Sci. USA* 88:6647.

Nilsson, J., Blackberg, L., Carlsson, P., Enerback, S., Hernell, O. and Bjursell, G., 1990, cDNA cloning of human milk bile-salt-stimulated lipase and evidence for its identity to pancreatic carboxylic ester hydrolase, *Eur. J. Biochem.* 192:543.

Reue, K., Zambaux, J., Wong, H., Lee, G., Leete, T.H., Ronk, M., Shively, J.E., Sternby, B., Borgstrom, B., Ameis, D. and Schotz, M.C., 1991, cDNA cloning of a carboxyl ester lipase from human pancreas reveals a unique proline-rich repeat unit, *J. Lipid Res.* 32:267.

Wang C.-H. and Hartsuck, J.A., 1993, Bile salt-activated lipase, a multiple function lipolytic enzyme, *Biochim. Biophys. Acta* 1166:1.

DRAMATIC CHANGES IN GAP JUNCTION EXPRESSION IN THE MAMMARY GLAND DURING PREGNANCY, LACTATION AND INVOLUTION

Paul Monaghan, Nina Perusinghe and W. Howard Evans[*]

Institute of Cancer Research, Haddow Laboratories,
Sutton, Surrey, SM2 5NG, U.K.
[*]Department of Medical Biochemistry, University of Wales
College of Medicine, Cardiff CF4 4XN, U.K.

Gap junctions constitute a widespread mechanism for intercellular communication. They are formed by a family of highly conserved connexin proteins which produce channels between adjacent cells permeable to molecules of approximately < 1kDa. Gap junctions are presumed to be important in the regulation of embryonic development, growth and tissue differentiation via ionic and second messenger molecules. In electrically excitable tissues e.g. cardiac muscle, they facilitate the propagation of electrical currents (Bennett *et al.*, 1991). Expression of gap junctions is developmentally and metabolically regulated, as seen in the uterus where there is a large increase in the levels of myometrial connexin43 immediately prior to parturition (Garfield *et al.*, 1977). The mechanisms regulating gap junction expression remain unclear. In mouse mammary gland, gap junctions have been reported by freeze-fracture electron microscopy (Pitelka *et al.*, 1973). Study of gap junctions has recently been facilitated by production of a range of site-directed anti-connexin antibodies (Evans *et al.*, 1992). Using a panel of these antibodies, expression of connexins in mammary tissue has been investigated during pregnancy, lactation and involution.

Female Parke's mice were mated and killed on day 0 (day of mating), and days 1, 4, 8, 12, 16, 18, of pregnancy, day 19 (parturition) days 20, 22, and 24 (lactating) and following removal of the litter on day 28, days 29,32,35 and 41 (involuting). The fourth mammary gland was frozen and stored in liquid N_2. Frozen sections were cut and immunolabelled using rabbit anti-connexin peptide antibodies recognising connexins 26, 32, 40 and 43. After incubation with primary antiserum, sections were washed and incubated with biotinylated sheep anti-rabbit IgG, then streptavidin-fluorescein. Sections were also double-labelled with connexin26 detected by fluorescein as above and mouse anti-keratin 14 antibody (LLOO2) detected by an anti-mouse rhodamine conjugate.

Connexin26 gap junctions were the only gap junctions detected in mammary tissue at any

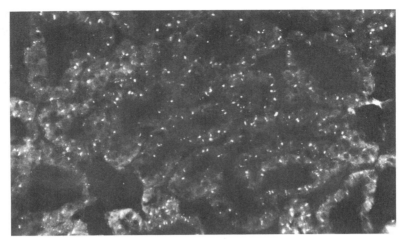

Figure 1. Fluorescence micrograph of connexin26 labelling of lactating mouse mammary gland. x 250.

stage of pregnancy or lactation. In the virgin gland, no gap junctions were detected, but by day 4 of pregnancy, small connexin26 immunoreactive gap junctions were detected in the duct epithelium. As pregnancy progressed, the developing alveolar structures contained clusters of gap junctions surrounding the developing lumina. Connexin26 immunolabelling reached a maximum in both quantity and intensity during lactation (Figure 1). Within 24 h of litter removal, gap junction immunolabelling was dramatically reduced. Double immunofluorescence labelling of gap junctions and cell-type specific anti-keratin antibodies indicated that the majority of gap junctions were present in the luminal cell population (Monaghan *et al.*, 1994).

These results demonstrate rapid modulation of connexin26 expression in mouse mammary gland during pregnancy, lactation and involution. Virgin mouse mammary gland showed no reactivity with any connexin antibody, but connexin 26 gap junctions in the luminal cell population increased rapidly during pregnancy and decreased rapidly at involution. The increase in gap junction expression during pregnancy may indicate an increased demand for intercellular communication during lactation.

ACKNOWLEDGEMENTS

This work is supported by the Cancer Research Campaign and Medical Research Council.

REFERENCES

Bennett, M.V.L., Barrio, L.C., Bargiello, T.A., Spray, D.C., Hertzberg, E.A. & Saez, J.C., 1991, Gap Junctions: new tools, new answers, new questions, *Neuron* 9:305.

Evans, W.H., Carlile, G., Rahman, S. & Torok, K., 1992, Gap junction communication channel: peptides and antipeptide antibodies as structural probes, *Biochem. Soc. Trans.* 20:856.

Garfield, R.E., Sims, S., and Daniel, E.E., 1977, Gap junctions: the presence and necessity in myometrium during parturition, *Science* 198:958.

Monaghan, P., Perusinghe, N.P., Carlile, G. and Evans W.H., 1994, Rapid modulation of gap junction expression in the mouse mammary gland during pregnancy, lactation and involution, *J. Histochem. Cytochem*: 42:931.

Pitelka, D.R., Hamamoto, S.T., Duafala, F.G. and Nemanic, M.K., 1973, Cell contacts in the mouse mammary gland, *J. Cell Biol.* 56:797.

AROMATASE INDUCTION IN STROMAL VASCULAR CELLS FROM HUMAN BREAST ADIPOSE TISSUE BY GROWTH FACTORS

Martin Schmidt and Georg Löffler

Institute of Biochemistry, Genetics and Microbiology,
University of Regensburg, D-93040 Regensburg, FRG

INTRODUCTION

Stromal vascular cells from human adipose tissue are able to produce large amounts of estrogens; aromatase induction in these cells by cAMP or glucocorticoids in the presence of as yet unidentified serum factors was described (reviewed by Simpson *et al.*, 1989). The finding of elevated aromatase activity in the adipose tissue near breast tumours (O'Neill *et al.*, 1988) implies that tumour cells produce aromatase inducing factors. To test whether there exists a paracrine mechanism for aromatase induction based on growth factors, which are released by breast cancer cells (Dickson *et al.*, 1990), we studied the regulation of aromatase activity in stromal cells of human breast adipose tissue by some of these factors. We showed that aromatase induction by cortisol depends on the addition of either PDGF-BB or bFGF together with db-cAMP (Schmidt and Löffler, 1994a). Here we present data which suggest that stromal vascular cells from breast adipose tissue are more susceptible to aromatase induction by PDGF than those from subcutaneous abdominal adipose tissue.

METHODS

Preparation and culture of cells (Schmidt and Löffler, 1994b) and aromatase induction and the [^{3}H]release assay (Schmidt and Löffler, 1994a) were performed as described earlier.

RESULTS

In subcultivated stromal vascular cells from breast adipose tissue, aromatase activity could be induced by the addition of cortisol and recombinant human PDGF-BB. The effect of db-cAMP in the presence of PDGF is strongly affected by cortisol: aromatase induction was suppressed by PDGF in the absence of cortisol, while it was potentiated in its presence. In primary cultures of stromal vascular cells from breast adipose tissue, cortisol alone led to a significant induction of aromatase activity within 1 day. This effect was potentiated by PDGF. When the induction period was extended to 3 days, only slight differences to the values obtained after one day could be observed. In cells from subcutaneous abdominal adipose tissue PDGF potentiated the cortisol effect after 1 day incubations, but the activities were significantly lower than in cells from breast tissue. PDGF was without any effect when cells were incubated for 3 days.

DISCUSSION

PDGF-BB induced aromatase activity in stromal cells from breast adipose tissue in the presence of cortisol. Addition of db-cAMP increased this effect. bFGF was unable to support the cortisol effect on aromatase activity in the same way as PDGF or serum (Schmidt and Löffler, 1994a), but it acted like PDGF in the presence of cortisol together with db-cAMP. In the absence of cortisol and at physiological concentrations of insulin, PDGF and bFGF inhibit aromatase induction by db-cAMP. The presence of glucocorticoid seems to be a prerequisite for a switch in the mode of action of growth factors, as their inhibitory effect in the absence of cortisol turns into a strong stimulatory effect in its presence.

Breast adipose tissue stromal cells respond to PDGF with significantly higher aromatase activity than stromal cells from subcutaneous adipose tissue. Beyond that they seem to retain this activity for markedly longer. This could be interpreted as follows: in cells from abdominal origin PDGF induces aromatase transcription for a very limited period, resulting in a short-lived peak of activity, in cells from breast tissue PDGF would drive transcription for a longer period. The fact that the effect of PDGF on aromatase induction in stromal cells depends on the source of adipose tissue used raises a number of pathological and physiological issues. The results presented here strongly support the hypothesis that breast tumour cells are able to induce elevated aromatase activity in surrounding adipose tissue by a paracrine mechanism, PDGF being a candidate inducing agent. In view of our results one can speculate that PDGF might be the (or one of the) serum component(s) responsible for glucocorticoid induction of aromatase. The physiological significance of our results remains to be elucidated. It would be tempting to speculate that even nonmalignant epithelial cells of the mammary gland react to elevated concentrations of circulating estrogens by release of growth factors. These in turn could induce local estrogen production in breast adipose tissue stromal cells, maintaining high concentrations of estrogens around the mammary gland for a limited time, but independent of oscillations of circulating estrogens.

REFERENCES

Dickson, R.B., Thompson, E.W. and Lippman, M.E., 1990, Regulation of proliferation, invasion and growth factor synthesis in breast cancer by steroids, *J. Steroid Biochem. Mol. Biol.* 37:305.

O'Neill, J.S., Elton, R.A. and Miller, W.R., 1988, Aromatase activity in adipose tissue from breast quadrants: a link with tumour site, *Br. Med. J.* 296:741.

Schmidt, M. and Löffler, G., 1994a, Induction of aromatase in stromal vascular cells from human breast adipose tissue depends on cortisol and growth factors, *FEBS Lett.* 341:177.

Schmidt, M. and Löffler, G., 1994b, The human breast cancer cell line MDA-MB231 produces an aromatase stimulating activity, *Eur. J. Cell Biol.* 63:96.

Simpson, E.R., Merrill, J.C., Hollub, A.J., Graham-Lorence, S. and Mendelson, C.R., 1989, Regulation of estrogen biosynthesis by human adipose cells, *Endocr. Rev.* 10:136.

DEVELOPMENTAL REGULATION OF HOMEOBOX GENES IN THE MOUSE MAMMARY GLAND

Deborah Phippard[1], Paul Sharpe[2] and Trevor Dale[1]

[1]Institute of Cancer Research
Haddow Laboratories, Sutton, Surrey SM2 5NG
[2]United Medical and Dental Schools
Guy's Hospital Campus
London Bridge, London SE1 9RT, UK

INTRODUCTION

Northern analysis has been used to demonstrate that *Msx-1*, *Msx-2* and *Msx-3*, which are expressed in areas of putative epithelial/mesenchymal interactions during embryogenesis, are expressed differentially in the mouse mammary gland.

RESULTS

Msx-2 is down-regulated earlier in pregnancy than *Msx-1* [8.5 days post coitus (p.c.) compared to 12.5 days p.c.]. *Msx-1* is re-expressed during involution. *Msx-3* is expressed at low levels in the virgin mammary gland, with a strong induction 2.5 days p.c. that declines by 4.5 days p.c.

mRNA extracted from virgin mammary fat pads prior to invasion by the epithelium was used to demonstrate that *Msx-1* and *Msx-3* are expressed in the mesenchyme, while *Msx-2* is either expressed in the epithelium itself, or expression is dependent upon the close proximity of epithelium.

A ~ 10-fold down-regulation of *Msx-2 in vivo* was observed as a consequence of ovariectomy. A potential role for estradiol in *Msx-2* regulation was investigated *in vitro* using an estrogen dependent sub-line of MCF-7 cells (see Figure 1). Cells withdrawn from estradiol showed a 3-fold down-regulation of the *Msx-2* 1.7 kb transcript and a 2-fold down-regulation of the 2.9 kb transcript, relative to control cells in 10^{-8}M estradiol. Whereas, pS2, a gene previously demonstrated to be estradiol dependent (Chalbos *et al.*, 1993), was reduced 4-fold relative to control. $1,25(OH)_2D_3$ did not affect *Msx-2* expression.

Intercellular Signalling in the Mammary Gland
Edited by C.J. Wilde *et al.*, Plenum Press, New York, 1995

DISCUSSION

Epithelial/mesenchymal interactions are known to be crucial for normal mammary gland development. It is interesting that *Msx* genes which are temporally expressed in either the mesenchymal or the epithelial compartment during embryogenesis, also retain this specificity during the development of the mammary gland.

Msx-3 has an interesting expression profile in the mammary gland as it is induced at the onset of pregnancy coincident with changes in phenotype. This is suggestive of a role in initiating the pregnant phenotype. *Msx-3* expression is currently being investigated in more detail. *Msx-2* appears hormonally regulated, as demonstrated both by the ovariectomy data and the fact that estradiol withdrawal causes down-regulation of *Msx-2* expression in a transformed cell line. *Msx-1*, while clearly differentially expressed in the mammary gland, does not have an expression profile which corresponds with an obvious change in phenotype. Although Hodgkinson *et al.* (1993) have shown up-regulation of *Msx-2* by $1,25(OH)_2D_3$ in human bone cell cultures, *Msx-2* expression was not up-regulated by $1,25(OH)_2D_3$ in our system. MCF-7 cells are competent to respond to $1,25(OH)_2D_3$ as Colston *et al.* (1992) have demonstrated growth inhibition on treatment with this vitamin. However, the induction of *Msx-2* may be a feature of the vitamin D_3 response that is specific for bone-derived cells.

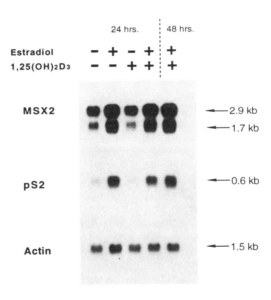

Figure 1. Effect of Estradiol and $1,25(OH)_2D_3$ on MSX2 Expression in the Human Breast Cancer Cell Line MCF-7.

REFERENCES

Chalbos, D., Philips, A., Galtier, F. and Rochefort, H., 1993, Synthetic antiestrogens modulate induction of pS2 and cathepsin-D messenger ribonucleic acid by growth factors and adenosine 3',5'-monophosphate in MCF7 Cells, *Endocrinology* 133:571.

Colston, K., Mackay, A.G., James, S.Y., Binderup, L., Chander, S. and Coombes, R.C., 1992, A new vitamin D analogue that inhibits the growth of breast cancer cells *in vivo* and *in vitro*, *Biochem. Pharm.* 44:2273.

Hodgkinson J.E., Davidson, C.L., Beresford, J. and Sharpe, P.T., 1993, Expression of a human homeobox-containing gene is regulated by $1,25(OH)_2D_3$ in bone cells. *Biochim. Biophys. Acta* 1174:11.

MURINE α-LACTALBUMIN GENE INACTIVATION AND REPLACMENT

Alexander Stacey and Angelika Schnieke

Pharmaceutical Proteins Ltd.
Roslin, Midlothian EH25 9PP, UK

We are using gene targeting to investigate the role of α-lactalbumin in lactation and the regulation of α-lactalbumin gene expression. Gene targeting in embryonic stem (ES) cells is now widely used to inactivate genes of interest. We have recently demonstrated a powerful technique, termed double replacement, whereby a wide variety of genetic alterations can be engineered at a chosen locus and the effects examined in mice (Stacey *et al.*, 1994) The procedure involves two consecutive rounds of gene targeting: the first creates a null allele of the gene of interest, which can then be replaced with DNA carrying experimental alterations in the second round. This process effects a clean replacement of DNA sequences, and no unwanted exogenous DNA (eg marker genes) are left at the target site. A most useful aspect of double replacement gene targeting is that multiple independent second step targeting experiments can be carried out, making it possible to produce and compare many experimental alterations at the same locus.

We have used double replacement to generate two novel alleles at the murine α-lactalbumin locus as a first step in the analysis of α-lactalbumin protein function and gene expression. The targetted alleles are:

1. Complete deletion of the murine α-lactalbumin coding region and promoter (0.57kb of upstream sequences).

2. Replacement of the deleted murine gene with the human α-lactalbumin gene coding region and promoter (0.772kb of upstream sequences).

Both alleles have been transferred to mice, and substrains established.

Experimental animals carrying a genetic deficiency of α-lactalbumin allow a definitive test of the role and importance of this protein in lactation and mammary gland physiology. In the lactating mammary gland, α-lactalbumin associates with ß(1-4) galactosyltransferase to form the lactose synthetase enzyme complex responsible for milk lactose synthesis. α-lactalbumin content varies widely between milks of different species, being the major whey protein in humans, but is almost absent in marine animals (Schmidt *et al.*, 1971). Milk α-lactalbumin concentration correlates with milk lactose concentration (Fitzgerald *et al.*, 1970). Lactose is thought to provide the major osmotic component of milk and therefore determines milk volume. We are studying the relationships between α-lactalbumin null allele in heterozygous and homozygous form. Preliminary studies indicate that homozygous animals produce thickened milk which fails to sustain pups beyond 8-10 days of life. A more comprehensive analysis is in progress. Mice homozygous for the human α-lactalbumin gene replacement provide a means of testing whether a functional interspecies hybrid can form between human α-lactalbumin and murine ß(1-4) galactosyltransferase. Preliminary results are that these animals produce apparently normal milk and rear offspring successfully, indicating a functional replacement of the mouse gene.

Human/mouse α-lactalbumin heterozygotes can provide information about the major determinants of α-lactalbumin expression levels in different species. α-lactalbumin is present at a significantly higher concentration in human milk ($\sim$ 2mg/ml) than in mouse milk ($\sim$ 0.2mg/ml). Analysis of the human gene placed at the mouse locus will inform us as to whether the level of α-lactalbumin expression is determined by properties of the gene itself or regulatory elements close to it, as other studies of α-lactalbumin transgenes have suggested (Soulier *et al.*, 1992), or whether expression level is primarily determined by the genetic locus. Comparison of RNA and protein expression by a single human and a single mouse gene present as alleles at the same locus in the same mouse will allow us to distinguish between these alternatives.

Embryonic stem cells carrying the mouse α-lactalbumin gene deletion provide an excellent tool with which to carry out a detailed study of the regulation of α-lactalbumin gene expression. Experimental alterations can be engineered *in vitro* in any region of the α-lactalbumin gene covered by the deletion, a mutant gene used to replace the deletion by gene targeting, and the effects studied in mice. Clearly, mutations introduced in this way are more informative than those contained in randomly integrated transgene constructs, because they are at the correct locus and subject to the same regulatory influences. It is also important to emphasise that mutations introduced by double replacement can be studied in an otherwise undisturbed genome, without the influence of exogenous marker genes.

Such investigations are now feasible because recent advances in ES cell technology have significantly improved the practicality and efficiency of this approach to genetic manipulation. Specifically these are: an ES cell line which contributes to the germ line of mice with high efficiency (Magin *et al.*, 1992), a simple practical method of gene replacement in ES cells (Stacey *et al.*, 1994) and a simplified method of chimera production (reviewed in Wood *et al.*, 1993). We suggest that the advantages of gene targeting as a means of studying molecular aspects of mammary physiology are now compelling.

REFERENCES

Fitzgerald, D.K., Brodbeck, U., Kiyosawa, I., Mawal, R., Colvin, B. and Ebner, K.E., 1970, α-lactalbumin and the lactose synthetase reaction, *J. Biol. Chem.* 245:2103.

Magin, T.M., McWhir, J. and Melton, D.W., 1992, A new mouse embryonic stem cell line with good germ line contribution and gene targeting frequency, *Nucl. Acids Res.* 14:3795.

Schmidt, D.V., Walker, I.E. and Ebner, K.E., 1971, Lactose synthetase activity in northern fur seal milk, *Biochim. et Biophys. Acta* 252:439.

Soulier, S., Vilotte, J.L., Stinnakre, M.G, and Mercier, J.C., 1992, Expression analysis of ruminant α-lactalbumin in transgenic mice: developmental regulation and general location of important cis-regulatory elements, *F.E.B.S. Letts.* 297:13.

Stacey, A.J., Schnieke, A., McWhir, J., Cooper, J., Colman, A. and Melton, D.W., 1994, Use of double replacement gene targeting to replace the murine α-lactalbumin gene with its human counterpart in embryonic stem cells and mice, *Mol. Cell Biol.* 14:1009.

DIFFERENTIAL EXPRESSION OF ß-LACTOGLOBULIN ALLELES A AND B IN DAIRY CATTLE

R.J. Wilkins[1,2], H.W. Davey[1], T.T. Wheeler[1] and C.A. Ford[2]

[1]AgResearch, Ruakura Agricultural Centre, Hamilton, New Zealand
[2]Centre for Research in Animal Biology University of Waikato, Hamilton, New Zealand

Two allelic variants of the whey protein ß-lactoglobulin gene are commonly found in dairy cattle. Interestingly, some 20 to 40% more of the A than the B variant protein is produced in milk. We report here our efforts to elucidate the basic molecular mechanisms responsible for the differential expression of the two ß-lactoglobulin alleles.

Samples of total RNA from lactating mammary tissue of A/B heterozygotes were run in duplicate and probed with either A or B allele-specific oligonucleotide probes (Wilkins, 1993) under stringent hybridisation conditions to determine the relative amounts of A and B allelic mRNA. Four animals were analysed in this way and found to give ratios of 1.66, 1.56, 1.37 and 1.87 to one for the ratios of A to B mRNA (Ford, 1993). One animal which consistently had a much higher ratio of B to A protein in its milk, 2.7 to one, was found to have a correspondingly high ratio of the mRNA alleles, 2.8 to one (Ford, 1993).

Because of problems in quantitating the relative hybridisation of probes to the A and B allelic mRNA, samples of total RNA were reverse transcribed, subjected to PCR, using a 5' oligonucleotide primer end-labelled with ^{33}P, cleaved with restriction enzyme *Hae* III and the fragments corresponding to the A and B allelic mRNAs separated by denaturing PAGE (Ford *et al.*, 1993). Relative levels of A and B allelic mRNA from two normal animals was 1.6 to one and for the animal with an abnormally large ratio of A to B variant protein, 2.4 to one. Whether the higher level of A allele mRNA is due to transcriptional or stability differences is uncertain.

Sequencing of 5' upstream DNA from semen of several AA and BB homozygotic bulls showed several consistent differences between the A and the B alleles (Figure 1), and two of these were within sequence motifs reported to bind nuclear proteins with putative gene regulation properties (Watson *et al.*, 1991). Hypothesising that these proteins might bind the A and B allele motifs with different affinities, oligonucleotides were synthesised for each region for each allele and used in gel shift assays with nuclear extracts prepared from lactating dairy cattle. No significant differences were seen in the gel shifting properties of the two sets of allelic oligomers, nor in competition assays were there any differences in the effectiveness of A and B allele oligomers in sequestering proteins bound to labelled oligomers.

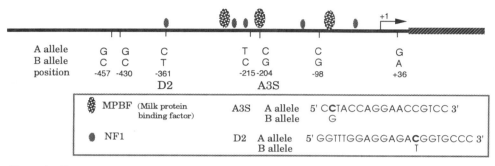

Figure 1. Upstream regions of the A and B alleles of ß-lactoglobulin. Few, if any, significant differences exist in the immediate upstream regions of the A and B alleles of the ß-lactoglobulin gene.

In order to measure the relative levels of ß-lactoglobulin A and B variant protein synthesis *in vitro*, mammary tissue samples were taken immediately after slaughter, diced into ~ 1 mm cubes and incubated for 3 h in the presence of ^{35}S-methionine. Protein samples were analysed by 2-D PAGE. Two experiments were done. In the first, tissue was stored on ice for 1 h, then dispersed and incubated. In this case ^{35}S incorporation was poor but equal amounts of the A and B variants proteins were synthesised - as evidenced by autoradiography. In the second case, tissue was immediately dispersed into pre-warmed media after slaughter. This resulted in a 10-fold increase in ^{35}S incorporation, with protein synthesis approaching *in vivo* levels. In this case about 1.4 times more A than B variant protein was synthesised, a ratio very similar to that observed in the bulk milk from these animals.

In conclusion, differences at the mRNA level appear to be, at least in part, responsible for the excess of ß-lactoglobulin A over B variant protein in milk. It is not yet clear whether these differences arise at the level of mRNA transcription or stability. Preliminary evidence suggests that under conditions of high metabolic activity the differences become accentuated, possibly due to differing affinities for a limited pool of mRNA transcription and processing factors.

REFERENCES

Ford, C.A., 1993, Gene expression of the A and B variants of ß-lactoglobulin, MSc Thesis, The University of Waikato, New Zealand.

Ford, C.A., Connett, M.B. and Wilkins, R.J., 1993, ß-lactoglobulin expression in bovine mammary tissue, *Proc. NZ Soc. Anim. Prod.* 53:167.

Watson, C.J., Gordon, K.E., Robertson, M. and Clark, A.J., 1991, Interaction of DNA-binding proteins with a milk protein gene promoter *in vitro*: identification of a mammary gland-specific factor, *Nucl. Acids Res.* 19:6603.

Wilkins, R.J., 1993, Allele specific oligonucleotide probes for mRNA, *Biotech.* 14:352.

DNA REPLICATION, ß-CASEIN GENE TRANSCRIPTION AND DNase-I SENSITIVITY OF ß-CASEIN GENE IN RABBIT MAMMARY CELL NUCLEI

Lech Zwierzchowski[1], Wanda Sokól-Misiak[1], Eve Devinoy[2]
Louis-Marie Houdebine[2], Wieslaw Niedbalski[1]
and Danuta Kleczkowska[1]

[1]Institute of Genetics and Animal Breeding
Polish Academy of Sciences, Jastrzrbiec,
05-551 Mroków, Poland
[2]National Institute for Research in Agronomy (INRA)
Laboratory of Cellular and Molecular Biology
78352 Jouy-en-Josas, France

INTRODUCTION

ß-casein is a major milk protein synthesized by the mammary gland during lactation. In the mammary gland of the rabbit, the expression of ß-casein mRNA changes markedly as a function of reproductive stage, being very low in virgin animals and fully expressed in the lactation stage.

RESULTS AND DISCUSSION

In this work, we studied the DNase-I sensitivity of the ß-casein gene locus in cell nuclei prepared from the mammary glands of rabbits at different physiological stages - pregnancy or early lactation. Changes in overall transcription rates, transcription of ß-casein gene, and replicative DNA synthesis were also measured in these nuclei and related to changing DNase-I sensitivity of the gene. Our results show that the replicative type of DNA synthesis predominates in mammary tissue during pregnancy and just after parturition when mammary epithelial cells proliferate extensively and the mass of the mammary gland increases significantly. On the other hand, maximum RNA-polymerase-I-dependent RNA synthesis and maximal transcription of rRNA genes was observed in early lactation. At this time the synthesis of rRNA attained a high level characteristic of terminally-differentiated mammary epithelial cells. Transcription carried out by RNA polymerase-II peaks in the mammary gland during the second half of pregnancy and correlates well with the morphological differentiation of the mammary tissue. On the other hand, the transcription rates of the ß-casein gene, which is expressed specifically in the mammary epithelial cells, increased mostly in late pregnancy and reach a maximum in early lactation.

Transcriptionally-active genes or genes being potentially active are often located in a chromatin fraction more sensitive to DNase-I digestion than the rest of the chromatin. Our experiments show that on day 15 of pregnancy the relative sensitivity of ß-casein gene (i.e. the gene which is active or at least potentially active in mammary gland cells) equals that of α-globin gene. In late pregnancy (day 25) or in early lactation (day 5) the sensitivity to DNase-I of the ß-casein gene substantially exceeded that of α-globin gene. In order to localize regions of DNase-I sensitivity in the rabbit ß-casein gene, Southern-blot analysis was performed on promoter region and coding sequences of the gene. Cell nuclei prepared from the mammary glands of pregnant or lactating rabbits were digested with DNase-I. Then, DNA was isolated from the nuclei, digested with restriction nucleases *EcoRI* or *BglII* and hybridized to ^{32}P-labelled ß-casein cDNA probe or to 0.38 probe (380-bp DNA fragment complementary to the rabbit ß-casein gene promoter; -324 bp to +56 bp region). Our results show that in the developing mammary gland, both coding sequences and promoter region become DNase-I sensitive in the actively-transcribed ß-casein gene. The increase in the sensitivity of ß-casein gene to DNase-I digestion parallels the induction of the gene transcription which occurs in the mammary gland between days 15 and 25 of pregnancy. Assuming that the increased level of DNase-I sensitivity reflects a more open structure of the transcriptionally-active chromatin, we conclude that conformational changes in the chromatin structure occur in mammary gland cells in the second half of pregnancy. Once activated, the chromatin region encompassing the milk protein gene remains open and transcriptionally-active during lactation.

AUTOCRINE CONTROL OF MILK SECRETION: DEVELOPMENT OF THE CONCEPT

Malcolm Peaker

Hannah Research Institute
Ayr, Scotland KA6 5HL, UK

INTRODUCTION

Over the past twenty years it has been established that the mammary gland is subject to more than one level of physiological control. The ultimate pattern of parental expenditure is obviously subject to natural selection and is, therefore, controlled genetically. The strategic level decisions, which may also involve interplay between mother and young, are signalled to the mammary gland by the endocrine system (see Peaker, 1989). This strategic level of control is then subject to environmental constraints, including nutrition, which can modify the rate of milk secretion systemically (see Loudon and Racey, 1987; Peaker, 1989). Such plasticity in the rate of milk secretion also appears to be mediated mainly, if not solely, by the endocrine system. However, these strategic and environmental forms of control, exerted systemically, are modulated by a tactical control mechanism operating within each mammary gland to set the actual rate of milk secretion. This local, intra-mammary mechanism responds to the frequency or completeness of milk removal by the young or by milking in dairy animals. The intra-mammary control of secretion has now been shown to operate in all mammals which have been studied - from dairy cows and goats to marsupials and man. This article describes the discovery of local control and its mechanism.

The intra-mammary level of control is possible because the mammary gland is a very unusual exocrine gland in that: (i) it secretes continuously during lactation whereas most glands (salivary glands, sweat glands, pancreas, for example) secrete intermittently in response to an acute stimulus, (ii) secretion is stored within its lumen until removed at intervals by the sucking young or, in dairy animals, by milking. Examples of exocrine glands which store their secretion are not common but snake venom glands are one (see Paine *et al.*, 1992). Indeed the suicide of Cleopatra in 30 B.C., by clasping an asp to her bosom must be one of the few occasions in history when two unusual exocrine glands were brought into apposition.

HISTORY

The history of the local, i.e. intra-mammary, control of milk secretion has three strands, one agricultural, the second physiological and the third in comparative reproductive physiology. Intercellular communication within the mammary gland as a whole and not just in the control of milk secretion has an additional strand from the study of cancer and the cellular products of cancerous cells.

Dairy Research

In dairying the importance of the frequency of milking and the interval between milkings on the rate of milk secretion has long been recognised. This earlier work, all done in the dairy cow, was reviewed by Elliott (1959a, b). In short-term experiments there were great problems in interpreting the findings because of the confounding effect of the variable amount of residual milk present after each milking. A small variation in the amount of milk remaining in the gland after each milking could make it appear that under a certain regimen the rate of secretion increased while under another it decreased when in fact there was no change in rate at all. For the secretion of fat, interpretation was even more complicated because the concentration of fat is higher in the milk stored in the alveoli and small ducts than in that stored in the large ducts and cistern of the gland. However, this problem did not interfere with the longer-term effects of increasing the frequency of milking and these studies clearly showed an increase in the rate of milk secretion when cows were milked three or four times each day instead of twice.

Elliott (1961) then went on to compare the effect of milking two glands (half the udder) thrice daily while the other two glands of the same animal continued to be milked twice daily. The rate of milk secretion increased rapidly but only in those glands milked thrice daily. Therefore, the physiological mechanism responsible was not systemic, because all glands would then have been affected equally, but local. Similar experiments were done later by Morag, (1973). Elliott concluded, '... *it is perhaps due to some local effect of the treatment on the secretory cells ...*'. Elliott (1961) also manipulated the milking process to leave behind an increased but relatively small amount of residual milk; in those animals the rate of milk secretion declined.

How an individual interpreted the quantitative determination of the rate of milk secretion seemed to depend on his or her background. At the time Elliott's (1961) experiment was done classical endocrinology was entering its heyday and determination of the rate of milk secretion was being interpreted by endocrinologists as being by hormones released in response to the milking stimulus and by nutritionists as being by the substrate supply to the mammary gland. However, the rate of milk secretion between milkings had been seen by dairy scientists, particularly by those in the U.S.A., as being controlled locally - by the increase in intra-mammary pressure brought about by milk accumulating in the mammary gland. According to this hypothesis, the rate of milk secretion decreased as milk accumulated and pressure rose. Frequent removal of milk simply relieved this pressure and the mean rate of secretion over a day was higher. This simple theory received a knock when it was shown in the dairy cow that between milkings the rate of milk accumulation is approximately linear with no tail-off until 16 h or more after the last milking (Elliott *et al.*, 1960). The difference between milking at 8-h or 12-h intervals could not be due to relief of inhibition by distension because there was no inhibition to relieve. These findings do not mean to say that mammary distension cannot or does not inhibit milk secretion; it does, eventually. There is very good direct evidence that long periods of milk accumulation do inhibit secretion by stretching the mammary secretory epithelium and that this process

194

occurs after the cessation of milk removal intentionally or accidentally in dairy animals (Fleet and Peaker, 1978; Peaker, 1980).

Physiological Research

The physiological investigations came to the same problem from a different direction since there were claims that oxytocin not only caused milk ejection but also caused an increase in the rate of secretion in cows, goats and sheep (see references in Linzell and Peaker, 1971). The late James Lincoln Linzell in developing his techniques for studying mammary function *in vivo* needed a technique for very frequent and virtually complete milk removal, particularly to investigate short-term effects on the rate of secretion and to obtain milk for isotopic studies using labelled precursors of milk constituents.

The milk yield of goats and cows milked every hour is variable because the milk ejection reflex cannot be elicited that frequently and sometimes more and sometimes less residual milk is obtained. Therefore, a number of workers had given oxytocin intravenously just before each milking in order to contract the myoepithelium and thereby obtain as much milk as possible; in that way more uniform yields were obtained. Linzell (1967) refined that procedure, in the goat. He used the amount of oxytocin normally released in response to milking in order to avoid any side effects of using the large and sometimes huge doses that had been used, and even in the 1990s still are used, by others.

It was at that point that the present author became involved. Hourly milking with the aid of even low doses of oxytocin causes, in some goats, changes in the concentrations of ions and lactose in the milk. When I joined Linzell at Babraham in 1968 to work on the mechanism of milk secretion, we realised that change in composition had to be explained. Looking through the results of earlier experiments we noticed two things. Firstly, there was a tendency for milk yield to increase during hourly milking. In Linzell's (1967) paper, oxytocin from two sources had been used, a natural extract and synthetic oxytocin; the increase in milk yield was more evident with the latter. Secondly, in some studies involving close-arterial infusion of isotopically-labelled substrates only the one gland had been milked hourly in order to obtain the labelled products; it was evident that only in that gland had milk yield increased. That anomaly was the key to uncovering the intra-mammary control of milk secretion.

While we both spotted the anomaly of the unilateral effect, we still needed a method of handling the data in order to distinguish between random changes in both milk output and composition in the two glands of the goat. For this I can claim the credit with the pride only a fellow inept mathematician could appreciate. As I gathered the data from the first few experiments I realised that day-to-day and milking-to-milking changes in yield and composition in the two glands varied not independently but in parallel. If yield, or the concentration of Na^+, increased by 15% in the left gland then it increased by 15% in the right. I devised a simple calculation, which we still use, to test for deviation from parallel behaviour of the two glands of goats or the four glands of cows.

We, therefore, studied not only the change in milk composition (and explained the mechanism) that occurs with frequent milking but also the increase in secretory rate (Linzell and Peaker, 1971). It was shown that the rate of milk secretion increased progressively over 10 h of hourly milking. When auto-transplanted glands were milked hourly without the aid of oxytocin and the volumes of milk obtained throughout the day were added to the normal afternoon yield, there was also a stimulation. Auto-transplanted glands are denervated and milking them does not evoke a milk ejection. Thus, it was shown that milk removal and not oxytocin is important in stimulating milk secretion. This view was reinforced by the finding that massaging an autotransplanted gland without actual milk removal resulted in no significant stimulation. Results that could be interpreted similarly had been obtained by Denamur and Martinet (1961) in sheep. They injected oxytocin after each milking and found an increase only when the residual milk was removed; oxytocin treatment without additional milking had no effect.

Unfortunately, despite these and other findings that the effect of exogenous oxytocin is solely to contract the myoepithelium and that milk removal is the key element in stimulating secretory rate, one still finds those investigating the effects of oxytocin without reference to the extensive studies on intra-mammary control of milk secretion (for example, Ballou *et al.*, 1993).

In our short-term experiments and from the mechanical behaviour of secretory alveoli when treated with oxytocin which Linzell had observed in his early studies on the myoepithelium, we did not consider it likely that the relief of mammary distension was involved and that the most likely explanation of the phenomenon was some sort of local negative feed-back mechanism within the gland itself, expressed as follows: '*Therefore, it is possible that some local humoral factor may be involved in the local regulation of the rate of secretion*' (Linzell and Peaker, 1971).

In 1971 we had a putative mechanism by which the mammary gland could adjust the rate of milk secretion to the size and nutritional requirements of the young at different stages of lactation. The energy and specific substrate demands of lactation are huge (see Loudon and Racey, 1987) and the selective advantages of a mechanism that conserves expenditure are obvious. We also appreciated that in a number of species from the pig (De Passille *et al.*, 1988, for example) to the hyraxes (Hoeck, 1977; Rudnai, 1984) each of the young feeds from the same teat at each suckling and that the local control mechanism may be involved in regulating the amount of milk secreted for young of different sizes in any one litter, whereas the central mechanism by which milk secretion is maintained by suckling and the release of adenohypophysial hormones would affect all glands equally.

Comparative Reproductive Physiology

The third strand in the history of intra-mammary control of milk secretion lay in the important and fascinating studies on the reproduction of macropod marsupials, led with great distinction by G.B. Sharman in the 1960s. I can do no better than to quote from his review of reproductive physiology in marsupials published in *Science* (Sharman, 1970):

> *Some marsupials, notably kangaroos, have overlapping periods of lactation - a young on foot being suckled on one of the four teats while another, much smaller, young is suckled in the pouch on another teat. The two milks, secreted concurrently by separate mammary glands of the same female, appear very different, and because the young on foot is over 200 days older than the pouch young the two milks apparently differ in composition... It is...clear that kangaroos commonly produce, at the same time from separate mammary glands, two milks which differ vastly in chemical composition and that these are produced under the same endocrine environment.*

ANOTHER CASE OF INTRA-MAMMARY CONTROL

Arising from further work on the cellular transport mechanisms responsible for the formation of the aqueous phase of milk another physiological situation involving local control of the mammary gland was found: the change in milk yield and composition that occurs at the onset of copious secretion in the peri-parturient period (lactogenesis stage II).

If milking is started in late pregnancy, the composition of the secretion changes from that of colostrum to that of normal milk. Milking pre-partum, therefore, accelerates the change that occurs naturally in the peri-parturient period. In terms of the aqueous phase of milk, there is a fall in the concentrations of sodium and chloride and a rise in potassium and lactose. In investigating the cellular mechanisms responsible for this change, goats were milked pre-partum, but only one gland was milked; the other was left unmilked until after parturition. Milk composition and permeability of the mammary epithelium changed only in the gland that was milked. Therefore, again, this was a clear case of local intra-mammary control compared with systemic control (Linzell and Peaker, 1974). The following hypothesis was proposed to account for these observations:

'... a substance is produced within, or accumulated by, the mammary gland in the hormonal milieu of pregnancy; this substance acts on the secretory epithelium to maintain a paracellular pathway. Near term as hormonal changes take place, the substance is no longer produced and that remaining in the gland is metabolized, so that the epithelium becomes "tight" and the composition changes to that of normal milk. In pre-partum milking the substance is removed at a greater rate than its rate of production, so that again milk composition would alter in the milked gland'.

Evidence was obtained later that the change in milk composition is related to the production and metabolism of prostaglandins by the mammary gland (Maule Walker and Peaker, 1980). During late pregnancy in the goat, prostaglandin $F_{2\alpha}$ is produced by the mammary gland; this output ceases near term. Although the prostaglandin continues to be produced up to the time of parturition, its breakdown to an inactive metabolite begins about 6 days pre-partum. In other words active prostaglandin starts to be mopped up near term. Since these changes coincided with the change in milk composition, the effect of unilateral, local administration of the prostaglandin $F_{2\alpha}$ analogue, cloprostenol, was investigated. Such treatment prevented the normal changes in milk composition that occur in late pregnancy, indicating the possibility that the local control of milk composition at this time is mediated by prostaglandin synthesis, metabolism and removal. Whether the local changes resulting in altered milk composition are related to the local control of milk yield in lactation is not known. In pre-partum milking, milk yield rises but there is still a major rise at parturition, suggesting that systemic factors still hold the rate of secretion in check.

Preliminary studies were made of the inhibitory content of the mammary secretion of non-lactating goats in early pregnancy (Blatchford *et al.*, 1985). Small quantities of mammary secretion were infused into one mammary gland of lactating goats; in all goats the permeability of the mammary epithelium was affected and, in two out of five, the rate of milk secretion decreased. Using culture techniques, the secretion also markedly inhibited lactose synthesis. Whether the secretion from pregnant goats contains the same inhibitory substance as that isolated later from milk proper or whether there are different or additional factors present is open to investigation.

LOCAL CHEMICAL INHIBITION: TESTING THE HYPOTHESIS

The hypothesis on the presence in milk of an inhibitor of milk secretion was tested by experiments in the goat *in vivo*.

Goat mammary glands respond to milking thrice instead of twice daily by increasing the rate of secretion (Henderson *et al.*, 1983). Henderson and Peaker (1984) showed that this local response is not due to relief from the pressure of stored milk. At one of the three milkings a volume of inert, isotonic solution, equal in volume to the milk removed at that milking, was infused into the lumen of one gland. The rate of milk secretion increased even with the degree of distension caused by 16 h of accumulation.

Further evidence that a chemical inhibitor is involved was obtained. When milk in the mammary gland was diluted with isotonic lactose or sucrose solutions, the rate of secretion increased (Fleet and Peaker, 1978; Henderson and Peaker, 1987). The isotonic solution was infused and the animal milked 7 h later. For example, with milk comprising 61% of the total volume at the time of removal, the rate of milk secretion between milkings increased significantly by a mean of 31%. Such an increase is compatible with the dilution of a chemical inhibitor present in milk.

Henderson and Peaker (1987) also drained milk every hour by catheter from the duct system of one gland; this procedure does not evacuate the secretory alveoli. These results were compared with a similar experiment in which oxytocin was given before hourly catheter milking. While in the latter the rate of secretion increased, in those milked frequently by catheter, there was no increase, showing that it is the secretory alveoli which are the site of the inhibitory action.

THE SEARCH FOR THE PUTATIVE INHIBITOR

At the time of the Linzell and Peaker (1971) paper proposing the presence of a chemical inhibitor of milk secretion in milk, there had been a suggestion of local chemical regulation of milk secretion during involution. Levy (1963, 1964) found that as milk accumulated in rat mammary glands following teat obstruction, the ability of the mammary tissue to synthesise fatty acids was impaired. He found that mammary acetyl CoA carboxylase was inhibited by rats' milk and by free fatty acids (C_{10}-C_{14}) and suggested this as a mechanism of feed-back control. Later studies showed that the non-esterified fatty acid fraction of rat milk can irreversibly inhibit acetyl CoA carboxylase purified from mammary tissue (Miller et al., 1970). Whether milk free fatty acid concentrations are reflected at the intracellular site of the enzyme is not known but in the longer-term this is a possibility and the idea has been resurrected by Heesom et al. (1992) as a mechanism to account for the decrease in lipid synthesis during periods of milk accumulation leading to involution.

In the early 1970s, the view that the putative feed-back inhibitor might be lipid soluble seemed attractive (Linzell and Peaker, 1971). A lipid soluble substance would be able to cross the luminal membrane of the secretory cell and raised concentrations in milk could be reflected inside the cell at a possible site of action. Indeed, evidence in favour of this view was forthcoming in 1973. Milk injected intra-peritoneally into lactating mice inhibited litter growth. Only milk containing milk-fat globules had this effect; skim-milk did not inhibit secretion (Sala et al., 1973). More recent evidence indicates that the feedback inhibitor is a protein and only recently has the inhibitory effect of milk fat i.p. been re-examined. Milk fat globules but not skim-milk were found to inhibit litter growth. However, none of the individual fractions of milk fat globules had an effect and when the concentration of fat globules was raised there was a markedly deleterious effect on the mother. It was concluded that milk fat globules administered intra-peritoneally by virtue of some physical property, have an adverse effect on health rather than a specific effect on the mammary gland (Peaker and Taylor, 1994).

In the 1970s culture systems for mammary tissue were relatively crude and the cells showed little secretory activity. We argued that an in-vitro system would be needed to search for the putative inhibitor in milk fractions since only a relatively small number of experiments could be done in goats and in those there was a high risk of intra-mammary infection by introducing mammary extracts into the gland. Therefore, the search for the inhibitor never got started in earnest until better culture systems for mammary tissue were developed.

More recently and by using mammary tissue explants which secrete milk constituents at a relatively high rate in culture, it has been possible to screen milk fractions for their ability to inhibit milk synthesis and secretion and to isolate active, inhibitory components from the milk of a number of species. The inhibitor has been dubbed FIL (Feedback Inhibitor of Lactation). The isolation and identity of FIL are covered elsewhere in this volume.

AUTOCRINE CONTROL

The term autocrine was coined, many years after the local control of the mammary gland by an inhibitor in milk was postulated, to describe the phenomenon in some cancerous cells by which growth promoters secreted by one type of cell have a stimulatory effect locally on the same type of cell (Sporn and Todaro, 1980). Since the mammary secretory epithelium is comprised of one type of cell, the term has been thought appropriate to describe local control within the mammary gland by a process which, in this case, is physiological and inhibitory.

Apart from FIL, milk has since been found to contain other substances that can affect mammary growth and differentiation. Examples are mammary-derived growth inhibitor which is associated with the milk fat-globule membrane (Böhmer et al., 1987; Brandt et al., 1988) and a substance from the same source that increases the longevity of mammary cells (Wendrinska et al., 1993); it has also been claimed that α-lactalbumin is a growth inhibitor (Thompson et al., 1992). It will be important in the future to not only be certain of chemical identity but also to distinguish between those substances in milk that can, say in vitro, affect some aspect of mammary function from those that actually are involved in physiological regulation. With the vast range of biologically-active substances in milk it is not a simple task to find whether they have any rôle in the young (as some now appear to do) (Peaker and Neville, 1991) let alone any autocrine or paracrine effects on the mammary gland.

KINETICS OF AUTOCRINE CONTROL

There is a very large gap in our knowledge of autocrine control. The problem arises because the secretory alveoli are never free of milk. Immediately after even complete emptying they pull milk back from the ducts by elastic recoil. Therefore, simply having a constant concentration of inhibitor in milk is not sufficient to explain the phenomenon. A number of hypotheses can be simulated numerically but hard evidence is needed to eliminate the improbables from the list of possibles. Much of this evidence will be obtained now that immuno-assays for the inhibitor, FIL, have been developed.

One possible mechanism is that a pro-inhibitor is secreted in milk and that with time the active inhibitor is formed, giving an increase in inhibitor concentration with time. Another possibility is that the inhibitor, once secreted, is subject to inactivation by a first-order process. During the accumulation of milk, the first-order catabolic mechanism would be saturated and the concentration of the inhibitor in the stored milk would rise. Changes in the receptivity of the secretory cell cannot be ruled out, and it is possible that as the luminal membrane of the cell increases in area during the accumulation of milk and expansion of the alveoli (Peaker and Blatchford, 1988; Knight *et al.*, 1994) receptors for the inhibitor might be exposed.

While the mechanism of action, kinetics of production and metabolism and the longer-term effects of FIL are being elucidated, and potential applications explored, it has to be remembered that if autocrine secretion by this or homologous milk components constitutes a ubiquitous method of controlling the rate of milk secretion in all mammals - as appears to be the case so far - then it must form, and be interpreted in terms of, the evolutionary strategy of lactation in a particular species. For example, it is known that a different hierarchy of mechanisms operates to arrest milk secretion when milk removal ceases. Thus, in the goat it would seem that the order in time is: autocrine control (FIL); then mammary distension acting to stretch the cells and inhibit secretion; finally, the effects of withdrawal of prolactational hormones released by the final suckling or milking (Fleet and Peaker, 1978; Peaker, 1980). By contrast, in the rat, there is strong evidence that the lack of hormonal stimulation is of major importance. Following separation from the young, the rate of milk accumulation in the glands decreases. However, when the young are not removed but removal of milk is prevented by sealing the teats, the rate of milk secretion remains high. Eventually, of course, the rate falls in the glands with sealed teats but clearly mammary distension is of secondary importance to the withdrawal of suckling-induced hormone release (Hanwell, 1972; Hanwell and Linzell, 1973). The time-course and influence of any autocrine control in the rat is not known. Clearly, investigation is needed of autocrine control in species which lack a capacious duct system for milk storage but which only suckle relatively infrequently either throughout or during a particular stage of lactation; such species include the rabbit (every 24 h) (Zarrow *et al.*, 1965), tree-shrews (every 48 h) (D'Souza and Martin, 1974) and the Antarctic fur-seal (4-5 days) (Bonner, 1984).

Another, related, problem is the time-course over which FIL acts. Does FIL modulate the amount of milk at the next suckling or the one after that? Does the time-course vary adaptively with the normal period between sucklings in different species? In other words, in molecular biological terms, the question is: how long does the message last?

CONCLUDING COMMENTS

The discovery of what is now called autocrine control of milk secretion broke the existing paradigm for the mechanism responsible for the quantitative control of the rate of milk secretion and has opened up a field of investigation from the whole mammalian organism in its environment to control of events at the molecular level. It is clearly a special mechanism in a special organ, for the mammary gland is the quintessence of mammalian life. But we should not be surprised when local interactions between cells are shown to be important; it is, after all, the primitive condition in phylogeny and the ontogeny of complex organisms (Wolpert, 1991).

On a final note, what happens in those other unusual glands that store their secretion, from the nectar-producing organs of flowering plants to snake-venom glands? There is good evidence that synthesis is stimulated by removal of nectar (Pyke, 1991) and by discharge of venom, respectively (Sobol-Brown *et al.*, 1975). As Cleopatra's asp fell to the ground was autocrine control taking charge of its next fatal dose?

REFERENCES

Ballou, L.U., Bleck, J.L., Bleck, G.T. and Bremel, R.D., 1993, The effects of daily oxytocin injections before and after milking on milk production, milk plasmin, and milk composition, *J. Dairy Sci.* 76:1544.

Blatchford, D.R., Neville, M.C., Peaker, M. and Wilde, C.J., 1985, Effects of mammary secretion from non-lactating goats on milk secretion *in vivo* and *in vitro*, *J. Physiol., Lond.* 361: 75.

Böhmer, F.-D., Krafts, T, Otto, A., Wernstedt, C., Hellman, U., Kurtz, A., Müller, T., Rohde, K., Etzold, G., Lehmann, W., Langen, P., Heldin, C.-H. and Grosse, R., 1987, Identification of a polypeptide growth inhibitor from bovine mammary gland, *J. Biol. Chem.* 262:15137.

Bonner, W.N., 1984, Lactation strategies in pinnipeds: problems for a marine mammalian group, *Symp. Zool. Soc. Lond.* 51: 253.

Brandt, R., Pepperle, M., Otto, A., Kraft, R., Boehmer, F.-D. and Grosse, R.A., 1988, A 13-kilodalton protein purified from milk fat globule membranes is closely related to a mammary-derived growth inhibitor, *Biochemistry* 27:1420.

D'Souza, F. and Martin, R.D., 1974, Maternal behaviour and the effects of stress in tree shrews, *Nature, Lond.* 251:309.

De Passille, A.M., Rushen, J. and Hartsock, T.G., 1988, Ontogeny of teat fidelity in pigs and its relation to competition at suckling, *Can. J. Anim. Sci.* 68:325.

Denamur, R. and Martinet, J., 1961, Action de l'ocytocine sur la sécrétion du lait de brebis, *Annal. Endocrinol* 22:777.

Elliott, G.M., 1959a, The direct effect of milk accumulation in the udder of the dairy cow upon milk secretion rate, *Dairy Sci. Abstr.* 21:435.

Elliott, G.M., 1959b, The effect on milk yield of the length of milking intervals used in twice a day milking, twice and three times a day milking and incomplete milking, *Dairy Sci. Abstr.* 21:482.

Elliott, G. M., 1961, The effect on milk yield of three times a day milking and of increasing the level of residual milk, *J. Dairy Res.* 28:209.

Elliott, G.M., Dodd, F.H. and Brumby, P.J., 1960, Variations in the rate of milk secretion in milking intervals of 2-24 hours, *J. Dairy Res.* 27:293.

Fleet, I.R. and Peaker, M., 1978, Mammary function and its control at the cessation of lactation in the goat, *J. Physiol., Lond.* 279:491.

Hanwell, A., 1972, Cardiovascular aspects of lactation in the rat, PhD Thesis, University of Cambridge.

Hanwell, A. and Linzell, J.L., 1973, The effects of engorgement with milk and of suckling on mammary blood flow in the rat, *J. Physiol., Lond.* 233:111.

Heesom, K.J., Souza, P.F.A., Ilic, V. and Williamson, D.H., 1992, Chain-length dependency of interactions of medium-chain fatty acids with glucose metabolism in acini isolated from lactating rat mammary glands. A putative feed-back to control milk lipid synthesis from glucose, *Biochem. J.* 281:273.

Henderson, A.J., Blatchford, D.R. and Peaker, M., 1983, The effects of milking thrice instead of twice daily on milk secretion in the goat, *Quart. J. Exp. Physiol.* 68:645.

Henderson, A.J. and Peaker, M., 1984, Feedback control of milk secretion in the goat by a chemical in milk, *J. Physiol., Lond.* 351:39.

Henderson, A.J. and Peaker, M., 1987, Effects of removing milk from the mammary ducts and alveoli, or of diluting stored milk, on the rate of milk secretion in the goat, *Quart. J. Exp. Physiol.* 72:13.

Hoeck, H.N., 1977, "Teat order" in hyrax (*Procavia johnstoni* and *Heterohyrax brucei*), *Z. Säugetier.* 42 (H2):112.

Knight, C.H., Hirst, D. and Dewhurst, R.J., 1994, Milk accumulation and distribution in the bovine udder during the interval between milkings, *J. Dairy Res.* in press

Levy, H.R., 1963, Inhibition of mammary gland acetyl CoA carboxylase by fatty acids, *Biochem. Biophys. Res. Commun.* 13:267.

Levy, H.R., 1964, Effects of weaning and milk on mammary fatty acid synthesis, *Biochim. Biophys. Acta* 84:229.

Linzell, J.L., 1967, The effect of very frequent milking and of oxytocin on the yield and composition of milk in fed and fasted goats, *J. Physiol., Lond.* 190:333.

Linzell, J.L. and Peaker, M., 1971, The effects of oxytocin and milk removal on milk secretion in the goat, *J. Physiol., Lond.* 216:717.

Linzell, J.L. and Peaker, M., 1974, Changes in colostrum composition and in the permeability of the mammary epithelium at about the time of parturition in the goat, *J. Physiol., Lond.* 243:129.

Loudon, A.S.I. and Racey, P.A. (eds), 1987, "Reproductive Energetics in Mammals", Symposia of the Zoological Society of London 57, Clarendon Press, Oxford.

Maule Walker, F.M. and Peaker, M., 1980, Local production of prostaglandins in relation to mammary function at the onset of lactation in the goat, *J. Physiol., Lond.* 309:65.

Miller, A.L., Geroch, M.E. and Levy, H.R., 1970, Rat mammary-gland acetyl-coenzyme A carboxylase. Interaction with milk fatty acids, *Biochem. J.* 118:645.

Morag, M., 1973, Two and three times-a-day milking of cows. II. Possible mechanisms for the increase in yield of cows milked thrice-daily, *Acta Agric. Scand.* 23:256.

Paine, M.J.I., Desmond, H.P., Theakston, R.D.G. and Crampton, J.M., 1992, Gene expression in *Echis carinatus* (carpet viper) venom glands following milking, *Toxicon* 30:379.

Peaker, M., 1980, The effects of raised intramammary pressure on mammary function in the goat in relation to the cessation of lactation, *J. Physiol., Lond.* 301:415.

Peaker, M., 1989, Evolutionary strategies in lactation:nutritional implications, *Proc. Nutr. Soc.* 48:45.

Peaker, M. and Blatchford, D.R., 1988, Distribution of milk in the goat mammary gland and its relation to the rate and control of milk secretion, *J. Dairy Res.* 55:41.

Peaker, M., Neville, M.C., 1991, Hormones in milk:chemical signals to the offspring, *J. Endocrinol.* 131:1.

Peaker, M. and Taylor, E., 1994, Inhibitory effect of milk fat on milk secretion in the mouse:a re-examination, *Exp. Physiol.* 79:561.

Pyke, G.H., 1991, What does it cost a plant to produce floral nectar?, *Nature, Lond.* 350:58.

Rudnai, J., 1984, Suckling behaviour in captive *Dendrohyrax arboreus* (Mammalia:Hyracoidea), *S. Afr. J. Zool.* 19:121.

Sala, N.L., Cannata, M.A., Luther, E., Arballo, J.C. and Tramezzani, J.H., 1973, Inhibition of milk secretion by intraperitoneal injections of milk into mice, *J. Endocrinol.* 56:79.

Sharman, G.B., 1970, Reproductive physiology of marsupials, *Science* 167:1221.

Sobol-Brown, R., Brown, M., Bdolah, A. and Kochva, E., 1975, Accumulation of some secretory enzymes in venom glands of *Vipera palaestinae*, *Amer. J. Physiol.* 229:1675.

Sporn, M.B. and Todaro, G.J., 1980, Autocrine secretion and malignant transformation of cells, *New Engl. J. Med.* 303:878.

Thompson, M.P., Farrell, H.M., Mohanam,S., Liu, S., Kidwell, W.R., Bansal, M.P., Cook, R.G., Medina, D., Kotts, C.E. and Bano, M., 1992, Identification of human milk α-lactalbumin as a cell growth inhibitor, *Protoplasma* 167:134.

Wendrinska, A., Addey, C.V.P., Orange, P.R., Boddy, L.M., Hendry, K.A.K. and Wilde, C.J., 1993, Effect of a milk fat globule membrane fraction on cultured mouse mammary cells, *Biochem. Soc. Trans.* 21:220S.

Wolpert, L., 1991, "The Triumph of the Embryo", Oxford University Press, Oxford.

Zarrow, M.X., Denenberg, V.H. and Anderson, C.O., 1965, Rabbit:frequency of suckling in the pup, *Science* 150:1835.

ENDOCRINE AND AUTOCRINE STRATEGIES FOR THE CONTROL OF LACTATION IN WOMEN AND SOWS

Peter E. Hartmann, Craig S. Atwood, David B. Cox
and Steven E.J. Daly

Department of Biochemistry
The University of Western Australia
Nedlands, WA 6009

INTRODUCTION

As understanding of the strategic control of lactation advances, it becomes increasingly clear that the lactational strategy of each mammalian species reflects the requirements of that species in close and complex ways. In this review, we choose two species - the human being and the domestic pig - in order to illustrate this with new findings concerning their lactational strategies. This comparison serves not only to depict the complexity of lactational strategies but also to demonstrate the differences that occur between two species with different lactational needs.

The central problem of the strategic control of lactation, which has been stated by Peaker and Wilde (1987), is that milk production must be balanced between the needs of the young, as the effect of insufficient milk on the suckling young may be devastating, and the fact that milk production is metabolically expensive to the mother. Linzell and Peaker (1971) similarly argued that multiparous species which exhibit establishment of a teat order may need a control mechanism to ensure that each gland produces just the right amount of milk for the attached young. Thus the problem of stategic control of lactation may be simply posed as "How is milk production balanced to the needs of the young?"

Traditionally, it has been understood that the mechanisms controlling lactation were endocrine in nature. However, it is becoming increasingly apparent that autocrine mechanisms are also important, particularly with respect to short-term and local spheres of influence. Therefore, in this review we have examined both endocrine and autocrine control mechanisms at the initiation of milk production after birth and during established lactation.

MAMMARY DEVELOPMENT AND LACTOGENESIS I

In mammalian species mammary development starts with the embryonic differentiation of ectodermal tissue. Early in human gestation, a two to four cell thick 'galactic band' develops from the axilla to the groin on either side of the body. This ectodermal structure differentiates and invaginates the surrounding mesenchymal tissue, continuing development until the formation of the primitive ducts by twenty to thirty-two weeks of gestation, followed by the juvenile lobulo-alveolar structures, which form up until birth.

During gestation in multiparous species, many bilaterally located mammary buds form along the embryonic 'galactic band'. Uniparous species have smaller numbers of mammary buds, although once again bilaterally located either anteriorly or posteriorly along the 'galactic band' (Vorherr, 1974). Throughout juvenile life the mammary structures remain quiescent, with allometric growth starting during puberty and continuing into early adulthood. The culmination of mammary growth occurs during pregnancy with the ductular and lobulo-alveolar development (Vorherr, 1974; Hartmann, 1991). In the sow, by day 90 of gestation mammogenesis is essentially complete (Kensinger et al., 1982). Lactogenesis I begins between day 90 and 105 of gestation and is associated with the 'initiation of structural and metabolic differentiation' within the mammary glands (Kensinger et al., 1982). However, the mammary glands of sows and women do not become fully functional until just prior to delivery, when colostrum becomes available. During pregnancy, progesterone and other sex steroid hormones stimulate ductular, lobular, and alveolar growth. Insulin, cortisol, thyroid and parathyroid hormone, placental lactogen, prolactin and possibly HCG also support this growth. In the sow similar hormone interactions occur, although placental lactogen has not been detected. While the lactogenic hormones assist in the growth and development of the mammary gland, their effect on the promotion of milk synthesis is suppressed until the removal of progesterone.

STRATEGIC CONTROL AT LACTOGENESIS II

At or around the time of delivery in women and sows, dramatic changes occur in their mammary glands. After the delivery of the neonate, copious milk secretion is initiated, in a process termed lactogenesis II and is characterised by the transition from colostrum, a sticky viscous solution high in protein, especially the immunoglobulins, and low in lactose, to mature milk, a high lactose, low protein, more fluid solution. Whereas in women, lactogenesis II occurs 31 ± 14 hours after delivery (Arthur et al., 1989b; Neubauer et al., 1993), in sows it occurs during farrowing (Atwood, 1993).

Endocrine Control

The increase in synthetic activity of the mammary gland during lactogenesis II is thought to be a balance between the withdrawal of an inhibitory hormone in the presence of lactation promoting hormones. Studies in rats (Kuhn, 1969; Nicholas and Hartmann, 1981a, b), cows (Hartmann, 1973), and ewes (Hartmann et al., 1973), suggest that the hormonal signal for the initiation of milk synthesis in the mammary glands of these species is decline in the concentration of progesterone in the blood at the end of pregnancy. There is growing evidence that this also is the case after birth in women (Vorherr, 1974; Kulski et al., 1977) and sows (Martin et al., 1978; Gooneratne et al., 1979; Hartmann et al., 1984a, b; Holmes and Hartmann, 1993).

In women during late pregnancy, the majority of the circulating progesterone is supplied by the placenta (Prichard *et al.*, 1985). Following the removal of the placenta, the concentration of progesterone declines, releasing its inhibition on the lactogenic hormones. In women, unlike other mammals, an abrupt increase in the rate of lactose synthesis occurs in the mammary gland approximately thirty hours after parturition (Arthur *et al.*, 1989b), consistent with the fall in the concentration of progesterone in blood and milk over the first day *postpartum* (Vorherr, 1974; Kulski and Hartmann, 1981; Cowie *et al.*, 1980). Consequently, women who retain viable placental fragments, fail to initiate lactation until the placenta fragments have been removed (Neifert *et al.*, 1981).

On the other hand, in sows, the majority of circulating progesterone is derived from the ovary and this progesterone falls to low levels by the time farrowing commences (Ellicott and Dziuk, 1973; Robertson and King, 1974). In the sow, a significant negative correlation was found between the concentration of lactose in the colostrum and the concentration of progesterone in the blood on the day of parturition (Martin *et al.*, 1978). Also, the concentration of lactose in the blood (an indicator of the synthetic activity of the mammary gland) increased soon after the commencement of the decline in the concentration of progesterone in the blood (Hartmann *et al.*, 1984b). The initiation of lactation in the sow was delayed by the administration of progesterone (Gooneratne *et al.*, 1979; Whitely *et al.*, 1990). Thus the sow's strategy is to undergo lactogenesis II concurrent with parturition.

Given that the pig has many glands with the potential for milk synthesis, synchronized timing of the initiation of lactation with the birth of the piglets allows for the efficient utilisation of maternal resources whilst providing adequate nourishment and immunological protection for the neonate. This timing of the initiation of lactation with farrowing is critical for piglet survival since the piglet is born with very low fat reserves (Mellor and Cockburn, 1986) and has a limited capacity for the *de novo* synthesis of fatty acids (Dividich *et al.*, 1989). Furthermore, glycogen reserves of piglets will last only 15 h in an 18-26°C environment, and as little as 2 h in cold environments, before hypoglycaemia and hypothermia become evident (Mellor and Cockburn, 1986). Therefore, if piglets do not obtain colostrum within the first few hours of life their chance of survival is greatly diminished. However, most piglets will consume between 50 and 100 ml of colostrum during the first 4 h *postpartum* (Atwood, 1993).

Lactogenic hormones are required for the increased rate of milk synthesis at the onset of lactation in the woman (Kulski *et al.*, 1978) and sow (Cowie *et al.*, 1980; Forsyth, 1986; Dodd *et al.*, 1993; de Passillé *et al.*, 1993). Recent evidence indicates that the lactogenic hormone, prolactin, plays an essential role in the initiation of lactose synthesis in both species. *In vivo* studies have shown that there is a parallel increase in the concentration of lactose and prolactin in the blood of sows during late pregnancy (Whitely *et al.*, 1990). Furthermore, plasma α-lactalbumin, the specifier protein in the lactose synthase complex, recently has been found to be negatively correlated with the fall in progesterone and positively correlated with the rise in prolactin (Dodd *et al.*, 1993). Indeed, the inhibition of prolactin secretion in sows (Taverne *et al.*, 1982) and women (Kulski *et al.*, 1978) with ergot alkaloids such as bromocryptine results in inhibition of the onset of lactation. *In vitro* studies have supported these findings, suggesting that prolactin has an important role in lactogenesis II of women and sows.

Insulin, cortisol and prolactin have been shown to increase glucose oxidation and glucose incorporation into lipids in mammary explants from pregnant gilts (Jerry *et al.*, 1989). The addition of insulin to mammary explants increased glucose oxidation by 59% and glucose incorporation into lipid by 150% compared to a low dose of insulin. The addition of prolactin to mammary explants cultured with insulin and cortisol increased the oxidation rate of glucose by 130% and fat synthesis by 400%, compared with mammary explants cultured with insulin and cortisol alone (Jerry *et al.*, 1989). In addition, a dose-dependent response to prolactin was demonstrated within the concentrations of prolactin normally found in the blood of pregnant gilts. This study demonstrated that the lactogenic complex of insulin, cortisol and prolactin induced metabolic activity in porcine mammary tissue from late pregnancy. Evidence in women would suggest a similar lactogenic complex (Jacobs, 1977; Kulski *et al.*, 1978; Arthur *et al.*, 1989b). In diabetic women the increase in milk citrate following birth was delayed with respect to normal women, suggesting that insulin may have a role in the synthesis of milk fats (Arthur *et al.*, 1989b). While these workers did not test the effect of cortisol alone on tissue metabolism, Martin *et al.* (1978) reported that changes in plasma corticosteriods during lactogenesis did not support a positive involvement of corticosteroids in the initiation of lactation in the sow.

In women, the increase in secretory activity of the mammary gland occurs bilaterally, independent of colostrum removal. This suggests that the initiation of milk secretion in women is hormonally controlled (Kulski, *et al.* 1978). With the withdrawal of progesterone in the presence of the lactogenic hormones (notably prolactin) resulting in an increase in the rate of milk synthesis around parturition in many species, including women (Hartmann, 1973; Fleet *et al.*, 1975; Arthur *et al.*, 1989b; Neubauer *et al.*, 1993), it is plausible that lactation could be artificially initiated. By modelling this mechanism, women wishing to relactate or initiate lactation for an adopted baby do so successfully. Women wishing to relactate are those who have had a previous pregnancy, and who after premature weaning wish to re-establish milk secretion. Since these women have undergone pregnancy induction of mammary development, re-establishment of milk secretion is uncomplicated, being initiated by increasing prolactin secretion with prolactin-releasing or dopamine-inhibiting medications, and increased nipple stimulation. For women wishing to lactate for an adopted baby, in the absence of a previous pregnancy, the initiation of lactation is more complex. First, the mammary gland must be primed with sex steroid treatment. This treatment is in the form of oestrogen supplementation, and leads to mammary gland development. Immediately prior to the arrival of the adopted infant, the oestrogen treatment is stopped, and prolactin secretion is enhanced following the mechanisms used for relactating mothers. These methods involve stimulating the body to secrete increased amounts of prolactin, either by TRH (Thyroliberin) or Metaclopramide treatment in the presence or absence of oestrogen priming (Lawrence, 1980; Phillips, 1993; Riordan and Auerbach, 1993).

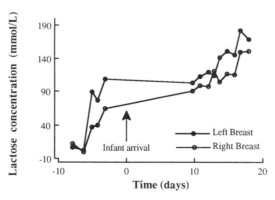

Figure 1. Lactose concentration of breast secretion, from a woman trying to relactate. The infant was born at day 0 and was fed using a supply line (D.B. Cox and P.E. Hartmann, unpublished observations).

In a single case study we were able to measure relactation in the absence of prolactin stimulatory drugs, or oestrogen therapy (Figure 1). The woman, 36 years of age and having had two children some years previously, expressed a desire to relactate for an adopted infant. In this case, to re-establish milk secretion the mother stimulated her breasts by frequent breast pumping without medications. We measured the concentration of lactose (by the method of Arthur et al., 1989b) in breast secretion samples from this woman before and after the arrival of the adopted infant (Figure 1), and concluded that lactose synthesis was established from negligible levels, to the levels found in lactating women, with nipple stimulation alone. Indeed Phillips (1993) reported that in relactating women, metaclopramide administration was "as much for the placebo effect as its prolactin-enhancing properties". Although this woman (Figure 1) did not fully breast feed her adopted infant, we know of a nulliparous woman, who by nipple stimulation established her lactation, and fully breastfed her adopted baby. Although hormonal changes are unknown in these women, clearly pregnancy is not a necessary prerequisite for lactogenesis II in women.

Autocrine Control

In sows, the removal of colostrum from the mammary glands is required for an increase in the rate of milk synthesis. (Atwood, 1993). In addition, in unsuckled mammary glands, the composition of the mammary secretion does not undergo the same changes seen in the secretion of suckled glands (Atwood, 1993, Figure 2). This observation strongly suggests the involvment of autocrine factors in the control of milk production at lactogenesis II for this species. Since the sow has a large number of glands with the potential for milk synthesis, a mechanism of autocrine regulation during lactogenesis II would allow only those glands that are required for milk synthesis to become functional, allowing unsucked glands to regress and thereby conserving maternal resources. Such a mechanism may have developed to cope with the variable number of young. This mechanism also confers advantages for the piglet. The piglet is born agammaglobulinaemic and, at birth, passes from a virtually pathogen-free environment to a microbiologically rich environment (Hartmann et al., 1984a). Therefore, the amount of immunoglobulins consumed within the first few hours, when macromolecules can be absorbed by the intestine, is important in establishing the piglets' serum immunity. Since piglets are born over a period of time (~ 4 h), the simultaneous 'switching on' of all the mammary glands at the commencement of farrowing would result in the loss of colostrum due to leakage from the glands, disadvantaging the younger piglets, due to the dilution of the milk immunoglobulins. Instead of this, it appears that individual glands are 'switched on' as and when required by the arrival of a newborn piglet.

In women, lactogenesis II also is marked by sudden compositional changes in the mammary secretion - increases in the concentrations of lactose and citrate in the mammary secretion are apparent from approximately 24 to 72 h after birth (Arthur et al. 1989a). These compositional changes are concurrent with the start of a marked increase in milk production, which may take over a week to plateau (Roderuck et al., 1946; Houston et al., 1983; Saint et al., 1984; Neville et al., 1988; Arthur et al., 1989b; Neville et al., 1991). At about this time mothers typically experience a feeling of sudden breast overfullness and their milk supply is said to be "coming in". Not uncommonly, the overfullness is quite severe and accompanied by swelling and inflammation of the breasts as well as considerable pain and discomfort. This experience is termed breast engorgement (Newton and Newton, 1951; Newton and Newton, 1962).

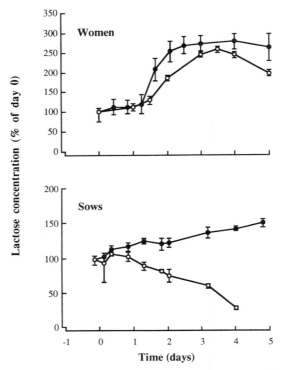

Figure 2. Changes in the lactose concentration of the secretion of the mammary glands of women and sows over the first 5 days after parturition. Closed circles represent data from suckled glands, open circles represent data from unsuckled glands. Each point represents the mean ± SEM (n = 4, for the human data; n = 13, for porcine data). These data are taken from Kulski *et al.* (1978) and Atwood (1993) for the two graphs, respectively.

It has long been assumed that "coming in" and engorgement were marking the beginning of lactogenesis II in women (Newton and Newton, 1962). However, Arthur *et al.* (1989b) found that the sensation of milk "coming in" varied considerably in timing between mothers (range 38 to 98 h after birth) and occurred well after the compositional changes discussed above. Therefore, the sensations of milk "coming in" and engorgement may mark changes that are distinct from lactogenesis II.

Engorgement is a phenomenon that has received very little attention in the medical literature despite the severe pain that can accompany it. It seems likely that breast engorgement is not merely fullness of milk but in fact an inflammatory reaction possibly as a consequence of the mismanagement of early breastfeeding (Newton and Newton, 1948; Newton and Newton, 1951; Auerbach *et al.*, 1993). The risk of engorgement may be lessened with avoidance of breastfeeding to a schedule and the adoption of demand breastfeeding practices (Hartmann, 1991; Auerbach *et al.*, 1993). On the other hand, breast fullness (of milk) should be seen, according to Auerbach *et al.* (1993), as a normal transitory state which marks the beginning of close regulation of milk synthesis to infant milk requirement and Hartmann (1991) refers to this transition state as lactogenesis III. That there is a change in the control of milk synthesis at lactogenesis III is evident from the fact that mothers who choose not to breastfeed still undergo breast-filling and engorgement with concurrent milk compositional changes in an identical fashion to breastfeeding mothers, before absence of breast emptying leads to involution (Kulski *et al.*, 1978; Kulski and Hartmann, 1981). Thus, in contrast to the sow, human breasts are initially 'switched on' whether needed or not, and only later are 'switched off' if not required. The secretion of unsuckled breasts undergoes much the same change in composition, at this time, as the secretion from suckled breasts (Figure 2). As will be discussed further below, the strategic control of lactation during established lactation, for women, appears to be under autocrine control. Therefore, lactogenesis III may be the transition from an endocrine-driven lactogenesis II to an autocrine-driven established lactation phase.

STRATEGIC CONTROL DURING ESTABLISHED LACTATION

Following lactogenesis II the production and composition of milk remains relatively stable until weaning occurs. Nevertheless the maintenance of established lactation and the less striking variations in both production and composition during this time are under both endocrine and autocrine control.

Endocrine Control

Both prolactin and insulin play a major role in the maintenance of milk secretion in lactating non-ruminants such as women (Kulski *et al.*, 1978; Arthur *et al.*, 1989b) and sows (Cowie *et al.*, 1980; Collier *et al.*, 1984; Forsyth, 1986; Vernon, 1989). Other hormones known to be galactopoietic in lactating ruminants, eg growth hormone, have varying effects in non-ruminants.

Prolactin is a hormone produced in the anterior pituitary gland, and released under hypothalamic control. In lactating mammals, prolactin is released following tactile nipple/teat stimulation. The sucking induced increase in the concentration of prolactin in the blood of women and sows is similar to the mechanism that has been observed in lactating rats (see Cowie *et al.*, 1980). The effect of sucking on the release of prolactin in sows may be mediated by endogenous opioid peptides (Mattioli *et al.*, 1986; Armstrong *et al.*, 1988), although the release of prolactin in women remains under dopaminergic control (Tay *et al.*, 1993).

In women, after the commencement of nursing, plasma prolactin increases to a 30-45 minute peak, and subsequently declines to the pre-nursing (basal) concentration by 180 minutes (Noel *et al.*, 1974). The nursing control on prolactin release has lead to a hypothesis that prolactin controls milk production. In women this appears not to be the case, with the prolactin response to nipple stimulation, declining over duration of lactation (Rattigan, 1987; Cox and Hartmann, 1993), with no associated decrease in milk production. In addition, for the first six months of lactation, the short-term rates of milk synthesis vary unilaterally, with these variations bearing no relationship to the preceding prolactin responses (Cox and Hartmann, 1993).

In contrast, the concentration of prolactin is maintained at a high level in the peripheral blood of sows throughout most of the lactation, due to the frequent sucking of the piglets (Threlfall *et al.*, 1974; van Landeghem and van de Wiel, 1978; Bevers *et al.*, 1978; Mulloy and Malvern, 1979; Dusza and Krzymowska, 1981; Taverne *et al.*, 1982; Plaut *et al.*, 1989; Holmes, 1991; Rojkittikhun *et al.*, 1991). Its concentration declines only late in lactation, after about 2-3 weeks (van Landeghem and van de Wiel, 1978), a change attributed to the tendency of piglets to suckle less frequently as they grow older (van Landeghem and van de Wiel, 1978; Mulloy and Malvern, 1979; Kirkwood *et al.*, 1984). Several workers have reported that circulating concentrations of prolactin are elevated during and immediately after periods of sucking (van Landeghem and van de Wiel, 1978; Taverne *et al.*, 1982; Kendall *et al.*, 1983; Mattioli *et al.*, 1986; Plaut *et al.*, 1989; Holmes, 1991). However, the concentrations of prolactin do not correlate with the number of piglets in the litter (Bevers *et al.*, 1978; Rojkittikhun *et al.*, 1991).

The lack of correlation between the number of piglets and the concentration of prolactin in the blood in the sow, and the constancy of milk production while the secretion of prolactin declined over the duration of lactation in women, are consistent with the finding in the rat that only a small proportion of the prolactin released at sucking is essential for the maintenance of milk secretion during lactation (Grosvenor *et al.*, 1975). From studies in laboratory rodents, it is thought that prolactin may have a regulatory role in the rate of lipogenesis in the mammary glands, adipose tissue and liver during lactation (Amenomori *et al.*, 1970; Zinder *et al.*, 1974; Romsos *et al.*, 1978; Agius *et al.*, 1979) and in the regulation of protein synthesis in the mammary glands (Williamson *et al.*, 1984). Also, prolactin has been shown to increase lipogenic enzymes in mammary explants of pseudopregnant rabbits (Martyn and Falconer, 1985). In contrast, arteriovenous differences in lactating rats fed bromocryptine showed that prolactin withdrawal for 24 h does not alter the uptake of glucose by the mammary glands (Robinson and Williamson, 1977).

In the sow, Plaut *et al.* (1989) have shown that the oxidation of either glucose or acetate to CO_2, and glucose incorporation into lipid, was low in mammary biopsy samples collected during pregnancy, but increased two- to five-fold on day 4 of lactation and thereafter paralleled the lactation curve for sows. These measures of mammary metabolism were positively correlated with prolactin binding to its receptor on mammary cell membranes. Also, during lactation negative correlations were observed between the concentrations of prolactin in plasma, prolactin binding and mammary metabolic rate. These results indicated that the binding of prolactin to its receptor on mammary cell membranes is an important effector of milk production in the sow. The number of prolactin receptors has been shown to increase from the 80th day of gestation in sows, with a more pronounced increased between parturition and day 20 of lactation (Berthon *et al.*, 1987).

A continual supply of insulin is important for the long-term maintenance of milk secretion in both ruminants and non-ruminants (Vernon, 1989). Insulin deficiency resulted in a reduction in milk secretion and enzymatic activity (Walters and McLean, 1968; Martin and Baldwin, 1971a; b). Consequently, insulin is thought to be required for the maintenance or survival of lactocytes during lactation (Baldwin and Louis, 1975). Insulin also has an important role in the short-term regulation of lipogenesis and protein synthesis but not lactose synthesis in the mammary glands of lactating rats (Jones *et al.*, 1984; Williamson *et al.*, 1984). In women insulin may have a role in the synthesis of lactose, as insulin dependent diabetic women show a delay in the initiation of lactose synthesis (Arthur *et al.*, 1989b; Neubauer *et al.*, 1993). Bolander *et al.* (1981) demonstrated that insulin was capable of stimulating the synthesis of casein and accumulation of casein mRNA in mouse mammary explants in the presence of cortisol and prolactin. This indicated that in addition to its role in cell maintenance, insulin may function in the expression of casein genes in mouse lactocytes.

The rate of glucose uptake by the mammary glands of the sow, as in other animals, does not appear to be under the influence of insulin. Holmes (1991) found no change in the mammary arteriovenous difference when the concentration of insulin in the blood was elevated during glucose infusion or after insulin administration. No acute effect of either insulin or glucagon on lactose synthesis has been observed in the goat or rat (Carrick and Kuhn, 1978; Wilde and Kuhn, 1979; Hove, 1978a; b). In women, on the other hand, insulin may have some control on the transport of glucose into the mammary gland. Arthur *et al.* (1994) showed that there is a delay in the influx of glucose into the mammary gland during lactogenesis II in insulin dependent diabetic women.

As in ruminants, growth hormone may have an important role in stimulating galactopoiesis in women. Breier *et al.* (1993) showed a significant milk production increase in women, following the injection of recombinant growth hormone. Although there is conflicting evidence, growth hormone does not appear to be important in controlling milk synthesis in the sow. Harkins *et al.* (1989) showed that the daily administration of recombinant porcine growth hormone from day 12 of lactation caused a 22% greater milk yield at day 28 of lactation in treated sows compared with untreated sows. However, Kveragas *et al.* (1986), Cromwell *et al.* (1989) and Toner *et al.* (1991) found that injections of recombinant porcine growth hormone daily into sows for 21 days prior to parturition, between day 108 of gestation and day 24 of lactation, and from day 8 to day 39 of lactation, respectively, had no effect on either milk yield or the concentration of fat, protein or lactose in the milk. Furthermore, sows injected with growth hormone prior to farrowing in a hot environment became more susceptible to heat stress (Cromwell *et al.*, 1989).

Another hormone likely to be involved in the strategic control of established lactation is oxytocin. This octapeptide is secreted from the posterior pituitary in response to a range of neural stimuli and released into the blood stream and is removed relatively swiftly (30 - 90 s). The mammary myoepithelium is responsive to oxytocin in the maternal blood stream, causing the alveoli to contract and the ducts to shorten and widen (Pitelka and Hamamoto, 1983). There is, as a result, the net movement of milk from the alveolar end of the glandular tissue to the nipple or teat end, increasing its availability to the suckling young. This is the functional end-point of the milk-ejection reflex.

Most commonly, the neural arc of the milk-ejection reflex is activated by tactile stimulation of the sensory nerve endings in the nipple during sucking (Cowie *et al.*, 1980; Cowie, 1984; Lincoln, 1984). This reflex is also controlled, in lactating mothers, by higher centres in the brain - apart from nipple stimulation, thinking of her infant, hearing an infant cry or other infant-centred cues may trigger a milk-ejection in a lactating mother (Hartmann, 1991). Presence of the milk-ejection reflex may be apparent to the breastfeeding mother through a "pins and needles" sensation in her breasts and the sudden release of milk from both her nipples as a dribble or stream (observable from the non-suckled nipple if the infant is already attached to a nipple when the reflex occurs). Alternatively, the mother may have no sensations associated with the presence of this reflex. Less commonly, this reflex may be associated with strong pain or uterine contractions.

As indicated above, the primary role of oxytocin is increasing the availability of stored milk to the infant. In species with large capacity to store milk outside of alveolar-lumen space, such as dairy species which possess large cisterns, much milk may be obtained by a milker without the milk-ejection reflex. If a milk-ejection reflex occurs after the milker has apparently emptied the udder, the same volume again may then be obtained (Cowie *et al.*, 1980). However, in species without such large storage spaces (eg. the rat, rabbit, dog, pig), the synthesized milk may be largely or completely unobtainable without an accompanying milk-ejection reflex. Women are considered part of this latter group (Cowie *et al.*, 1980), however, it is possible that women vary with respect to their non-alveolar storage capacity and with respect to their need for a milk-ejection reflex when breastfeeding (Lucas *et al.*, 1980).

Oxytocin is considered to be galactopoietic. Serial injections of oxytocin increase milk production in a number of species (Linzell and Peaker, 1971). The use of oxytocin nasal sprays is highly effective in stimulating milk production in mothers in very early lactation who are expressing their milk rather than breastfeeding. Ruis, (1981) recorded that the cumulative amount of milk expressed by these mothers on days 2 and 5 *postpartum* was, on average, 3.5 times greater than those mothers given a placebo. In contrast, no evidence for a relationship between oxytocin secretion and milk production by breastfeeding mothers in established lactation has been found (Lucas *et al.*, 1980). How oxytocin stimulates milk production is uncertain. High concentrations of oxytocin in the blood may cause the release of other hormones, such as prolactin (Cowie *et al.*, 1980). More likely is that the expulsion of newly synthesized milk from the alveolar lumen, in response to a milk-ejection reflex, releases inhibition of milk synthesis by the accumulation of milk within the alveolar lumen. Therefore, simply the removal of milk from the alveolar-lumenal space may sponsor an increase in the rate of milk synthesis. Oxytocin may, therefore, have a [minor] role in the regulation of milk production and this regulation may chiefly depend on the frequency of breastfeeds. Thus this hormone may be a link between endocrine and autocrine control mechanisms.

Another strategy for the survival of piglets may be the development of tight control over milk ejection. This is established within the first few hours *postpartum* in the sow, the frequency at which milk can be removed from the glands decreasing from intervals of 15 to 20 min to about 40 min by 24h *postpartum*. The removal of milk from the alveoli and ductal system of the mammary glands requires the release of oxytocin from the posterior pituitary, in response to the activation of neural receptors within the teats by the nuzzling and sucking of piglets (Cowie *et al.*, 1980; Ellendorf *et al.*, 1982). The release of oxytocin and milk ejection reflex will only occur if the majority of the piglets are stimulating the teats by their nuzzling and sucking actions. This requirement that all piglets must suck at the same time in order to obtain milk may prevent any one or a smaller number of piglets from dominating and removing more milk, thus depriving the other weaker piglets. Furthermore, the short duration (15 to 20 sec) of milk ejection makes it difficult for a dominant piglet to suck from more than one gland at each milk ejection. It has been proposed that relaxin may have a regulatory role during sucking, and may control the release of oxytocin or the action of oxytocin on the myoepithelial cells (Summerlee *et al.*, 1984; Jones and Summerlee, 1986). This tight control of milk ejection, as illustrated by the requirement for oxytocin for release of milk and perhaps including the reversal of oxytocin action by relaxin, therefore may be of evolutionary importance in allowing for the equal distribution of milk amongst all piglets whilst allowing adequate time for the synthesis of milk. This is further supported by the suggestion that the formation of a teat order confers an advantage on the piglets by minimising fights for teats during each sucking period (Hartstock and Graves, 1976; De Passillé *et al.*, 1988). The consumption of milk (and postulated regulatory factors within milk e.g. caseinomorphins) also provides a mechanism for the synchronisation and modulation of behaviour which may be important during the early stages of life of the suckled mammal.

While the 'design' of the tight milk ejection allows for the survival of more piglets, these same strategies do however limit the amount of milk that can be removed by the suckling piglets. If an autocrine control mechanism regulates the rate of milk secretion in the mammary glands of the sow during established lactation the short duration of milk flow at a milk ejection may not only prevent the removal of all the available milk from the mammary gland, but also the removal of the inhibitory compounds thus preventing an increase in the rate of milk secretion. Furthermore, the selection of sows for large litter sizes results in less milk being available for each piglet since total milk yield decreases with increasing litter size (King *et al.*, 1989; Atwood, 1993). In addition, increasing the food intake of the sow does not necessarily increase milk yield (Lellis and Speer, 1983). These lactational strategies of the sow on the one hand may maximise the number of piglets surviving, but on the other hand are restrictive to maximising piglet growth during the lactation period. While this tight control over milk ejection would seem to be essential for the survival of a maximum number of piglets, increasing the duration of milk ejection may allow for an increased removal of milk by the piglets. Ultimately however, with an increase in the time of milk ejection, the maximum volume of milk removed will be determined by the capacity of the sow to synthesise milk.

Autocrine control

By the early 1970's it was understood that a variety of factors could potentially influence the amount of milk produced by a breastfeeding mother during established lactation. Extrapolating from studies of dairy species, it was felt that the nutritional intake of a mother may play a key role in determining the amount of milk produced by that mother. However, by 1986 (Prentice et al., 1986) it had been well demonstrated that maternal nutrition had little influence on the daily milk production of women both in developing and developed countries and that the most important determinant of maternal milk production in both groups of countries was the optimisation of the breast feeding relationship between the mother and the infant. Indeed, it has become clear over the last decade that, in demand breastfeeding during established lactation, milk supply is controlled in such a way as to match the demand (normally the infant's appetite) for milk (Dewey and Lonnerdal, 1986; Neville and Olivia-Rasbach, 1987; Dewey et al., 1991). The next question, which remained unanswered, was to determine how the breast measured the infant's appetite.

Given the close relationship that can potentially exist between a mother's milk production and the appetite of her infant, methods were needed that could measure milk synthesis as opposed to milk production. The term "milk production" is used in the literature to refer to volumes of milk removed from the breast (either by the infant or by expression). Thus, milk production may refer to the volume of milk removed from the breast at a breastfeed/expression or the average volume of milk removed per unit of time, e.g. ml/24h. Milk production is traditionally measured by test weighing, but may also be determined by expression techniques, Doplar milk flow measurement, infant swallow counting or isotope dilution techniques (Coward et al., 1979; Hartmann and Saint, 1984; Woolridge et al., 1985; Lucas et al., 1987; Lau and Henning, 1989). Alternatively, milk synthesis refers to the accumulation of milk within the breast. "Rate of milk synthesis" refers to the rate at which newly synthesized milk is accumulating within the breast, e.g. ml/h.

Our laboratory has sought to develop methods to measure the short-term rates of milk synthesis in breastfeeding mothers through the measurement of breast volume (Arthur et al., 1989a; Daly et al., 1992). With these methods we are able to measure short-term rates of milk synthesis, between breastfeeds, by measuring the rate of increase in breast volume between those breastfeeds.

In our laboratory, several different methods of measuring breast volume were tried, including stereography and water displacement. We have published the details of two methods of measuring breast volume. The first involved Moiré topography (Arthur et al., 1989a). This method, while accurate, was limited to women with a particular breast shape. Also, the computation of breast volume by this method was restrictively slow. In response to these two problems, a rapid, computerized technique was developed that was named Computerized Breast Measurement (CBM; Daly et al., 1992) (Figure 3). Both techniques have been validated in the same way - the change in breast volume at a breastfeed was compared to the amount of milk consumed by the infant, as measured by test-weighing. The close relationship observed between the change in breast volume and milk removal (Daly et al., 1992) indicates that changes in breast volume can be used to make accurate determinations of changes in the volume of milk within the breast.

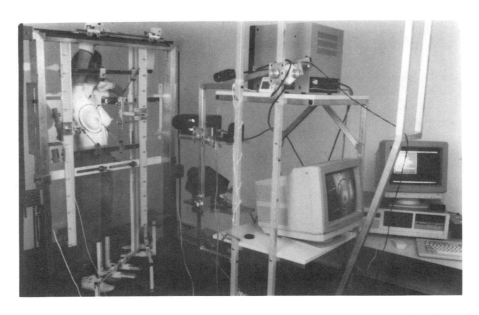

Figure 3. The CBM system in operation: A positioning frame allows the volunteer to position either of her breasts in line with two projectors that project horizontal light stripes. CCD cameras, above and below the two projectors, relay video images to the computer. Analysis by the computer of the distortion of the light stripes on the chest surface allows calculation of the volume of tissue within the black circle painted around the breast. The CBM system is based on the Shape © Measurement System (Alexander and Ng, 1987).

Our studies to date using breast volume measurement methods have illustrated a number of previously unstudied or unquantifiable aspects of human lactation. (1) We observed marked variability in the rates of milk synthesis both between women as well as between days for individual breasts (Daly *et al.*, 1992). In addition, rates of milk synthesis were observed to be highly variable both between breasts and between interfeed intervals (Daly *et al.*, 1993a; Figure 4). This shows that the breast can rapidly change its rate of milk synthesis from one interfeed interval to the next. Also, it was seen that the average rate of milk synthesis during a 24 h period for an individual breast was on average only approximately 64% of the highest rate of milk synthesis observed for that breast (Daly *et al.*, 1993a). Thus, our results suggest that the breasts of the mothers we studied had the capacity to synthesize much more milk than their infants usually required. The between-breast variability in rates of milk synthesis was also of interest. Clearly, milk synthesis is controlled independently between the two breasts to a considerable degree.

(2) The storage capacity of a breast could be determined by measuring breast volume changes over 24 h periods, that is, by calculating the difference between the maximum and minimum breast volumes observed over the 24 h period (Daly *et al.*, 1993a). We observed that there was no relationship between total milk storage capacity of a mother's two breasts and 24-h milk production (Daly *et al.*, 1993a) thus confirming the conventional wisdom that small breast size does not restrict a mother's ability to provide milk for her infant. However, it did appear that increasing storage capacity, between mothers, allowed greater flexibility with regard to breastfeeding patterns. Women with larger storage capacities appeared capable of greater flexibility in terms of intervals between breastfeeds and the amount of milk transferred at each breastfeed.

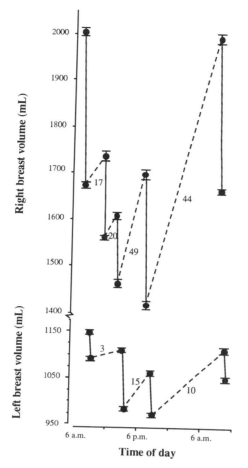

Figure 4. The right and left breast volume changes of a breastfeeding woman over a period of 28 h. This mother was fully breastfeeding her second infant at approximately 2 months *postpartum*. Each point represents the mean ± SEM of replicate breast volume measurements. Lines link pre- and post-feed mean breast volumes. Dashed lines link postfeed mean breast volume of a breastfeed to the pre-feed mean breast volume of the next breastfeed; their slope thus indicating rate of milk synthesis between the two breastfeeds. Rate of milk synthesis also given by the number (in ml/h) accompanying each dashed line. This figure is taken from Daly *et al.*, 1993a.

(3) It was seen that the breasts of the mothers contained a demonstrable amount of available milk after most breastfeeds (Figure 4). Indeed, on average, the breasts of the women were only 76 % empty of their storage capacity after a breastfeed (Daly *et al.*, 1993a). Therefore, infants do not usually empty the breast at a breastfeed. This implies that the infants were regulating their own milk intake.

(4) There was a positive relationship between the degree of breast emptying after a breastfeed and the rate of milk synthesis observed after that breastfeed. This relationship was significant for 6 of 13 breasts studied (Daly *et al.*, 1993a). The degree of emptying explained from 32% to 95% of the variability in postfeed rate of milk synthesis for these six breasts. Thus for these six breasts, the greater the degree of emptying at a breastfeed, the greater the rate of milk synthesis after that breastfeed.

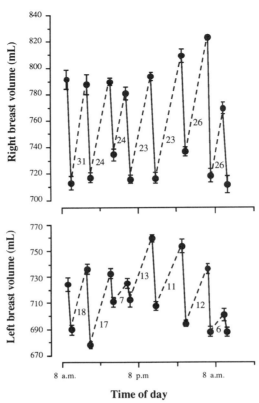

Figure 5. The right and left breast volume changes of a mother with a small storage capacity over a period of 24 h. See Figure 3 for description of symbols. This figure is taken from Daly *et al.*, 1993a.

(5) Despite the variability observed in short-term rates of milk synthesis, there was a close relationship between the total milk synthesized in a 24 h period and the total milk withdrawn in that period, confirming the hypothesis that the breast does balance supply to demand over the 24 h period (Daly *et al.*, 1993a). Indeed, one subject who had a particularly small total storage capacity (approx. 200 ml) was able to produce large volumes of milk per 24 h (approx. 900 ml/24 h) with a regime of more frequently and uniformly spaced breastfeeds (Figure 5).

(6) When the fat content of fore and hind milk samples was compared to changes in breast volume, it was seen that the fat content of milk increased as the breast was emptied (Daly *et al.*, 1993b; Figure 6). Overall, almost 70 % of the variation in the fat content of the breast milk samples could be described by the degree to which the breast was emptied when each sample was taken. However, for the nine individual breasts, between 41 - 95 % of the variance of the fat content of milk was explained by this factor. This relationship appears to explain circadian variation in the fat content of breast milk and also is highly consistent with the existence of a gradient in the fat content of the milk between nipple and alveoli (Daly *et al.*, 1993b).

Whilst some of these observations are not consistent with an endocrine mechanism for the short-term control of milk synthesis in women (eg, 1), all of these observations are consistent with an autocrine control mechanism. Our observations suggest that the breast

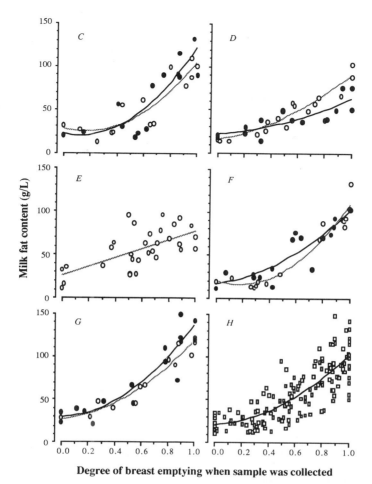

Figure 6. The relationship between the fat content of the milk samples (abscissa) and the degree to which the breast was emptied when the sample was taken (ordinate). Graphs C to G correspond to subjects C to G respectively; from Daly *et al.*, 1993b. Graph H displays the relationship between the two factors for the data for five subjects. For graphs C to G, open circles represent right breast data, closed circles represent left breast data. The regression equation relating the two factors has been calculated separately for right and left breasts for each subject and the line of best fit is given as a hatched (right breast) or solid (left breast) line. This figure is taken from Daly *et al.*, 1993b.

measures the infant's appetite by gauging the degree to which it is emptied by the infant at the end of each breastfeed. The research of other laboratories indicates that the human breast may be amply endowed with mechanisms that could potentially be involved in the communication of infant appetite, through degree of breast emptying, to the epithelial cells. There is some evidence for the existence of a feedback-inhibitor of lactation protein in human milk. Prentice *et al.*, (1989) prepared a sample of whey protein of molecular weight > 10 KDa from human milk samples. This fraction significantly inhibited the synthesis of lactose (by approximately 26 %) and casein (by > 10%) in rabbit mammary tissue culture.

A number of studies suggest the possibility that milk fat synthesis is regulated by an autocrine, feedback inhibitory mechanism. However, in this situation the putative inhibitory compounds are not polypeptides but rather free fatty acids. Heesom *et al.* (1992) reported that the medium chain fatty acids hexanoic acid (6:0), octanoic acid (8:0) and decanoic acid (10:0) decreased glucose conversion into lipid in isolated acini from lactating rat mammary glands. This work is in line with earlier studies showing that short-chain fatty acids inhibit acetyl-CoA carboxylase activity (Levy, 1963; Levy, 1964). In addition, extracts from rat mammary glands that have been unsuckled for extended periods (9 h), demonstrate only 10 % of the [^{14}C]acetate incorporation into fatty acids of extracts from suckled glands (Levy, 1964). These various authors have concluded that trace free fatty acids may act in a feed-back mechanism to control milk lipid synthesis. The presence of both a nipple-alveoli gradient in the fat content of milk (Daly *et al.*, 1993b) and lipases in milk (Hernall and Blèckberg, 1985) makes this hypothesis highly attractive. Also, the apical membrane of the epithelial cell plays a dynamic role in the secretion of milk constituents into the lumen of the alveolus. Membrane material is added to the apical membrane through the exocytosis of Golgi and other vesicles. Membrane material is lost from the apical membrane throught the secretion of fat globules and double membrane-bound vesicles, budding microvilli and other vesicles (Kanno, 1990). In addition, as the lumen fills with milk the epithelial cells change in shape from columnar to cuboidal (Mayer and Klein, 1961). Thus the apical membrane must either increase in surface area as milk accumulates in the lumen but then decrease as milk is withdrawn or change dramatically with respect to the degree of its convolution.

The apical membrane budget of an epithelial cell is, therefore, complicated. A further complication is introduced by the fact that it is not known whether all the components of milk are synthesized in unison. The work of Molenaar *et al.* (1992) suggests that alveoli synthesizing some constituents of milk (eg, α-lactalbumin) may not be synthesizing others (e.g. fat). These observations can be interpreted as showing that either milk secretion is heterogeneous across the alveoli of a mammary gland or that milk secretion is heterogeneous across time. According to the former idea, different alveoli may be "responsible" for different constituents of milk. This idea requires mechanisms that act across the gland to coordinate the synthesis of different alveoli, so that milk composition is constant. According to the latter idea, each epithelial cell synthesizes and secretes each constituent but not at the same time, rather secretion of the different constituents may occur in a cyclical fashion. The latter idea presents fewer obstacles to its incorporation with existing ideas about milk secretion.

The hypothesis that the constituents of milk are secreted in a cyclical manner may be helpful in explaining how the apical membrane budget is managed. Molenaar *et al.* (1992) reported that the alveoli expressing α-lactalbumin were characterized as collapsed (ie, having little lumenal space) with columnar-type epithelial cells. Such morphology is characteristic of "filling" alveoli (Mayer and Klein, 1961). These alveoli were characterized by Molenaar *et al.* (1992) as being "non-fatty", that is the lumens did not contain many fat globules. "Fatty" alveoli, on the other hand, tended to contain many fat globules and to have large lumenal spaces. According to Salazar and Tobon (1974) such alveoli are "resting". In addition, the protein α_{s1}-casein appeared to be co-expressed with α-lactalbumin, whereas the protein lactoferrin appeared to be expressed in the fatty alveoli and not co-expressed with α-lactalbumin. From these observations it is possible to construct models of possible short-term secretory cycles. For example, after an alveolus is emptied, protein and lactose synthesis and secretion commence, but fat globule secretion is delayed. Fat globule secretion might be inhibited by the restricted apical membrane surface area and the requirement for this membrane to expand as the volume of milk in the lumen increases due to lactose secretion. However, exocytosis of the Golgi vesicles that occurs with protein and lactose secretion may provide the membrane material necessary both for apical membrane expansion and for fat globule secretion (Kanno, 1990). As fat globule secretion continues, protein and lactose synthesis may become inhibited through the action of the FIL protein (Wilde and Peaker, 1990). Similarly, the accumulation of fat globules, as the alveolus becomes very full and distended, might inhibit further fat synthesis and secretion through the inhibition of fat synthesis by free fatty acids. At the next breastfeed, collapse of the alveolus might relieve these inhibitions through the removal of the lumenal contents, and also precipitate the final budding of milk fat globules from the apical membrane as this membrane collapses.

Molenaar *et al.* (1992) note that the expression of α-lactalbumin appears to be very finely controlled, with adjacent lobules expressing either high or zero levels of this protein. Such a pattern of gene expression, and indeed the model outlined above, would require "exquisitely local control over mammary gene expression" (Molenaar *et al.* 1992). Immunochemical studies in rats (Nolin, 1979; Nolin and Bogdanove, 1980; Nolin, 1985) have shown that epithelial cells cycle through "secretory" and "resting" phases with a cycle time of hours. This cycle corresponds to the filling and emptying of the alveolus. In addition, prolactin bound only to the cells in the "secretory" phase, when the alveoli were empty and the epithelial cells were tall and columnar, and not in "resting" cells associated with full and distended alveoli (Nolin, 1979). These findings appear to be consistent with a putative action of the FIL inhibitor, that being the modulation of the responsiveness of the epithelial cell to prolactin (McKinnon *et al.*, 1988; Bennett *et al.*, 1990; Wilde *et al.*, 1990).

Therefore, the inhibitory nature of certain milk constituents together with the possible existence of a secretory cycle and the changes that this cycle might involve in terms of apical membrane budget form an impressively large array of mechanisms whereby milk synthesis and milk secretion might be precisely geared to milk removal, particularly the degree to which the breast is emptied at a breastfeed by an infant.

It is possible that the human species may be a better model for the study of autocrine control than dairy species. Dairy species have been bred for maximum milk production, possibly resulting in a breeding-out of sensitivity to this mechanism of control of milk secretion (Henderson and Peaker, 1987). In contrast to the fact that dairy species may often be milked so as to be close to their maximum possible synthetic capacity, women breastfeeding singleton infants may be several fold below their maximum potential for synthesising milk. In contrast with the fact that dairy species and most other mammalian species may have their glands emptied at regular and predictable intervals, the human species is characterised by large variation in feeding frequency both between and within women (Hartmann and Saint, 1984; Hartmann *et al.*, 1984c). Furthermore, the development of breast volume measurement techniques make it possible for the short-term responsiveness of the human mammary gland to be described in terms of actual rates of milk synthesis within the gland, rather than in terms of changes in the amount of milk leaving the gland (ie milk production) as found in studies of dairy animals.

In conclusion, that the milk production of the human breast in established lactation, and in a demand feeding regime, matches the infants need for milk is clear. Our work has demonstrated that the human breast is capable of responding to the needs of the infant for milk at a variety of levels. Our current picture of how the human breast responds to infant demand for milk requires that the breast is able to gauge the degree to which it is emptied. It is very clear that there are ample potential mechanisms that could be involved in such a precise communication of infant need to the epithelial cells. The challenge is now to discern how these potential mechanisms are actually integrated to the control of the synthesis of human milk.

REFERENCES

Agius, L., Robinson, A.M., Girard, J.R. and Williamson, D.H., 1979, Alterations in the rate of lipogenesis *in vivo* in maternal liver and adipose tissue on premature weaning of lactating rats. A possible regulatory role of prolactin, *Biochem J.* 180:689.

Alexander, B.F. and Ng, K.C., 1987, 3-D Shape Measurement by active triangulation using an array of coded light stripes, *SPIE, Optics, Illumination and Image Processing for Machine Vision II* 850:199.

Amenomori, Y., Chen, C.L. and Meites, J., 1970, Serum prolactin levels in rats during different reproductive states, *Endocrinololology* 86:506.

Armstrong, J.D., Kraeling, R.R. and Britt, J.H., 1988, Effects of naloxone or transient weaning on secretion of LH and prolactin in lactating sows, *J. Reprod. Fert.* 83:301.

Arthur, P.G., Jones, T.R., Spruce, J. and Hartmann, P.E., 1989a, Measuring short-term rates of milk synthesis in breast-feeding mothers, *Quart. J. Exp. Physiol.* 74:419.

Arthur, P.G., Kent, J.C. and Hartmann, P.E., 1989b, Milk lactose, citrate and glucose as markers of lactogenesis in normal and diabetic women, *J. Pediatr. Gastroenterol. Nutr.* 9:488.

Arthur, P.G., Kent, J.C. and Hartmann, P.E., 1994, Metabolites of lactose synthesis in milk from diabetic and non-diabetic women during lactogenesis II, *J. Pediatr. Gastroenterol. Nutr.* 19:100.

Atwood, C.S., 1993, "The measurement of fat, protein, lactose and cellular metabolites in milk for the assessment of lactocyte function during the lactation cycle of the sow," PhD Thesis, The University of Western Australia, Perth.

Auerbach, K.G., Riordan, J. and Countryman, B.A., 1993, The breastfeeding process, *in*: "Breastfeeding and Human Lactation", Riordan, J. and Auerbach, K.G., eds., Jones and Bartlett, Boston.

Baldwin, R.L. and Louis, S., 1975, Hormonal actions on mammary metabolism, *J. Dairy Sci.* 58:1033.

Bennett, C.N., Knight, C.H. and Wilde, C.J., 1990, Regulation of mammary prolactin binding by secreted milk proteins, *J. Endocrinol.* 127, suppl:141.

Berthon, P., Katoh, M., Dusanter-Fourt, I., Kelly, P.A. and Dijane, J., 1987, Purification of prolactin receptors from sow mammary gland and polyclonal antibody production, *Mol. Cell. Endocrinol.* 51:71.

Bevers, M.M., Willemse, A.H. and Kruip, A.M., 1978, Plasma prolactin levels in the sow during lactation and the postweaning period as measured by radioimmunoassay, *Biol. Reprod.* 19:628.

Bolander jr, F.F., Nicholas, K.R., Van Wyk, J.J. and Topper, Y.J., 1981, Insulin is essential for accumulation of casein mRNA in mouse mammary epithelial cells, *Proc. Nat. Acad. Sci. USA* 78:5682.

Breier, B.H., Milsom, S.R., Blum, W.F., Schwander, J., Gallaher, B.W. and Gluckman, P.D., 1993, Insulin-like growth factors and their binding proteins in plasma and milk after growth hormone-stimulated galactopoiesis in normally lactating women, *Acta Endocrinol.* 129:427.

Carrick, D.T. and Kuhn, N.J., 1978, Diurnal variation and response to food withdrawal of lactose synthesis in lactating rats, *Biochem. J.* 174:319.

Collier, R.J., McNamara, J.P., Wallace, C.R. and Dehoff, M.H., 1984, A review of endocrine regulation of metabolism during lactation, *J. Anim. Sci.* 59:498.

Coward, W.A., Sawyer, M.B., Whitehead, R.G. and Prentice, A.M., 1979, New method for measuring milk intakes in breast-fed babies, *Lancet* II:13.

Cowie, A.T., 1984, Lactation, *in*: "Hormonal control of reproduction", Austin, C.R. and Short, R.V., eds., Cambridge.

Cowie, A.T., Forsyth, I.A. and Hart, I.C., 1980, "Hormonal Control of Lactation", Springer-Verlag, Berlin.

Cox, D.B. and Hartmann, P.E., 1993, Plasma prolactin and milk synthesis in women, *Neuroendocrinology* 58(Supp 1):47

Cromwell, G.L., Stahly, T.S., Edgerton, L.A., Monegue, H.J., Burnell, T.W., Schenck, B.C. and Elkins, T., 1989, Recombinant porcine somatotropin in lactating sows, *J. Dairy Sci.* 72(Supp 1):257.

Daly, S.E.J., Di Rosso, A., Owens, R.A. and Hartmann, P.E., 1993b, Degree of breast emptying explains changes in the fat content, but not fatty acid composition, of human milk, *Exp. Physiol.* 78:741.

Daly, S.E.J., Kent, J.C., Huynh, D.Q., Owens, R.A., Alexander, B.F., Ng, K.C. and Hartmann, P.E., 1992, The determination of short-term breast volume changes and the rate of synthesis of human milk using computerized breast measurement, *Exp. Physiol.* 77:79.

Daly, S.E.J., Owens, R.A. and Hartmann, P.E., 1993a, The short-term synthesis and infant-regulated removal of milk in lactating women, *Exp. Physiol.* 78:209.

De Passillé, A-M.B., Rushen, J. and Hartstock, T.G., 1988, Ontogeny of teat fidelity in pigs and its relation to competition at suckling, *Can. J. Anim. Sci.* 68:325.

De Passillé, A.M.B., Rushen, J., Foxcroft, G.R., Aherne, F.X. and Schaefer, A., 1993, Performance of young pigs: relationships with periparturient progesterone, prolactin, and insulin of sows, *J. Anim. Sci.* 71:179.

Dewey, K.G., Heinig, M.J., Nommsen, L.A. and Lonnerdal, B., 1991, Maternal versus infant factors related to breast milk intake and residual milk volume: The DARLING study, *Pediatr.* 87:829.

Dewey, K.G. and Lonnerdal, B., 1986, Infant self-regulation of breast milk intake, *Acta Pediatr. Scand.* 75:893.

Dividich, J. le, Esnault, T. and Lynch, B., 1989, Influence de la teneur en lipides du colostrum sur l'accetion lipidique et la regulation de la glycemie chez le porc nouveau-né, *J. Rech. Porcine France* 21:275.

Dodd, S.C., Buttle, H.L., Forsyth I.A. and Dils, R.R., 1993, Milk whey protein as markers of mammary gland activity in the plasma of intact and mastectomised pregnant sows, *J. Endocrinol.* 121(Supp 1):312.

Dusza, L. and Krzymowska, H., 1981, Plasma prolactin levels in sows during pregnancy, parturition and early lactation, *J. Reprod. Fert.* 61:131.

Ellendorf, F., Forsling, M.L. and Poulain, D.A., 1982, The milk ejection reflex in the pig, *J. Physiol.* 333:577.

Ellicott, A.R. and Dziuk, P.J., 1973, Minimum daily dose of progesterone and plasma concentration for maintenance of pregnancy in ovariectomized gilts, *Biol. Reprod.* 9:300.

Fleet, I.R., Goode, J.A., Harmon, M.H., Laune, M.S., Linzell, J.L. and Peaker, M., 1975, Secretory activity of goat mammary glands during pregnancy and the onset of lactation, *J. Physiol.* 333:577.

Forsyth, I.A., 1986, Variation among species in the endocrine control of mammary growth and function: the roles of prolactin, growth hormone, and placental lactogen, *J. Dairy Sci.* 69:886.

Gooneratne, A., Hartmann, P.E., McCauley, I. and Martin, C.E., 1979, Control of parturition in the sow using progesterone and prostaglandin, *Aust. J. Biol. Sci.* 32:587.

Grosvenor, C.E., Whitworth, N. and Mena, F., 1975, Milk secretory response of the conscious lactating rat following intravenous injections of rat prolactin, *J. Dairy Sci.* 58:1803.

Hacker, R.R. and Hill, D.L., 1972, Nucleic acid content of mammary glands of virgin and pregnant gilts, *J. Dairy Sci.* 55:1295.

Harkins, M., Boyd, R.D. and Bauman, D.E., 1989, Effect of recombinant porcine somatotropin on lactational performance and metabolite patterns in sows growth of nursing pigs, *J. Anim. Sci.* 67:1997.

Hartmann, P.E., 1973, Changes in the composition and yield of the mammary secretion of cows during the initiation of lactation, *J. Endocrinol.* 59:231.

Hartmann, P.E., 1991, The breast and breast-feeding, *in*: "Scientific foundations of obstetrics and gynaecology", Phillip, E., Setchell, M. and Ginsburg, M., eds., 4th edition, Butterworth Heinemann, Oxford.

Hartmann, P.E. and Arthur, P.G., 1986, Assessment of lactation performance in women, *in*: Human Lactation 2: Maternal and Environmental Factors", Hamosh, M. and Goldman, A.S., eds., Plenum Press, New York and London.

Hartmann, P.E., McCauley, I., Gooneratne, A.D. and Whitely, J.L., 1984a, Inadequacies of sow lactation: survival of the fittest, *in*: "Physiological Strategies in Lactation" (Symposia of the Zoological Society of London, Number 51), Peaker, M., Vernon, R.G. and Knight, C.H., eds., Academic Press, London.

Hartmann, P.E., Rattigan, S., Prosser, C.G., Saint, L. and Arthur, P.G., 1984c, Human lactation: back to nature, *in*: "Physiological Strategies in Lactation" (Symposia of the Zoological Society of London, number 51), Peaker, M., Vernon, R.G. and Knight, C.H., eds., Academic Press, London.

Hartmann, P.E. and Saint, L., 1984, Measurement of milk yield in women, *J. Pediatr. Gastroenterol. Nut.* 3:270.

Hartmann, P.E., Trevethan, P. and Shelton, J.N., 1973, Progesterone and oestrogen and the initiation of lactation in ewes, *J. Endocrinol.* 59:249.

Hartmann, P.E., Whitely, J.L. and Willcox, D.L., 1984b, Lactose in plasma during lactogenesis, established lactation and weaning in sows, *J. Physiol.* 347:453.

Hartstock, T.G. and Graves, H.B., 1976, Neonatal behaviour and nutrition-related mortality in domestic swine, *J. Anim. Sci.* 42:235.

Healy, D.L., Rattigan, S., Hartmann, P.E., Herington, A.C. and Burger, H.G., 1980, Prolactin in human milk: correlation with lactose, total protein, and α-lactalbumin levels, *Amer. J. Physiol.* 238:E83.

Heesom, K.J., Souza, P.F.A., Ilic, V. and Williamson, D.H., 1992, Chain-length dependency of interactions of medium-chain fatty acids with glucose metabolism in acini isolated from lactating rat mammary glands, *Biochem. J.* 281:273.

Henderson, A.J. and Peaker, M., 1987, Effects of removing milk from the mammary ducts and alveoli, or of diluting stored milk, on the rate of milk secretion in the goat., *Quart. J. Exp. Physiol.* 72:13.

Hernell, O. & Blèckberg, L., 1985, Lipase and esterase activities in human milk, *in*: "Human Lactation: Milk Components and Methodologies", Jensen, R.G. and Neville, M.C., eds., Plenum Press, New York and London.

Holmes, M.A., 1991, "Biochemical Investigations into Milk Secretion and Milk Removal in the Lactating Pig," Ph.D. Thesis, The University of Western Australia, Perth.

Holmes, M.A. and Hartmann, P.E., 1993, The concentration of citrate in the mammary secretion of sows during lactogenesis and established lactation, *J. Dairy Res.* 60:319.

Houston, M.J., Howie, P.W. and McNeilly, A.S., 1983, Factors affecting the duration of breastfeeding: 1 measurement of breast milk intake in the first week of life, *Early Hum. Devel.* 8:49.

Hove, K., 1978a, Effects of hyperinsulinemia on lactose secretion and glucose uptake by the goat mammary gland, *Acta Physiol. Scand.* 104:422.

Hove, K., 1978b, Maintenance of lactose secretion during acute insulin deficiency in lactating goats, *Acta Physiol. Scand.* 103:173.

Jacobs, L.S., 1977, The role of prolactin in mammogeneis and lactogenesis, *in*: "Comparitive Endocrinology of Prolactin", Dellmann, H-D., Johnson, J.A. and Klachko, D.M., eds., Plenum Press, New York.

Jerry, D.J., Stover, R.K. and Kensinger, R.S., 1989, Quantitation of prolactin-dependent responses in porcine mammary explants. *J. Anim.Sci.* 67:1013.

Jones, S.A., Ilic, V. and Williamson, D.H., 1984, Physiological significance of altered insulin metabolism in the conscious rat during lactation, *Biochem. J.* 220:455.

Jones, S.A. and Summerlee, A.J.S., 1986, Relaxin acts centrally to inhibit oxytocin release during parturition: an effect that is reversed by naloxone, *J. Endocrinol.* 111:99.

Kanno, C., 1990, Secretory membranes of the lactating mammary gland, *Protoplasma* 159:184.

Kendall, J.Z., Richards, G.E. and Shih, LI-N., 1983, Effect of haloperidol, suckling, oxytocin and hand milking on plasma relaxin and prolactin concentrations in cyclic and lactating pigs, *J. Reprod. Fert.* 69:271.

Kensinger, R.S., Collier, R.J., Bazer, F.W., Ducsay, C.A. and Becker, H.N., 1982, Nucleic acid, metabolic and histological changes in gilt mammary tissue during pregnancy and lactogenesis, *J. Anim. Sci.* 54:1297.

King, R.H., Toner, M.S. and Dove, H., 1989. Pattern of milk production, *in*: "Manipulating Pig Production II", Barnett, J.L. and Hennessey, D.P., eds., Australasian Pig Science Association, Werribee.

Kirkwood, R.N., Lapwood, K.R., Smith, W.C. and Anderson, I.L., 1984, Plasma concentrations of LH, prolactin, oestradiol-17ß and progesterone in sows weaned after lactation for 10 or 35 days, *J. Reprod. Fert.* 70:95.

Knight, C.H., 1987 Compensatory changes in mammary development and function after hemimastectomy in lactating goats, *J. Reprod. Fert.* 79:343.

Kuhn, N.J., 1969, Progesterone withdrawal as the lactogenic trigger in the rat, *J. Endocrinol.* 44:39.

Kulski, J.K. and Hartmann, P.E., 1981, Change in human milk composition during the initiation of lactation, *Aust. J. Exp. Biol. Med. Sci.* 59:101.

Kulski, J.K., Hartmann, P.E., Martin, J.D. and Smith, M., 1978, Effects of bromocriptine mesylate on the composition of the mammary secretion in non-breast-feeding women, *Obstet. Gynaecol.* 52:38.

Kulski J.K., Smith, M. and Hartmann, P.E., 1977, Perinatal concentrations of progesterone, lactose and α-lactalbumin in the mammary secretion of women, *J. Endocrinol.* 74:509.

Kveragas C.L., Seerley, R.W., Martin, R.J. and Vandergrift, W.L., 1986, Influence of exogenous growth hormone and gestational diet on sow blood and milk characteristics and on baby pig blood, body composition and performance, *J. Anim. Sci.* 63:1877.

Lau, C. and Henning, S.J., 1989, A noninvasive method for determining patterns of milk intake in the breast-fed infant, *J. Pediatr. Gastroenterol. Nut.* 9:481.

Lawrence, R.A., 1980, "Breastfeeding- a Guide for the medical profession," The C.V. Mosby Company, St. Louis.

Lellis, W.A, and Speer, V.C., 1983, Nutrient balance of lactating sows fed supplemental tallow, *J. Anim. Sci.* 56:1334.

Levy, H.R., 1963, Inhibition of mammary gland acetyl CoA carboxylase by fatty acids, *Biochem. Biophys. Research Comm.* 13:267.

Levy, H.R., 1964, The effects of weaning and milk on mammary fatty acid synthesis, *Biochim. Biophys. Acta* 84:229.

Lincoln, D.W., 1984, The posterior pituitary, *in*: "Hormonal control of reproduction", Austin, C.R. and Short, R.V., eds., Cambridge University Press, Cambridge.

Linzell, J. L. and Peaker, M., 1971, The effect of oxytocin and milk removal on milk secretion in the goat, *J. Physiol.* 216:717.

Lucas, A., Drewett, R.B. and Mitchell, M.D., 1980, Breast-feeding and plasma oxytocin concentrations, *Br. Med. J.* 281:834.

Lucas, A., Ewing, G., Roberts, S.B. and Coward, W.A., 1987, How much energy does the breast fed infant consume and expend? *Br. Med. J.* 295:75.

Martin, C.E., Hartmann, P.E. and Gooneratne, A., 1978, Progesterone and corticosteroids in the initiation of lactation in the sow, *Aust. J. Biol. Sci.* 31:517.

Martin, R.J. and Baldwin, R.L., 1971a, Effects of alloxan diabetes on lactational performance and mammary tissue metabolism in the rat, *Endocrinology* 88:863.

Martin, R.J. and Baldwin, R.L., 1971b, Effects of insulin and anti-insulin serum treatments on levels of metabolites in rat mammary glands, *Endocrinology* 88:868.

Martyn, P. and Falconer, I.R., 1985, The effect of progesterone on prolactin stimulation of fatty acid synthesis, glycerolipid synthesis and lipogenic-enzyme activities in mammary glands of pseudopregnant rabbits, after explant culture or intraductal injection, *Biochem. J.* 231:321.

Mattioli M., Conte, F., Galeati, G, and Seren, E., 1986, Effect of naloxone on plasma concentrations of prolactin and LH in lactating sows, *J. Reprod. Fert.* 76:167.

Mayer, G. & Klein, M., 1961, Histology and cytology of the mammary gland, *in*: "Milk: The Mammary Gland and its Secretion", Kon, S.K. and Cowie, A.T., eds., Academic Press, New York and London.

Mellor, D.J. and Cockburn, F., 1986, A comparison of energy metabolism in the new-born infant, piglet and lamb, *Q. J. Exp. Physiol.* 71:361.

McKinnon, J., Knight, C.H., Flint, D.J. and Wilde, C.J., 1988, Effect of milking frequency and efficiency on goat mammary prolactin receptor number, *J. Endocrinol.* 119, suppl:167.

Molenaar, A.J., Davis, S.R. and Wilkins, R.J., 1992, Expressions of a-lactalbumin, α-$_{s1}$-casein, and lactoferrin genes is heterogeneous in sheep and cattle mammary tissue, *J. Histochem. Cytochem.* 40:611.

Mulloy, A.L. and Malvern, P.V., 1979, Relationships between concentrations of porcine prolactin in blood serum and milk of lactating sows, *J. Anim. Sci.* 48:786.

Neifert, M.R., McDonough, S.L. and Neville, M.C., 1981, Failure of lactogenesis associated with placental retention, *Amer. J. Obstet. Gynecol.* 140:477.

Neubauer, S.H., Ferris, A.M., Chase, C.G., Fanelli, J. Thompson, C.A., Lammi-Keefe, C.J., Clark, R.M., Jensen, R.G., Bendel, R.B. and Green, K.W., 1993, Delayed lactogenesis in women with insulin-dependent diabetes mellitus, *Amer. J. Clin. Nutr.* 58:54.

Neville, M.C., Allen, J.C., Archer, P.C., Casey, C.E., Seacat, J., Keller, R.P., Lutes, V., Rasbach, J. and Neifert, M., 1991, Studies in human lactation: milk volume and nutrient composition during weaning and lactogenesis, *Amer. J. Nut.* 54:81.

Neville, M.C., Keller, R.P., Seacat, J., Lutes, V., Neifert, M., Casey, C.E., Allen, J.C. and Archer, P., 1988, Studies in human lactation: milk volumes in lactating women during the onset of lactation and full lactation, *Amer. J. Clin. Nut.* 48:1875.

Neville, M.C. and Olivia-Rasbach, J., 1987, Is maternal milk production limiting for infant growth during the first year of life in breast-fed infants? *in*: "Human Lactation 3: The effects of human milk on the recipient infant", Goldman, A.S., Atkinson, S.A. and Hanson, L., eds., Plenum Press, New York and London.

Newton, M. and Newton, N.R., 1947, The let-down reflex in human lactation, *J. Paedr.* 33:684.

Newton, M. and Newton, N.R., 1951, *Postpartum* engorgement of the breast, *J. Obstet. Gynaecol.* 61:664.

Newton, M. and Newton, N.R., 1962, The normal course and management of lactation, *Clin. Obstet. Gynaecol.* 61:664.

Nicholas, K.R. and Hartmann, P.E., 1981a, Progesterone control of the initiation of lactose synthesis in the rat, *Aust. J. Biol. Sci.* 34:435.

Nicholas, K.R. and Hartmann, P.E., 1981b, Progressive changes in plasma progesterone, prolactin and corticosteroid levels during late pregnancy and the initiation of lactose synthesis in the rat, *Aust. J. Biol. Sci.* 34:445.

Noel, G.L., Suh, H.K. and Frantz, A.G., 1974, Prolactin release during nursing and breast stimulation in *postpartum* and non-*postpartum* subjects, *J. Clin. Endocrinol. Metabol.* 38:413.

Nolin, J.M., 1979, The prolactin incorporation cycle of the milk secretory cycle, *J. Histochem. Cytochem.* 27:1203.

Nolin, J.M., 1985, Target cell prolactin II, *Int. Rev. Cytol.* 95:45.

Nolin, J.M. and Bogdanove, E.M., 1980, Effects of estrogen on prolactin (PRL) incorporation by lutein and milk secretory cells and on pituitary PRL secretion in the *postpartum* rat: correlations with target cell responsiveness to PRL, *Biol. Reprod.* 22:393.

Peaker, M. and Wilde, C.J., 1987, Milk secretion: autocrine control, *News Physiol. Sci.* 2:124.

Phillips, V., 1993, Relactation in mothers of children over 12 months, *J. Trop. Pediatr.* 39:45.

Pitelka, D.R. and Hamamoto, S.T., 1983, Ultrastructure of the mammary secretory cell, *in*: "Biochemistry of Lactation", Mepham, T.B., ed., Elsevier, New York.

Plaut K.I., Kensinger, R.S., Griel, L.C. and Kavanaught, J.F., 1989, Relationships among prolactin binding, prolactin concentrations in plasma and metabolic activity of the porcine mammary gland, *J. Anim. Sci.* 67:1509.

Prentice, A., Addey, C.V.P. & Wilde, C.J., 1989, Evidence for local feedback control of human milk secretion, *Biochem. Soc. Trans.* 17:122.

Prentice, A.M., Paul, A., Prentice, A., Black, A., Cole, T. and Whitehead, R. G., 1986, Cross-cultural differences in lactational performance, *in*: "Human Lactation 2: Maternal and Environmental Factors", Hamosh, M. and Goldman, A.S., eds., Plenum Press, New York and London.

Prichard, J.A., MacDonald, P.C. and Gant, N.F., 1985, "Williams Obstetrics-17th Edition," Appleton-Century-Crofts, Norwalk.

Rattigan, S., 1987, "Yield and composition of human milk: nutritional and hormonal aspects", Ph.D. Thesis, The University of Western Australia, Perth.

Riordan, J. and Auerbach, K.G., 1993,"Breastfeeding and human lactation," Jones and Bartlett Publishers, Boston.

Robinson, A.M. and Williamson, D.H., 1977, Comparison of glucose metabolism in the lactating mammary gland of the rat *in vivo* and *in vitro*. Effects of starvation, prolactin and insulin deficiency, *Biochem. J.* 164:153.

Robertson, H.A. and King, G.J., 1974, Plasma concentrations of progesterone, oestrone, oestradiol-17ß and of oestrone sulphate in the pig at implantation, during pregnancy and at parturition, *J. Reprod. Fert.* 40:133.

Roderuck, C., Coryell, M.N., Williams, H.H. and Macy, I.G., 1946, Metabolism of women during the reproductive cycle IX. The utilisation of riboflavin during lactation, *J. Nut.* 322:267.

Rojkittikhun, T., Uvnès-Moberg, K., Einarsson, S. and Lundeheim, N., 1991, Effects of weaning on plasma levels of prolactin, oxytocin, insulin, glucagon, glucose, gastrin and somatostatin in sows, *Acta Physiol. Scand.* 141:295.

Romsos, D.R., Muiruri, K.L., Lin, P-Y. and Leveille, G.A., 1978, Influence of dietary fat, fasting, and acute premature weaning on in vivo rates of fatty acid synthesis in lactating mice, *Proc. Soc. Exp. Biol. Med.* 159:308.

Ruis, H., Rolland, R., Doesburg, W., Broeders, G. and Corbey, R., 1981, Oxytocin enhances onset of lactation among mothers delivering prematurely, *Br. Med. J.* 283:340.

Saint, L., Smith, M. and Hartmann, P.E., 1984, The yield and nutrient content of colostrum and milk of women from giving birth to 1 month *post-partum*, *Br. J. Nut.* 52:87.

Salazar, H. and Tobon, H., 1974, Morphologic changes of the mammary gland during development, pregnancy and lactation, *in*: "Lactogenic Hormones, Fetal Nutrition, and Lactation", Josimovich, J.B., Reynolds, M. and Cobo, E., eds., John Wiley & Sons, New York, London, Sydney and Toronto.

Summerlee, A.J.S., O'Byrne, K.T., Paisley, A.C., Breeze, M.F. and Porter, D.G., 1984, Relaxin affects the central control of oxytocin release, *Nature* 309:372.

Taverne, M., Bevers, M., Bradshaw, J.M.C., Dieleman, S.J., Willemse, A.H. and Porter, D.G., 1982, Plasma concentrations of prolactin, progesterone, relaxin and oestradiol-17ß in sows treated with progesterone, bromocriptine or indomethacin during late pregnancy, *J. Reprod. Fert.* 65:85.

Tay, C.C.K., Glasier, A.F. and McNeilly, A.S., 1993, Effect of antagonists of dopamine and opiates on the basal and GnRH-induced secretion of luteinizing hormone, follicle stimulating hormone and prolactin during lactational amenorrhoea in breastfeeding women, *Hum. Reprod.* 8:532.

Threlfall, W.R., Dale, H.E. and Martin, C.E., 1974, Porcine blood and hypophyseal prolactin values, *Amer. J. Vet. Res.* 35:1491.

Toner M.S., King, R.H., Dove, H., Hartmann, P.E. and Atwood, C., 1991, The effect of growth hormone on milk production in first litter sows, *in*: "Recent Advances in Animal Nutrition in Australia 1991", Farrell, D.J., ed., University of New England, Armidale.

Van Landeghem, A.A.J. and Van De Wiel, D.F.M., 1978, Radioimmunoassay for porcine prolactin: plasma levels during lactation, suckling and weaning and after TRH administration, *Acta Endocrinol.* 88:653.

Vernon, R.G., 1989, Endocrine control of metabolic adaptation during lactation, *Proc. Nut. Soc.* 48:23.

Vorherr, H., 1974, "The Breast:Morphology, Physiology and Lactation," Academic Press, New York.

Walters, E. and McLean, P., 1968, Effect of alloxan-diabetes and treatment with anti-insulin serum on pathways of glucose metabolism in lactating rat mammary gland, *Biochem J.* 109:407.

Whitely, J.L., Hartmann, P.E., Willcox D.L., Bryant-Greenwood, G.D. and Greenwood, F.C., 1990, Initiation of parturition and lactation in the sow: effects of delaying parturition with medroxyprogesterone acetate, *J. Endocrinol.* 124:475.

Wilde C.J., Henderson, A.J., Knight, C.H., Blatchford, D.R., Faulkner A. and Vernon, R.G., 1987, Effects of long-term thrice-daily milking on mammary enzyme activity, cell population and milk yield in the goat, *J. Anim. Sci.* 64:533.

Wilde, C.J., Knight, C.H., Addey, C.V.P., Blatchford, D.R., Travers, M., Bennett, C.N., and Peaker, M., 1990, Autocrine regulation of mammary cell differentiation, *Protoplasma* 159:112.

Wilde, C.J. and Kuhn N.J., 1979, Lactose synthesis in the rat, and the effects of litter size and malnutrition, *Biochem. J.* 182:287.

Wilde, C.J. and Peaker, M., 1990, Autocrine control in milk secretion, *J. Agric. Sci. Cambridge* 114:235.

Williamson, D.H., Munday, M.R. and Jones, R.G., 1984, Biochemical basis of dietary influences on the synthesis of the macronutrients of rat milk, *Fed. Proc.* 43:2443.

Woolridge, M.W., Butte, N., Dewey, K.G., Ferris, A.M., Garza, C. and Keller, R.P., 1985, Methods for the measurement of milk volume intake of the breast-fed infant, *in*: "Human Lactation: Milk Components and Methodologies", Jenson, R.G. and Neville, M.C., eds., Plenum Press, New York and London.

Zinder O., Hamosh, M., Fleck, T.R.C. and Scow, R.O., 1974, Effect of prolactin on lipoprotein lipase in mammary gland and adipose tissue of rats, *Amer. J. Physiol.* 226:744.

AUTOCRINE CONTROL OF MILK SECRETION: FROM CONCEPT TO APPLICATION

Colin J. Wilde, Caroline V.P. Addey, Lynn M. Boddy-Finch
and Malcolm Peaker

Hannah Research Institute
Ayr, KA6 5HL
UK

INTRODUCTION

The rate of milk secretion is regulated acutely by frequency and completeness of milk removal from the mammary gland. Numerous experiments in dairy animals over more than 30 years (see Elliot, 1961; Linzell and Peaker, 1971) have shown that this acute regulation is exerted locally within each gland rather than by circulating galactopoietic hormones. Experiments in lactating goats also demonstrated that local regulation of milk secretion by milk removal was not related to physical distension of the gland by accumulated milk (Henderson and Peaker, 1984), but was, on the other hand, absolutely dependent on milk removal from the gland (Linzell and Peaker, 1971). This pointed to the involvement of a chemical mechanism, and suggested that local regulation of milk secretion was the result of changes in the degree of feedback inhibition exerted by a secreted milk constituent. Evidence for feedback inhibition of milk secretion is reviewed elsewhere in this volume (Peaker, 1995). Here we describe work which has identified and characterised the autocrine inhibitor in goat's milk.

IDENTIFICATION OF THE FEEDBACK INHIBITOR

We have used a mammary tissue bioassay to search in goat's milk for factors able to feedback-inhibit synthesis of other major milk constituents. Milk fractions were tested for their ability to inhibit lactose or casein synthesis in tissue explants from mid-pregnant rabbits, which synthesise milk constituents when cultured in the presence of lactogenic hormones. Initial results with relatively crude milk fractions indicated the presence of factors able to inhibit synthesis of milk protein and carbohydrate, and in both cases inhibitory activity in defatted and dialysed milk fractioned with the whey proteins. Co-fractionation of the two inhibitory activities persisted during further purification. Within the whey, inhibition was associated with constituents of M_r 10,000-30,000, based on their passage through or retention by ultrafiltration membranes, and a single bioactive protein was subsequently isolated from this fraction by a combination of anion exchange chromatography and chromatofocusing. By this procedure, lactose and casein bioactivity were each co-purified up to 40,000-fold compared with unfractionated whey (Wilde *et al.*, 1995a). Co-purification of bioassay activity, and the apparent homogeneity of protein obtained by chromatofocusing, suggested that a single milk protein was competent to regulate synthesis of both lactose and casein by feedback inhibition.

The inhibitory protein has an N-terminal sequence which contains no homology with other milk proteins, nor indeed any recorded sequence. In view of this novelty and its biological activity we have, accordingly, named the protein FIL, Feedback Inhibitor of Lactation. FIL is a glycoprotein. Positive reaction in assays for hexose and sialic acid was supported by lectin binding studies, and by chemical deglycosylation with trifluoromethanesulphonic acid (TFMS), which reduced its M_r of 7,600 on gel filtration by approximately 1,000. This oligosaccharide content appears to cause the protein's unusual behaviour on SDS-polyacrylamide gel electrophoresis (PAGE; Figure 1). Native protein migrated with an apparent M_r of 66,000, but after TFMS treatment, deglycosylated protein migrated at a position consistent with its M_r by gel filtration. Anomalous migration on SDS-PAGE is not without precedent amongst milk glycoproteins. Bovine and murine κ-casein show similar, albeit less extreme, behaviour, migrating at approximately twice their M_r determined by structural analysis. Significantly, this idiosyncratic behaviour is also shared by an inhibitory protein in cow's milk identified by the same combination of anion exchange chromatography, chromatofocusing and bioassay of whey proteins. The structural as well as functional similarity of bovine and caprine inhibitory proteins was also shown by specific and high-affinity binding of caprine FIL by a polyclonal antiserum raised against the bovine inhibitor. Indeed, the anti-bovine FIL antiserum was more specific for the goat inhibitor than antisera raised using goat FIL as antigen. Polyclonal and monoclonal antibodies against goat FIL each recognised other whey proteins of similar M_r but no bioassay activity, which may be precursors or products of the active constituent (see below).

AUTOCRINE CONTROL

The site of FIL synthesis was identified by immunohistochemistry using anti-bovine FIL antiserum. We adopted an approach in which mammary epithelial cells from late-pregnant goats were cultured on a reconstituted basement membrane termed EHS matrix. On this matrix, preparations highly enriched for epithelial cells by density gradient centrifugation form mammospheres, multicellular structures enshrouded in matrix material. Visualisation of nuclei within the mammospheres by DAPI staining showed the cells arranged peripherally around a central lumenal space, in a manner reminiscent of alveoli in the lactating tissue. As with differentiated epithelial cells *in vivo*, mammosphere cells also became polarised, and secreted proteins vectorially.

Casein and α-lactalbumin were detected by immunofluorescence microscopy and by radiolabelling in the lumenal space, whence they were extracted by EGTA treatment of the cells *in situ*. Therefore, goat mammary epithelial cells, like mouse cells on EHS matrix (Barcellos-Hoff *et al.*, 1989; Hurley *et al.*, 1994) were orientated such that apically-secreted milk were found in the lumenal i.e. EGTA-extractable fraction whereas basolaterallysecreted proteins accumulated in culture medium. Immunoblotting of EGTA extract and culture medium demonstrated that FIL secretion followed the same pattern as casein and α-lactalbumin, that is, FIL was detected in lumenal secretion, and was barely detectable in basal secretion (Figure 1). From this we have concluded that feedback inhibition is an autocrine mechanism: the inhibitory protein is synthesised by the secretory epithelial cells on which it acts, and accumulates preferentially, *in vivo* perhaps exclusively, in milk. Clearly, it is the unusual, indeed almost unique, ability of the mammary gland to store its

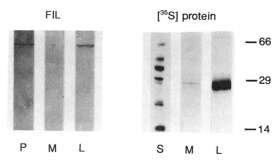

Figure 1. Vectorial secretion of FIL in goat mammary epithelial cells on EHS matrix. Purified protein (P) was detected by immunoblotting as a high M_r band in lumenal secretion (L) but not culture medium (M). [^{35}S]methionine-labelled casein detected by SDS-PAGE and fluorography showed a similar distribution. S, molecular weight standards.

its secretion extracellularly which confers this additional local level of control on the rate of milk secretion. However, autocrine feedback in an exocrine gland is not without precedent. "Milking" the venom glands - the only other exocrine gland in which secretion is stored extracellularly - of the carpet viper *Echis carinatus* increases expression of the genes encoding venom constituents (Paine *et al.*, 1992). Feedback inhibition by extracellular product has also been demonstrated in the flowers of *Blandiflora nobilis*, where daily removal of nectar stimulates the plant's nectar production (Pyke, 1991).

MANIPULATION OF AUTOCRINE INHIBITION

The ability of the protein identified by tissue culture bioassay to inhibit milk protein and lactose secretion *in vivo* was tested by injecting protein back into glands of lactating goats through the teat canal. Intra-ductal injection of microgram quantities of anion exchange-purified FIL produced a temporary, concentration-dependent inhibition of milk secretion. The effect appeared to be specific to the bioassay-positive protein, since neither carrier solution nor much larger quantities of other whey proteins had any effect on milk yield (Wilde *et al.*, 1988a). Both the degree and duration of inhibition depended on the amount injected. With higher doses of 500 μg and 750 μg of inhibitory protein, milk secretion was inhibited for up to 3 days (Figure 2), whereas the effect of 100 μg and 250 μg was relieved within 24 to 48 hours of injection, which took place immediatedly after the afternoon milking. Reversible, concentration inhibition of milk secretion by FIL confirmed the protein's ability to inhibit milk protein and lactose synthesis *in vitro*. On the other hand, it also identified an anomaly between the protein's effect *in vivo* and *in vitro* which is as yet unresolved. Irrespective of the amount of protein injected, FIL inhibited milk

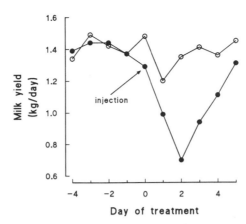

Figure 2. Effect of FIL on milk yield in the lactating goat. FIL (750 μg) was administered by injection through the teat canal into one gland of a lactating goat (●). The other gland (○) was injected with the same volume of carrier solution.

secretion *in vivo* without affecting gross milk composition i.e. protein, lactose or lipid content, suggesting that it was acting on all milk constituents equally. In contrast, partly-purified inhibitor had no effect on lipid synthesis from acetate in rabbit mammary explants (Wilde *et al.*, 1987a) or in suspension cultures of lactating mouse mammary acini (Rennison *et al.*, 1993). Whatever the reason for this dichotomy, the inhibitory protein's competence to regulate milk secretion coordinately in the lactating tissue is clearly consistent with its proposed role in mediating the response to changes in milk frequency, which also affects milk secretion but not milk composition (Henderson *et al.*, 1983).

FIL's ability to regulate milk secretion but not milk composition was supported by other experiments *in vivo*, which adopted an alternative strategy involving immuno-manipulation of autocrine inhibition. The theory behind this approach was that antibody against FIL produced by active immunisation, if secreted in milk, should act to neutralise autocrine inhibition and increase milk yield. When this was tested in goats after peak lactation, three immunisations of hapten-conjugated protein and a suitable adjuvant at 8-weekly intervals produced a circulating titre of antibody against FIL. The immune response was apparent after the second immunisation, and increased after the third, at which time anti-FIL antibody was also detected in milk. The goats were milked twice daily, and as expected at this stage of lactation, milk yield in sham-immunised animals (which showed no immune response) declined progressively throughout the experimental period. The change in milk yield of immunised goats was initially similar to that of the controls, but after the second immunisation, the decline in yield had slowed, and after the third immunisation there was little or no fall in milk yield for a short period. In immunised goats, the decrease in milk yield with once daily milking was also significantly lower than in controls, an observation consistent with partial immuno-neutralisation of the increase in autocrine feedback (Figure 3). Therefore, it appeared that the presence in milk of antibodies specific for FIL had conferred some protection against autocrine inhibition.

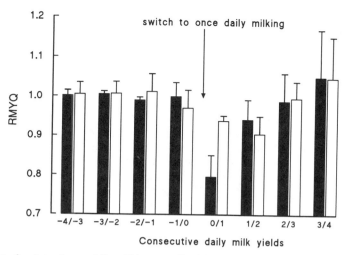

Figure 3. Effect of unilateral once daily milking on milk yield in lactating goats immunised against FIL (□) and in sham-immunised controls (■). In each group, the unilateral change in milk yield on introduction of once-daily milking is expressed as a relative milk yield quotient (Linzell and Peaker, 1971). Values < 1.0 indicate that milk yield on consecutive days had fallen only in the gland milked once daily.

INTRACELLULAR MECHANISM

Initially, it was difficult to envisage how a single milk protein could act in concerted fashion on all milk constituents. The efficacy of FIL administered by intra-ductal injection in lactating goats indicated that the protein's action was indeed one of feedback inhibition after secretion into the alveolar lumen. This suggested a receptor-mediated mechanism, and preliminary binding experiments with $[^{125}I]$-FIL have supported the presence of a specific receptor on the apical membrane. Intracellular signal transduction of autocrine feedback is as yet uncharacterised, but study of mammary protein secretion has suggested a mechanism of action which may explain the multipotent actions of the inhibitor. FIL treatment of suspension cultures of mouse mammary cells produced dramatic but fully-reversible dispersion of the Golgi apparatus and vesiculation of the endoplasmic reticulum (Rennison *et al.*, 1993; Burgoyne *et al.*, 1995). These ultrastructural changes were accompanied by a 50% inhibition of protein secretion, which again was fully-reversibly within one hour after FIL removal (Rennison *et al.*, 1993).

FIL's effect on protein secretion is at present without physiological parallel, inasmuch as it inhibits a constitutive secretory pathway, one in which exocytosis occurs without temporary intracellular storage of secretion. However, treatments in other secretory cells which disrupted membrane trafficking and caused accumulation of secretory protein in the endoplasmic reticulum (ER) also blocked protein synthesis (Kuznetsov *et al.*, 1992). Furthermore, FIL's effects can be mimicked by brefeldin A, a fungal drug which causes retrograde transport of *cis*, medial, and *trans*-Golgi elements (but not *trans*-Golgi network enzymes) back to the ER, with consequent inhibition of protein transport through the secretory pathway (Lippincott-Schwartz *et al.*, 1989; Orci *et al.*, 1991). Brefeldin A, like FIL, also inhibited mammary protein synthesis. This suggests that any agent which disrupts membrane trafficking between ER and Golgi may be competent to inhibit protein translation. An effect of FIL on vesicular transport could then explain its coordinated effects on synthesis and secretion of non-lipid milk constituents, and could also explain the increase in intracellular casein degradation induced by autocrine inhibition *in vivo* or *in vitro* (Stewart *et al.*, 1988; Wilde *et al.*, 1989a). Whereas in lactating goat mammary tissue, there is normally little or no intracellular degradation of secretory proteins (Wilde and Knight, 1986), incomplete milking prior to tissue explanting induced the degradation of a proportion of newly-synthesised caseins (Wilde *et al.*, 1989a), and this effect which was reproduced by culture in the presence of partly-purified FIL (Stewart *et al.*, 1988). Increased secretory protein degradation within the ER was also observed in CHO and COS cell lines and in mouse L-cell cultures when protein transport between ER and Golgi was inhibited (Wileman *et al.*, 1991).

As the mechanism of intracellular lipid translocation is as yet poorly characterised, it is difficult to predict the likely consequences of FIL's disruption of Golgi and ER integrity. On the one hand, in view of the ER's rôle as microlipid droplet progenitor (Keenan *et al.*, 1992; Keon *et al.*, 1993), it seems possible that disruption of ER integrity would also affect milk lipid synthesis and secretion. Clearly, this would explain FIL's concerted inhibition of all milk constituents when administered by intra-ductal injection in lactating goats (Figure 2). On the other hand, and in contrast to its effect *in vivo*, FIL had no effect on lipid synthesis in tissue or cell culture (Wilde *et al.*, 1987a; Rennison *et al.*, 1993). One explanation is that these cultures lack some factor mediating FIL's effect on lipid synthesis and secretion *in vivo*. For instance, in rabbit explant cultures and lactating mouse mammary cell suspensions where FIL was tested, *de novo* fatty acid synthesis takes place in the absence of triacylglycerol synthesis from exogenous fatty acids. *In vivo,* exogenous non-esterified fatty acids extracted from chylomicrons, VLDL and non-esterified fatty acids in the bloodstream (Hawkins and Williamson, 1972) normally account for a high proportion of mammary triacylglycerol synthesis, and regulate *de novo* lipid synthesis through an inhibitory mechanism similar to that described for medium chain fatty acids (Robinson and Williamson, 1978; Heesom *et al.*, 1992; Williamson *et al.*, 1995). Without the modulating influence of exogenous non-esterified fatty acids, *de novo* triacylglycerol synthesis may be rendered insensitive to autocrine feedback by FIL.

KINETICS OF AUTOCRINE INHIBITION

If, as the evidence indicates, FIL is a physiological regulator of the rate of milk secretion, reversible, concentration-dependent inhibition in lactating goats and in tissue and cell culture (Wilde *et al.*, 1987a; Rennison *et al.*, 1993) suggests that its concentration in milk increases as milk accumulates, so that the rate of milk secretion is reduced as the gland fills. Conversely, FIL's concentration should decrease after milking, in order to re-establish a high rate of milk secretion. Thus, frequent milking stimulates milk secretion by limiting the accumulation of inhibitory protein, whereas infrequent milking reduces milk yield by increasing the amplitude of changes in inhibitor concentration. Intra-ductal injection of FIL then mimics the effect of infrequent milking. We have yet to explore fully the kinetics of autocrine inhibition, but preliminary results support these predictions. Measurement of FIL's milk concentration in dairy cows suggests that milk storage in the gland causes FIL to accumulate in milk, but when milk secretion is not accompanied by milk storage, there is no increase in autocrine feedback (P Irving, K. Stelwagen, S.R. Davis & C.J. Wilde, unpublished work).

The kinetics of autocrine inhibition are complicated by the presence of residual milk left behind after milking which, according to our predictions, should contain a high concentration of FIL. This high concentration in residual milk cannot persist indefinitely, otherwise it would prevent cyclical changes in autocrine inhibition and milk secretion rate as milk accumulates and is removed. It is not yet known how autocrine inhibition in residual milk is relieved, nor how changes in FIL concentration are achieved during milk acumulation, but it is unlikely that they occur through a change in the rate of inhibitor secretion, since this would require that it be regulated independently of other milk constituents. A more plausible explanation is that FIL is the result of, or is susceptible to, processing after secretion. Computer modelling indicates that in either case, first order metabolism in the alveolar lumen would bring about an increase in active inhibitor during milk accumulation, even though FIL was being secreted at a constant rate relative to other milk constituents. The same mechanism could also neutralise inhibitor in residual milk. A mechanism involving FIL inactivation would rapidly reverse the effect of a small pool of residual inhibitor. In the event of pro-inhibitor activation, neutralisation of residual inhibitor would be slower, since it would depend on dilution of residual bioactive protein. Processing of secreted FIL may explain the presence in milk of other whey proteins which, based on their recognition by polyclonal and monoclonal antibodies raised against bioactive protein, are structurally-related to FIL, but inactive in the bioassay.

As FIL acts directly on the secretory epithelial cells of the tissue, for it to be effective it must be in the alveolar lumen of the gland. In other words, cisternal milk exerts no feedback inhibition. It follows that animals with relatively large cisterns are less subject to autocrine feedback than those which, by dint of a small cistern, store a high proportion of milk within the alveolar space. The cisternal to alveolar storage ratio should also influence the response to frequent or infrequent milking. Large-cisterned cows should be less responsive to frequent milking and more tolerant of infrequent milking than those those with small cisterns, because they are less subject to autocrine inhibition. This prediction has been confirmed in lactating goats (Knight *et al.*, 1989) and cows (Dewhurst and Knight, 1992; Knight and Dewhurst, 1992). It is also possible that milking frequency may up- or down-regulate its own longer-term effect on the rate of milk secretion by altering the anatomy of the mammary gland. Long term once-daily milking is reported to increase udder volume and residual milk volume compared with goats milked twice daily (Capote *et al.*, 1994). Another recent study has also shown that lactating goats become more tolerant of once-daily milking in successive treatment periods, apparently because a greater proportion of stored milk was accommodated in the cistern of the gland - where it exerts no feedback inhibition (Henderson and Peaker, 1987) - rather than in the alveolar lumen (C.H. Knight, personal communication).

From these observations it would also be predicted that other species which do not possess a mammary cistern, and so store their milk in the ducts and alveolar lumena of the mammary gland, would exhibit a rate of milk secretion that is especially dependent on the degree of gland emptying. A recent study of breastfeeding mothers supports this prediction (Daly *et al.*, 1993). A clear relationship was demonstrated between degree of breast emptying and the rate of milk secretion in the next interfeed period, suggesting that milk removal, and by extrapolation autocrine inhibition, may be the predominant factor regulating milk secretion in the breastfeeding mother.

LOCAL REGULATION OF MAMMARY DEVELOPMENT

Sustained changes in milking frequency or completeness of milking elicit a sequence of local adaptations in the mammary cell population, which act to accommodate the increased or decreased rate of milk secretion. Initially, this is seen as a change in the degree of secretory cell differentiation. For example, an increase from twice to thrice daily milking stimulated secretory cell differentiation, assessed by key mammary enzyme activities, within 10 days (Wilde *et al.*, 1987b), and four times daily milking of dairy cows produced a similar effect after four weeks (Hillerton *et al.*, 1990). Stimulation of cell differentiation has also been demonstrated by increased concentrations of messenger RNAs (mRNA) encoding key lipogenic enzymes (Wilde *et al.*, 1990) and milk proteins (JM Bryson, CVP Addey and CJ Wilde, unpublished work) after several weeks of thrice rather than twice or once daily milking. Conversely, unilateral once daily milking of goats for 4 weeks decreased secretory cell differentiation compared with contralateral glands milked twice daily (Wilde and Knight, 1990). Even a small perturbation of milk removal imposed by incomplete milking decreased secretory cell differentiation when sustained for several months (Wilde *et al.*, 1989b). It seems likely that this regulation of secretory cell differentiation is also by FIL. Partly-purified inhibitor prevented full differentiation of mouse mammary epithelial cells on floating collagen gels (Wilde *et al.*, 1991), and decreased cell differentiation in rabbit mammary glands treated by intra-ductal injection (Wilde *et al.*, 1988b).

Autocrine modulation of cell differentiation may be a consequence of FIL's primary effect on membrane trafficking, which may act to up- or down-regulate galactopoietic hormone receptors. Short-term treatment of lactating mouse mammary epithelial cells with partly-purified FIL decreased cell-surface prolactin binding (Bennett *et al.*, 1990), an effect recently confirmed using anion exchange-purified protein (Bennett *et al.*, 1994a). FIL also decreased total prolactin binding in rabbit mammary glands 24 h after injection through the teat canal (Bennett *et al.*, 1994b), mimicking the effect of milk accumulation (Bennett *et al.*, 1992). Autocrine modulation of endocrine control is also supported by the effects of milking frequency: thrice instead of twice daily milking increased total prolactin binding in goat mammary tissue after 4 weeks (McKinnon *et al.*, 1988) and conversely, incomplete milking reduced prolactin binding in concert with a decrease in secretory cell differentiation (Wilde *et al.*, 1989b).

Autocrine regulation of mammary epithelial differentiation is likely to be the long term consequence of FIL's primary effect on membrane trafficking, and its resultant effect on galactopoietic hormone receptor distribution. In other words, down-regulation of mammary hormone receptors is likely to be, in chronological terms, an effect secondary to FIL's primary, acute effect on milk secretion, and not the causitive factor. Although prolactin is reported to regulate milk secretion acutely in rabbit mammary explants (Ollivier-Bousquet, 1978), milk protein synthesis and secretion in short-term cultures of murine mammary epithelial cells was unaffected by removal of extracellular prolactin (Bennett *et al.*, 1994a), but was nevertheless regulated acutely by FIL (Rennison *et al.*, 1993). Thus, FIL's novel effect on the secretory pathway may serve to integrate autocrine control of milk secretion with endocrine effects on mammary development, the latter being necessary to sustain the tissue's response during long term manipulation of milking frequency or efficiency.

Whether such a mechanism may also acccount for the long term effect of milking frequency on mammary growth is not known at present. As in other cases, this growth response was elicited by a local mechanism. Secretory cell number increased unilaterally when one gland was milked three times daily for a period of months (Wilde *et al.*, 1987b). The increase in cell number was associated with a higher rate of DNA synthesis in freshly-prepared explants, suggesting that frequent milking stimulated cell proliferation. However, in late lactation when these measurements were made, cell number in goat mammary tissue is declining progressively (Wilde and Knight, 1989), and it may be that a greater cell number in the thrice-milked gland was due in part to a reduced rate of cell death. Recent work (Quarrie *et al.*, 1994, 1995) showed that apoptosis can be induced unilaterally by milk stasis in one gland, indicating that local mechanisms sensitive to milk accumulation do regulate mammary cell death. We have at this stage no reason to implicate FIL in this local control of cell proliferation and apoptosis. However, an understanding of this local regulation of mammary development may lead to new methods for manipulating lactation: maintenance of cell number after peak lactation should act to prevent the progressive decline in milk yield, and so increase the persistency of lactation.

CONCLUSION

A combination of experiments in lactating dairy animals and tissue or cell culture show that FIL, a novel milk protein, acts as an autocrine inhibitor of milk secretion. This review has focused on the characterisation of FIL proteins in goat's and cow's milk, but the cross-species activity of these ruminant proteins, and partial characterisation of similar proteins in the milk of breastfeeding mothers (Prentice *et al.*, 1989; Wilde *et al.*, 1995b) and the tammar wallaby (Hendry *et al.*, 1992; Nicholas *et al.*, 1995) raises the possibility that autocrine control of milk secretion may be a mechanism that, like lactation itself, is ubiquitous in mammals.

ACKNOWLEDGEMENTS

The authors are indebted to Crispin Bennett, David Blatchford, Jane Bryson, Shirley Connor, Marian Kerr, Kay Hendry and Chris Knight of the Director's Research Group at HRI who have contributed to this work. We are also grateful to those who have collaborated on aspects of the research programme, particularly Bob Burgoyne and colleagues (University of Liverpool), Pat Barker (Babraham Institute, Cambridge), Steve Davis and Kerst Stelwagen (Ruakura Agricultural Centre, New Zealand), Peter Hartmann and Steven Daly (University of Western Australia), Kevin Nicholas (CSIRO, Australia) and Anne Prentice (MRC Dunn Nutrition Unit, Cambridge). This work was funded by the Scottish Office Agriculture and Fisheries Department.

REFERENCES

Barcellos-Hoff, M.H., Aggeler, J., Ram. T.G. and Bissell, M.J., 1989, Functional differentiation and alveolar morphogenesis of primary mammary cultures on reconstituted basement membrane, *Development* 105: 223.

Bennett, C.N., Knight, C.H. and Wilde, C.J., 1990, Regulation of mammary prolactin binding by secreted milk proteins, *J. Endocrinol.* 127:S141.

Bennett, C.N., Knight, C.H. and Wilde, C.J., 1994, Down-regulation of prolactin receptors in lactating mouse mammary cells by FIL (Feedback Inhibitor of Lactation), a secreted milk protein, submitted for publication.

Bennett, C.N., Wilde, C.J. and Knight, C.H., 1992, Changes in prolactin binding with milk accumulation in the lactating rabbit, *Proc. Brit. Soc. Endocrinol.*, April, 1992:141.

Bennett, C.N., Wilde, C.J. and Knight, C.H., 1994, Milk accumulation inhibits hormone binding to lactating rabbit mammary gland through an action of FIL, the Feedback Inhibitor of Lactation, submitted for publication.

Blatchford, D.R. and Peaker, M., 1983, Effects of decreased feed intake on the response of milk secretion to frequent milking in goats, *Quart. J. Exp. Physiol.* 68:315.

Bryson, J.M , Wilde, C.J. and Addey, C.V.P., 1993, Effect of unilateral changes in milking frequency on mammary mRNA concentrations in the lactating goat, *Biochem. Soc. Trans.* 21:294S.

Burgoyne, R.D., Handel, S.E., Sudlow, A.W., Turner, M.D., Kumar, S., Simons, J.P., Blatchford, D.R. and Wilde, C.J., 1995, The secretory pathway for milk protein secretion and its regulation, *In* "Intercellular Signalling in the Mammary Gland", Wilde, C.J., Knight, C.H. and Peaker, M., eds., Plenum Publishing Company, New York.

Capote, J., Lopez, J.L., Darmanin, N., Caja, G., Peris, S. and Such, X., 1994, Once-a-day milking effects on lactation peformance and udder traits in the first lactation of Canarian dairy goats, *J. Dairy Sci.*, in press.

Daly, S.E.J., Owens, R.A. and Hartmann, P.E., 1993, The short-term synthesis and infant-regulated removal of milk in lactating women, *Exp. Physiol.*, 78:209.

Dewhurst, R.D. and Knight, C.H., 1992, The response to thrice daily milking and its relationship to cisternal storage capacity in dairy cows, *Anim. Prod.* 54:459.

Elliot, G.M., 1961, The effect on milk yield of three times a day milking and of increasing the level of residual milk, *J. Dairy Res.* 28:209.

Hawkins, R.A. and Williamson, D.H., 1972, Measurements of substrate uptake by mammary gland of the rat, *Biochem. J.* 129:1171.

Heesom, K.J., Souza, P.F.A., Ilic, V. and Williamson, D.H., 1992, Chain-length dependency of interactions of medium-chain fatty acids with glucose metabolism in acini isolated from lactating rat mammary glands. A putative feedback to control milk lipid synthesis, *Biochem. J.* 281:273.

Henderson, A.J., Blatchford, D.R. and Peaker, M., 1983, The effects of milking thrice instead of twice daily on milk secretion in the goat, *Quart. J. Exp. Physiol.* 68:645.

Henderson, A.J. & Peaker, M., 1984, Feedback control of milk secretion in the goat by a chemical in milk, *J. Physiol., Lond.* 351:39.

Henderson, A.J. & Peaker, M., 1987, Effect of removing milk from the mammary ducts and alveoli, or of diluting stored milk, on the rate of milk secretion in the goat, *Q. J. Exp. Physiol.*, 72:13.

Hendry, K.A.K., Wilde, C.J., Nicholas, K.R. and Bird, P.H, 1992, Evidence for an inhibitor of milk secretion in the tammar wallaby, *Proc. Aust. Soc. Biochem. Mol. Biol.* 24:POS-2-3.

Hillerton, J.E., Knight, C.H., Turvey, A., Wheatley, S.D. and Wilde, C.J., 1990, Milk yield and mammary function in dairy cows milked four times daily, *J. Dairy Res.* 57:285.

Hurley, W.L., Blatchford, D.R., Hendry, K.A.K. and Wilde, C.J., 1994, Extracellular matrix and mouse mammary cell function: comparison of substrata in culture, *In Vitro Cell Dev. Biol.* 30A:529.

Keenan, T.W., Dylewski, D.P., Ghosal, D. and Keon, B.H., 1992, Milk lipid globule precursor release from endoplasmic reticulum reconstituted in a cell-free system, *Eur. J. Biochem.* 57:21.

Keon., B.H., Ghosal, D. and Keenan, T.W., 1993, Association of cytosolic lipids with fatty acid synthase from lactating mammary gland, *Int. J. Biochem.* 25:533.

Knight, C.H., Brosnan, T., Wilde, C.J. and Peaker, M., 1989, Evidence for a relationship between gross mammary anatomy and the increase in milk yield obtained during thrice daily milking in goats, *J. Reprod. Fertil. (abst. series)*, 5:30.

Knight, C.H. and Dewhurst, R.D., 1992, Relationship between cisternal storage capacity and tolerance of once daily milking in dairy cows, *Anim. Prod.* 54:476.

Kuznetsov, G., Brostrom, M.A. and Brostrom, C.O., 1992, Demonstration of a calcium requirement for secretory protein processing and export. Differential effects on calcium and dithiothreitol, *J. Biol. Chem.* 267:3932.

Linzell, J.L. and Peaker, M., 1971, The effects of oxytocin and milk removal on milk secretion in the goat. *J. Physiol., Lond.* 216:717.

Lippincott-Schwartz, J., Yuan, L., Bonifacino, J.S. and Klausner, R.D., 1989, Rapid redistribution of Golgi proteins into the ER in cells treated with brefeldin A: evidence for membrane cycling from Golgi to ER, *Cell* 56:801.

McKinnon, J., Knight, C.H., Flint, D.J. and Wilde, C.J., 1988, Effect of milking frequency and efficiency on goat mammary prolactin receptor number, *J. Endocrinol.* 119 (Suppl.):167.

Nicholas, K.R., Bird, PH., Hendry, K.A.K. and Wilde, C.J., 1995, *In* "Intercellular Signalling in the Mammary Gland", Wilde, C.J., Knight, C.H. and Peaker, M., eds., Plenum Publishing Company, New York.

Olivier-Bousquet, M., 1978, Early effects of prolactin on lactating rabbit mammary gland, *Cell Tissue Res.* 187:25.

Orci, L., Tagaya, M., Amherdt, M., Perrelet, A., Donaldson, J.G., Lippincott-Schwartz, J., Klausner, R.D. and Rothman, J.E., 1991, Brefeldin A, a drug that blocks secretion, prevents the assembly of non-clathrin-coated buds on Golgi cisternae, *Cell* 64:1183.

Paine, M.J.I., Desmond, H.P., Theakston, R.D.G. and Crampton, J.M., 1992, Gene expression in *Echis carnatus* (carpet viper) venom glands following milking, *Toxicon*, 30:379.

Peaker, M., 1995, Autocrine control of milk secretion: development of the concept, *In* "Intercellular Signalling in the Mammary Gland", Wilde, C.J., Knight, C.H. and Peaker, M., eds., Plenum Press, New York.

Prentice, A., Addey, C.V.P. and Wilde, C.J., 1989, Evidence for local feedback control of human milk secretion, *Biochem. Soc. Trans.* 15:122.

Pyke, G.H., 1991, What does it cost a plant to produce floral nectar? *Nature* 350:58.

Quarrie, L.H., Addey, C.V.P. and Wilde, C.J., 1994, Local regulation of mammary apoptosis in the lactating goat. *Biochem. Soc. Trans.* 22:178S.

Quarrie, L.H., Addey, C.V.P. and Wilde, C.J., 1995, Local control of mammary apoptosis by milk stasis, *In* "Intercellular Signalling in the Mammary Gland", Wilde, C.J., Knight, C.H. and Peaker, M., eds., Plenum Publishing Company, New York.

Rennison, M.E., Kerr, M.A., Addey, C.V.P., Handel, S.E., Turner, M.D., Wilde, C.J. and Burgoyne, R.D., 1993, Inhibition of constitutive protein secretion from lactating mouse mammary epithelial cells by FIL (feedback inhibition of lactation), a secreted milk protein, *J. Cell. Sci.* 106:641.

Robinson, A.M. and Williamson, D.H., 1978, Control of glucose metabolism in isolated acini of the lactating mammary gland of the rat. Effects of oleate on glucose utilization and lipogenesis, *Biochem. J.* 170:609.

Stewart, G.M., Addey, C.V.P., Knight, C.H. and Wilde, C.J., 1988, Autocrine regulation of casein turnover in goat mammary explants, *J. Endocrinol.* 118:R1.

Wilde, C.J., Addey, C.V.P., Boddy-Finch, L.M. and Peaker, M., 1995a, Autocrine regulation of milk secretion by a protein in milk, *Biochem. J.*, in press.

Wilde, C.J., Addey, C.V.P. and Knight, C.H., 1989a, Regulation of intracellular casein degradation by secreted milk proteins, *Biochim. Biophys. Acta* 992:315.

Wilde, C.J., Addey, C.V.P., Casey, M.J., Blatchford, D.R. and Peaker, M., 1988a, Feedback inhibition of milk secretion: the effect of a fraction of goat milk on milk yield and composition, *Quart. J. Exp. Physiol.* 73:391.

Wilde, C.J., Blatchford, D.R., Knight, C.H. and Peaker, M., 1989b, Metabolic adaptations in goat mammary tissue during long-term incomplete milking, *J. Dairy Res.* 56:7.

Wilde, C.J., Blatchford, D.R. and Peaker, M., 1991, Regulation of mouse mammary cell differentiation by extracellular milk proteins, *Exp. Physiol.* 76:379.

Wilde, C.J., Calvert, D.T., Daly, A. and Peaker, M., 1987a, The effect of goat milk fractions on synthesis of milk constituents by rabbit mammary explants and on milk yield *in vivo*, *Biochem. J.* 242:285.

Wilde, C.J., Calvert, D.T. and Peaker, M., 1988b, Effect of a fraction of goat milk serum proteins on milk accumulation and enzyme activities in rabbit mammary gland, *Biochem. Soc. Trans.* 15:916.

Wilde, C.J., Henderson, A.J., Knight, C.H., Blatchford, D.R., Faulkner, A. and Vernon, R.G., 1987b, Effect of thrice daily milking on mammary enzyme activity, cell population and milk yield in the goat, *J. Anim. Sci.* 64:533.

Wilde, C.J. and Knight, C.H., 1986, Degradation of newly-synthesised casein in mammary explants from pregnant and lactating goats, *Comp. Biochem. Physiol.* 84B:197.

Wilde, C.J. and Knight, C.H., 1989, Metabolic adaptations in mammary gland during the declining phase of lactation, *J. Dairy Sci.* 72:1769.

Wilde, C.J. and Knight, C.H., 1990, Milk yield and mammary function in goats during and after once daily milking, *J. Dairy Res.* 57:441.

Wilde, C.J., Knight, C.H., Addey, C.V.P., Blatchford, D.R., Travers, M., Bennett, C.N. and Peaker, M., 1990, Autocrine regulation of mammary cell differentiation, *Protoplasma* 159:12.

Wilde, C.J., Prentice, A. and Peaker, M., 1995b, Breastfeeding: matching supply with demand in human lactation, *Proc. Nutr. Soc.*, in press.

Wileman, T., Kane, L.P., Carson, G.R. and Terhorst, C., 1991, Depletion of cellular calcium accelerates protein degradation in the endoplasmic reticulum, *J. Biol. Chem.* 266:4500.

Williamson, D.H., Ilic, V., and Lund, P., 1995, A rôle for medium-chain fatty acids in the regulation of lipid synthesis in milk stasis? *In* "Intercellular Signalling in the Mammary Gland", Wilde, C.J., Knight, C.H. and Peaker, M., eds., Plenum Publishing Company, New York.

A ROLE FOR MEDIUM-CHAIN FATTY ACIDS IN THE REGULATION OF LIPID SYNTHESIS IN MILK STASIS?

Dermot H. Williamson, Vera Ilic and Patricia Lund

Metabolic Research Laboratory
Nuffield Department of Clinical Medicine
Radcliffe Infirmary
Woodstock Road
Oxford OX2 6HE, U.K.

INTRODUCTION

It is well-established from farming practice that frequent milking increases milk yield in cow and goat. As milk is stored within the lumen of the mammary gland after secretion from the epithelial cells, it implies that once the lumen is full there is a feed-back mechanism to control further milk synthesis and secretion. The finding that the secretory rate increased when milk stored in autotransplanted glands of lactating goats was diluted with an inert solution provided strong evidence for the existence of an inhibitor(s) in milk (Henderson and Peaker, 1984; 1987). A milk fraction has been identified that rapidly, but reversibly, inhibits lactose and casein synthesis in a concentration dependent way. The inhibitor is a heat-labile constituent of whey proteins and has been shown to be effective in goat and rabbit *in vivo* (Wilde *et al.*, 1987; reviewed in Wilde and Peaker, 1990). Interestingly , the active material did not inhibit fatty acid synthesis (lipogenesis) from acetate (Wilde *et al.*, 1987). This implies that inhibition of protein and lactose synthesis results in a linked inhibition of lipid synthesis. The evidence is somewhat ambiguous on this point in that inhibition of rat mammary gland protein synthesis *in vivo* with cycloheximide did not decrease the rate of lipogenesis in the gland for at least 2h after administration, but then it declined markedly (Roberts *et al.*, 1982). Alternatively, it is possible that milk contains a specific inhibitor(s) of milk lipid synthesis that acts independently of the whey fraction protein inhibitor. Some 30 years ago a fraction from rat milk was described which inhibited fatty acid synthesis *in vitro*, and in particular, the activity of acetyl-CoA carboxylase (Levy, 1964), a key enzyme in the lipogenic pathway. Later work by Levy

(Miller *et al.*, 1970) identified the inhibitory components of this fraction as free fatty acids. The aim of this contribution is to assess whether medium-chain fatty acids (MCFA), present in milk at low concentration (Miller *et al.*, 1970), play a regulatory role in mammary gland lipid metabolism of non-ruminants and in particular the rat.

MILK LIPID SYNTHESIS

In most species lipid in the form of triacylglycerols is a major component of milk; the content (g/100 ml) varies from 3.8 g in the human to 33.1 g in the polar bear (Mepham, 1976). There are two sources of this lipid: (a) the synthesis *de novo* of fatty acids, mainly MCFA, within the mammary gland and (b) uptake of exogenous lipid from plasma lipoproteins (chylomicrons or very-low-density lipoproteins, VLDL) or non-esterified fatty acids (NEFA) bound to albumin. In non-ruminants the main substrate for mammary gland lipogenesis is glucose which provides both the acetyl-CoA for fatty acids and the glycerol moiety of the triacylglycerols (Figure 1; see Bauman and Davis, 1974). The products of lipogenesis are the medium-chain fatty acids of chain length C_6 to C_{14} (Dils, 1983). The longer chain fatty acids are provided from chylomicrons or VLDL. The former are derived from the absorbed dietary lipid and are secreted by the intestinal epithelial cells via the lymphatic system into the peripheral blood (Tso and Balint, 1986). VLDL are secreted from the liver and the fatty acid moiety of the triacylglycerols is formed by synthesis *de novo* or from NEFA released from stored triacylglycerol in adipose tissue (Gibbons, 1990). A portion of the NEFA released by adipose tissue can be extracted directly by mammary gland, but in quantitative terms it is less important than fatty acids derived from the triacylglycerols of chylomicrons and VLDL (Hawkins and Williamson, 1972). The key step for the uptake of exogenous lipid by the gland is the hydrolysis of the triacylglycerols to the constituent NEFA by lipoprotein lipase (Figure 1; see Scow and Chernick, 1987).

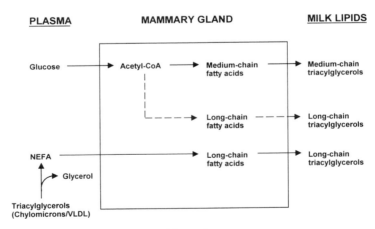

Figure 1.

The relative importance of the two pathways for milk lipid synthesis can be determined *in vivo* by a number of methods, including measurement of arteriovenous difference for substrates across the gland (Hawkins and Williamson, 1972). For a quantitative picture this method requires information on blood flow. In the chow-fed rat (high carbohydrate diet), approximately 43% of the milk fat can arise from glucose and 45% from fatty acids derived from triacylglycerols taken up by the gland (Williamson, 1990). Other, less invasive, methods are the measurement of the rate of lipogenesis *in vivo* (from all substrates) with ^3H_2O (Robinson *et al.*, 1978) and of the capacity for uptake of chylomicrons by feeding [^{14}C] lipid (Oller do Nascimento and Williamson, 1986) or intravenous injection of radioactive chylomicrons (Scow *et al.*, 1977).

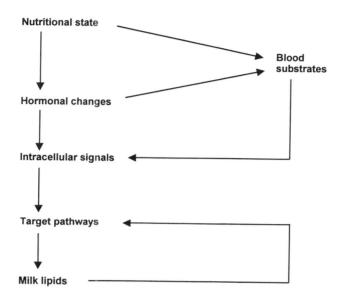

Figure 2.

The regulation of the production of milk lipid is complex and involves a number of factors including nutritional state, substrate availability and hormonal milieu (Figure 2, Table 1; see Williamson and Da Costa, 1993). The present contribution is specifically concerned with the possibility that in milk stasis (see next section) a component of the milk might inhibit the further production of milk lipid via autocrine control, analogous to the inhibition of casein and lactose synthesis (see Wilde and Peaker, 1990). In terms of substrate selection, this would be particularly advantageous in the case of lipid arising from the *de novo* synthesis pathway because this process consumes substantial amounts of glucose (some 40 mmol/24 h in the rat). If milk stasis does indeed have a regulatory effect on milk lipid synthesis, it is likely to occur relatively rapidly after onset of milk accumulation in order to spare further glucose (and other substrates) utilization. Potential candidates in milk for this form of regulation include MCFA which are the primary end-products of the pathway of lipogenesis (Knudsen *et al.*, 1976; Libertini and Smith, 1978) and therefore can be viewed as feed-back inhibitors. The finding in this laboratory that an oral load of medium-chain triacylglycerols (chain length: 56% C_8; 40% C_{10}) inhibited lipogenesis (by 80%) in the lactating mammary gland provided preliminary evidence for a possible interaction (Agius and Williamson, 1980). The present contribution addresses two questions: (a) is there evidence for a short-term inhibition of milk lipid synthesis during milk stasis; (b) if so, are MCFA feed-back regulators?

Table 1. Regulation of milk lipid synthesis

Effector	Lipogenesis	Exogenous lipid uptake
Insulin	Stimulation	Stimulation
Prolactin	No effect	No direct effect
Ketone bodies	Inhibition	?
Long-chain fatty acids	Inhibition	?

MILK STASIS

As already mentioned, the mammary gland has the capacity to adjust milk production to demand. In the extreme case of premature removal of the young in mid-lactation, the volume of milk continues to be produced at the normal rate for several hours. In the rat, this period is 4 h, with accumulation progressively decreasing thereafter (Grigor *et al.*, 1986).

The effects of milk stasis on the regulation of synthesis of milk constituents can be studied either by removal of the litter or by sealing the teats unilaterally whilst allowing suckling. Teat-sealing is achieved either by adhesive tape or by sealing the individual teat ducts; the latter allows suckling of the sealed teats to continue. The two experimental models (litter-removal versus teat-sealing) are fundamentally different: in the former, blood flow to the mammary gland progressively decreases after 4 h, whereas in the latter, blood flow is unaffected for at least 8 h (Hanwell and Linzell, 1973). Thus short-term teat-sealing excludes the possibility that decreased substrate and hormone supply *per se* are responsible for the inhibition of synthesis of milk constituents. It also indicates that accumulated milk does not restrict blood flow, at least in the short-term, and the unsealed gland acts as a control for the sealed gland (see below). Obviously, milk secretion in suckled, teat-sealed rats cannot continue at the same rate indefinitely since the glands have a limited capacity for storage.

There is considerable evidence that milk stasis in the teat-sealed model for 24 h and longer leads to loss of activity of key enzymes, including glucose-6-phosphate dehydrogenase (McLean, 1964), acetyl-CoA carboxylase, ATP citrate lyase and phosphofructokinase (Jones, 1968), and lipoprotein lipase (Scow and Chernick, 1987; Oller do Nascimento and Williamson, 1989). It is therefore not surprising that the rates of both lipogenesis and chylomicron extraction *in vivo* are considerably decreased in teat-sealed mammary glands after 24 h (Oller do Nascimento and Williamson, 1989; Scow *et al.*, 1977). The picture that emerges is of a gradual cessation of milk production after 4 h with the eventual loss of the capacity to produce the key components of milk. The question is whether there is a single signal or a multiplicity of signals that bring about the short-term inhibition of synthesis of milk components without altering the synthetic capacity of the gland. Other signals will eventually lead to a switch-off of gene expression and/or increased destruction of key enzymes or pre-formed milk constituents. The dramatic and rapid effects of hypophysectomy on the activities of, for example, acetyl-CoA carboxylase and ATP citrate lyase (Jones, 1967) and lipoprotein lipase (Scow and Chernick, 1987) indicate hormonal involvement, the primary candidate being prolactin (although other peptides secreted by the pituitary may be involved) in maintaining enzyme activity. This assumes that the operative procedure for hypophysectomy had no other effects. To identify the possible role of effectors within milk as opposed to humoral factors it is therefore important to study the early period of milk stasis (4-8 h), preferably in the teat-sealed model, before any significant change in mammary gland blood flow has occurred (Hanwell and Linzell, 1973).

In a detailed study in the rat involving pup-removal, Grigor and colleagues (1986) have shown that milk stasis results in expressed milk with a decreased lipid content (50% by 6 h), whereas protein and lactose remained constant. At first sight this suggests that triacylglycerol synthesis is more inhibited than the synthesis of the other major constituents. In fact, Grigor et al., 1986, found that lipid accumulated in the gland after removal of the pups. Moreover, the composition of the accumulated lipid had the same relative proportions of MCFA and long-chain fatty acids (LCFA) as normal rat milk. Calculations from these data show that synthesis of protein and lactose are inhibited to a similar, but considerably greater, extent than is lipid synthesis (Table 2). The number and size of lipid droplets in the

Table 2. Effects of short-term (6 h) litter removal on synthesis of milk macronutrients.

| Macronutrient | Total synthesis (g) | | Inhibition (%) |
	Control	Litter-removed (6 h)	
Lipid	2.0	1.5	25
Protein	1.04	0.58	44
Lactose	0.40	0.22	45

Control rates calculated for 6 h on the basis of 50g milk per day, assuming 50g = 50ml. 'Litter-removed' is the increment in milk macronutrients in the gland over 6 h of stasis. Calculated from Grigor et al., 1986.

epithelial cells increased during stasis as did accumulation of lipid in the alveolar lumen, which implies that transfer of lipid from the lumen to the milk ducts is inhibited by stasis whereas transfer of protein and lactose is not. Grigor et al. (1986) present evidence that oxytocin may have a specific effect on some stage of lipid transfer. Alternatively, there may be an involvement of prolactin in the transfer process, because prolactin-deficiency induced by bromocryptine treatment (24 h) resulted in decreased transfer of lipid and smaller milk clot in the stomachs of the suckling pups (Oller do Nascimento and Williamson, 1989; Da Costa and Williamson, 1993, 1994). Maternal plasma prolactin is decreased by more than 50% after removal of the pups for 12 h (Grigor et al., 1984).

The rate of mammary gland lipogenesis *in vivo* was also measured during stasis by Grigor and colleagues (1986). The mean rate decreased by 50% after removal of the pups for 6 h, but the inhibition was not statistically significant because of considerable variability in the values. The failure to have a control group of lactating rats to take into account the appreciable diurnal changes in the rate of lipogenesis (Munday and Williamson, 1983) may have contributed to the variability.

In view of the uncertainties concerning the existing data we have measured the rates of lipogenesis with ^3H_2O and the uptake over 15 min of injected (tail vein) [^3H]-chylomicrons, 30 μmol per rat, in the short-term (6 h) teat-sealed model (Table 3). To ensure a similar nutritional state, all rats were given an oral load of glucose (4 mmol) before administration of ^3H_2O. The mass and lipid content of the teat-sealed gland increased by 50% and 44% respectively over 6 h. The rate of lipogenesis decreased by 20% in the sealed gland, whereas there was no significant decrease in chylomicron uptake or lipoprotein

Table 3. Effects of short-term (6 h) unilateral teat-sealing on lipid metabolism in rat mammary gland

Measurement	Unsealed gland	Sealed gland
Wt of tissue (g)	10.5 ± 1.6	15.8 ± 2.8[***]
Wt of lipid (g)	1.23 ± 0.14	1.77 ± 0.089[***]
Lipogenesis (μmol ^{3}H$_2$0 incorporated per h)	980 ± 91	777 ± 44[*]
Chylomicron uptake (% administered dose per 15 min)	28.7 ± 3.7	33.5 ± 5.4
Lipoprotein lipase activity (μmol NEFA released per min)	9.3 ± 2.9	9.1 ± 3.1

The mean values ± S.D. are for at least seven rats. Values for sealed glands that are significantly different from unsealed by Student paired t test are shown: [*]$P < 0.05$; [***]$P < 0.001$.

lipase activity (Table 3). In contrast, the rate of hepatic lipogenesis increased from 323 ± 81 μmol ^{3}H$_2$O incorporated /h per liver in control rats to 543 ± 182 ($P < 0.015$) in rats with unilateral teat-sealing, indicating that the inhibition in teat-sealed mammary gland was a tissue-specific phenomenon. In control rats (neither side sealed) no statistical difference was found between the left and right glands in any of the parameters measured. Thus in the short-term (6 h) there appears to be a specific inhibition of the lipogenic pathway of milk lipid synthesis in teat-sealed glands.

The greater sensitivity of the lipogenic pathway compared with the exogenous pathway to inhibition by milk stasis is analogous to the findings on starvation of lactating rats (Oller do Nascimento and Williamson, 1988). As blood flow does not alter within 8 h of teat-sealing (Hanwell and Linzell, 1973; see also unchanged uptake of chylomicrons by sealed glands, Table 3) and there appears to be no change in the *in vitro* capacity for fatty acid synthesis over this period (Levy, 1964) it is reasonable to assume that the observed inhibition of lipogenesis (Table 3) is due to a signal generated within mammary tissue.

MEDIUM-CHAIN FATTY ACIDS *IN VIVO*

Two lines of evidence have led us to consider MCFA as possible feed-back inhibitors of lipogenesis in the mammary gland of the lactating rat. Firstly, the demonstration that an oral load of medium-chain triacylglycerols (MCT) rapidly inhibits lipogenesis in the lactating gland but not in the liver (Agius and Williamson, 1980). Secondly, the finding of Miller and colleagues (1970) that a complex mixture of non-esterified fatty acids isolated from rat milk could inhibit acetyl-CoA carboxylase. This mixture contained the wide spectrum of fatty acids found in milk triacylglycerols, and about 24% of the total were present as MCFA (C$_6$ to C$_{14}$); of the MCFA laurate (C$_{12}$) and myristate (C$_{14}$) were the most effective inhibitors.

Under normal circumstances the lactating mammary gland of the rat is not supplied with dietary MCFA by the blood stream, because they are efficiently extracted from the portal blood by the liver (Greenberger and Silkman, 1969; Bach and Babayan, 1982). Presumably on oral loading (Agius and Williamson, 1980) the liver's capacity to remove

MCFA is exceeded. Consequently, we have studied the chronic effects of increasing the dietary content of MCT (200 g/kg incorporated into chow diet; 56% C_8; 40% C_{10}) and fed for 8 d (Souza and Williamson, 1993); the effects with MCT were compared with a similar diet containing triolein.

Table 4. Chronic effects of feeding medium-chain triacylglycerols (MCT) on lipogenesis *in vivo*

Diet[1]	Lipogenesis (μmol ^3H_2O incorporated/h per tissue)	
	Mammary gland	Liver
Chow	138 ± 13.9	13.5 ± 2.3
Triolein	$23.9 \pm 5.0^{***}$	7.9 ± 1.3
MCT[2]	$60.1 \pm 9.3^{**}$	$38.1 \pm 4.2^{**}$

[1]Triolein or MCT (200g/kg incorporated into chow and fed 8 d). [2]MCT consisted of (g/kg) C_8, 560; C_{10}, 400. Mean values significantly different from chow: $^{**}P < 0.01$; $^{***}P < 0.001$.

Both high-fat diets decreased the rate of mammary gland lipogenesis (triolein, by 83%; MCT, by 57%) but the MCT increased the rate of hepatic lipogenesis by 182% (Table 4). Parallel with the decrease in lipogenesis there was a decrease in the arterio-venous differences for glucose and lactate across the mammary gland with both the triolein and MCT-containing diets; the decrease in uptake of C_3 units (glucose x 2 plus lactate) was around 52% for both substrates (Table 5).

Table 5. Chronic effects of feeding medium-chain triacylglycerols (MCT) on A-V differences for glucose and lactate across rat mammary gland

Diet	A-V difference (μmol/ml)		
	Glucose	Lactate	Total C_3[1]
Chow	2.35 ± 0.28	0.72	5.42
Triolein	$1.03 \pm 0.30^{**}$	0.37^{**}	2.43
MCT	$1.27 \pm 0.12^{**}$	0.08^*	2.62

[1]Glucose x2 plus lactate
Values $\pm$ SEM that are significantly different from chow: $^*P < 0.05$; $^{**}P < 0.01$.

Previous work from this laboratory has shown that lactation results in decreased oxidation and conservation for the mammary gland of [1-^{14}C]-triolein administered by oral intubation, whereas no difference in the high rate of oral [1-^{14}C]-octanoate (C_8) oxidation to $^{14}CO_2$ was observed between virgin and lactating rats (Oller do Nascimento and Williamson, 1986). Feeding a meal (5 g) of chow containing either [1-^{14}C]-octanoate or [1-^{14}C]-triolein to starved (24 h) lactating rats resulted in the oxidation of 65% of the

[1-^{14}C]-octanoate but only 38% of the [1-^{14}C]-triolein within 5 h (Souza and Williamson, 1993). Only 4% of the [1-^{14}C]-octanoate was found in the mammary gland lipid compared with 58% of the [1-^{14}C]-triolein. The low conversion of [1-^{14}C]-octanoate to mammary gland lipid observed in these experiments is in contrast to the much higher value reported by Lossow and Chaikoff (1958). We have no explanation for this discrepancy, but feel confident of our findings.

To see if the chronic effects of MCFA on glucose metabolism could be reproduced acutely, [1-^{14}C]-hexanoate (500 μmol) or [1-^{14}C]-octanoate (300 μmol) were injected into the tail vein of anaesthetized rats and blood collected for arterio-venous (AV) difference measurements 5 min later. Hexanoate (C$_6$) had no effect on the AV difference for glucose or lactate (Table 6), but it did increase the plasma glucose (saline-injected, 6.29 $\pm$ 0.43 μmol/ml, n=5, mean $\pm$ S.D.; hexanoate-injected, 8.11 $\pm$ 0.85 μmol/ml; $P < 0.002$). Consequently, the extraction of glucose decreased from 52% to 33%. In contrast, octanoate did not alter the glycaemia, but decreased the AV difference for glucose by 50% and that for lactate by 80% (Table 6). As in the chronic feeding experiments (Souza and Williamson, 1993), less than 10% of the administered dose of [1-^{14}C]-MCFA accumulated in the mammary gland lipid, although 95% of the label had disappeared from the plasma pool within 5 min. The low accumulation of MCFA in the lactating mammary gland is likely to be due to more effective removal by the liver rather than a limitation of esterification in mammary tissue.

Although these experiments *in vivo* indicate that the presence of increased MCFA in the plasma results in decreased lipogenesis and decreased uptake of glucose and lactate they cannot be taken as evidence for a direct link between MCFA and the changes in glucose metabolism. For example, there is no information on possible changes in blood flow to the mammary gland induced by MCFA. Availability of MCFA increases the concentration of blood ketone bodies (acetoacetate and hydroxybutyrate) which are potential inhibitors of glucose utilization and hence lipogenesis (Robinson and Williamson, 1977a). In the experiments reported in Table 6, the arterial blood ketone body concentration increased approximately three-fold after injection of hexanoate or octanoate (saline, 0.21 μmol/ml; hexanoate, 0.78 μmol/ml; octanoate, 0.72 μmol/ml). It is, however, noteworthy that although injection of the two MCFA increased blood ketone bodies to the same extent, octanoate had the more pronounced effects on glucose metabolism (Table 6).

Table 6. Acute effects of MCFA on A-V differences for glucose and lactate across mammary gland

Injection	A-V difference (μmol/ml)		
	Glucose	Lactate	Total C$_3$[1]
Saline	3.17 $\pm$ 0.88	0.61 $\pm$ 0.18	6.95
Hexanoate (C$_6$)	2.67 $\pm$ 1.32	0.57 $\pm$ 0.60	5.91
Octanoate (C$_8$)	1.60 $\pm$ 0.49[*]	0.12 $\pm$ 0.15[***]	3.32

[1]Glucose x2 plus lactate
Mean values $\pm$ SD for 5 experiments. Values significantly different from saline: [*]P < 0.05; [***]P < 0.001.

MEDIUM-CHAIN FATTY ACIDS IN VITRO

To circumvent the ambiguities inherent in the interpretation of experiments *in vivo* the effects of a range of MCFA (C_6 to C_{12}; 2 mM) on [1-^{14}C]-glucose metabolism have been studied in acini isolated from lactating rat mammary glands (Heesom *et al.*, 1992). Hexanoate (C_6) or octanoate (C_8) decreased glucose utilization by 50% and 33% respectively (Table 7). In contrast, decanoate (C_{10}) increased it by a small amount (16%), whereas laurate (C_{12}) had no effect. In the absence of MCFA the percentage of glucose (x 2) utilized which accumulated as lactate was 5.2%, and this value progressively increased with increasing chain length to C_{10}: hexanoate, 13.2%; octanoate, 19.6%; decanoate, 31.3%; laurate, 9.1%. The production of $^{14}CO_2$ from the [1-^{14}C]-glucose (an index of the pentose phosphate pathway) was decreased by MCFA; the degree of inhibition decreased with increasing chain length (Heesom *et al.*, 1992). As would be expected from the changes in glucose utilization and lactate formation, all the MCFA, except laurate, inhibited conversion of [1-^{14}C]-glucose to [^{14}C]-lipid (Table 7). The absence of an inhibitory effect of laurate (Table 7) is in agreement with the finding that feeding a diet rich in coconut oil (49% C_{12}) for 4 days does not inhibit lipogenesis *in vivo* in mammary tissue (Grigor and Warren, 1980).

The large accumulation of lactate in the presence of decanoate and to a lesser extent with octanoate (Table 7) suggests inactivation of pyruvate dehydrogenase as occurs in mammary glands of 24 h starved rats (Kankel and Reinauer, 1976; Baxter and Coore, 1978). This inactivation can be reversed *in vitro* by dichloroacetate (Munday and Williamson, 1981), an inhibitor of pyruvate dehydrogenase kinase (Whitehouse and Randle, 1973), which phosphorylates and inactivates pyruvate dehydrogenase. Dichloroacetate had no effect on glucose utilization in the absence or presence of MCFA, but as expected decreased pyruvate and, in the case of octanoate and decanoate, also lactate (by 50%) (Heesom *et al.*, 1992). Here the activation of pyruvate dehydrogenase by dichloroacetate increased the percentage of [1-^{14}C]-glucose utilized appearing as lipid, and reversed the inhibitory effect of these two MCFA. Accumulation of acetyl-CoA or MCFA-acyl-CoA, or the resulting depletion of CoA may explain the inhibition of pyruvate dehydrogenase by octanoate and decanoate. For these mechanisms to be operative there must be a

Table 7. Effects of MCFA on glucose metabolism in rat mammary gland acini

	Measurement (μmol/min per 100mg dry wt)		
Addition	Glucose removal	Lactate formation	Glucose to lipid
None	1.07 ± 0.031	0.11 ± 0.010	0.13 ± 0.012
Hexanoate (C_6)	0.54 ± 0.017[***]	0.15 ± 0.031	0.06 ± 0.005[***]
Octanoate (C_8)	0.72 ± 0.019[**]	0.29 ± 0.020[***]	0.11 ± 0.007[***]
Decanoate (C_{10})	1.24 ± 0.052[**]	0.68 ± 0.075[**]	0.25 ± 0.017[*]
Laurate (C_{12})	1.10 ± 0.010	0.20 ± 0.015[***]	0.34 ± 0.013

The results are mean values ± SEM of at least 10 separate experiments. Values significantly different from control: [*]$P < 0.05$; [**]$P < 0.01$; [***]$P < 0.001$.

mitochondrial MCFA acyl-CoA synthetase in addition to that in the cytosol. Inhibition of pyruvate dehydrogenase activity by octanoate and hexanoate would agree with depressed lactate extraction by mammary gland *in vivo* in MCT-fed rats (Table 5) and after octanoate injection (Table 6).

An inhibition of flux through pyruvate dehydrogenase, however, cannot fully explain the large inhibition of glucose utilization by hexanoate or octanoate (Table 7). Studies *in vivo* (Jones *et al.*, 1984; Mercer and Williamson, 1987; Burnol *et al.*, 1988) and *in vitro* (Williamson *et al.*, 1975; Robinson and Williamson, 1977b) have suggested that phosphofructokinase and hexokinase may be key sites of regulation of glucose utilization in the gland. All four MCFA increased G6P (Table 8) implying an inhibition of

Table 8. Effects of MCFA on G6P in rat mammary gland acini

Addition	G6P (μmol/100 mg dry wt)
None	0.073 ± 0.0035
Hexanoate (C$_6$)	0.180 ± 0.011[***]
Octanoate (C$_8$)	0.161 ± 0.007[***]
Decanoate (C$_{10}$)	0.148 ± 0.017[***]
Laurate (C$_{12}$)	0.099 ± 0.009[*]

Results are mean values $\pm$ SEM for 8 experiments.
Values significantly different from control: [*]$P < 0.05$; [***]$P < 0.001$.

phosphofructokinase I and/or G6P dehydrogenase, possibly brought about by an increase in MCFA acyl-CoA or citrate in the cytosol. Although the inhibition of glucose utilization by hexanoate and octanoate (Table 7) can be ascribed to the raised G6P and possible allosteric inhibition of hexokinase by its product G6P (Kuhn, 1989), this does not hold for decanoate which increases glucose utilization in the face of increased G6P (Tables 7 and 8). A regulatory role for citrate is difficult to define in this tissue in view of its dual function as an intermediate in lipogenesis and as an important component of milk.

Insulin, which can relieve the inhibition of glucose utilization by ketone bodies *in vitro* (Williamson *et al.*, 1975) and appears to stimulate glycolytic flux *in vivo* in the mammary gland (Jones *et al.*, 1984; Mercer and Williamson, 1987; Burnol *et al.*, 1988) also partially reverses the inhibitory effects of octanoate and decanoate (but not hexanoate) on conversion of glucose to lipid (Heesom *et al.*, 1992). Although relative hypoinsulinaemia is the physiological norm in lactation (see Williamson and Da Costa, 1993), increases in plasma insulin can modulate the interactions of MCFA with glucose metabolism, as shown by the absence of an inhibition of mammary gland lipogenesis *in vivo* when oral MCT and insulin are administered simultaneously (Agius and Williamson, 1980).

CONCLUDING REMARKS

There is now good evidence from experiments *in vivo* and *in vitro* that MCFA, and in particular octanoate and decanoate, can inhibit the conversion of glucose to lipid in mammary gland. At least three sites appear to be involved: hexokinase, phosphofructokinase and pyruvate dehydrogenase. Whether acetyl-CoA carboxylase is inhibited or inactivated has not been addressed; the *in vitro* experiments of Levy and colleagues (Miller *et al.*, 1970) suggest that it should be inhibited by MCFA. This might be particularly important during milk stasis in the ruminant. At present there is no definitive information on the mechanisms involved although it is reasonable to assume that the initial inhibition is brought about by MCFA or their respective acyl-CoA derivatives or by the resulting shortage of CoA. The finding that MCFA stimulate lipogenesis *in vivo* in liver, whereas they inhibit it in mammary gland (Agius and Williamson, 1980; Souza and Williamson, 1993) highlights a fundamental difference between the two tissues in the regulation of lipogenesis and/or the metabolism of MCFA. The major fate of MCFA in liver is partial oxidation to ketone bodies; little is directly converted to esterified products. In contrast, mammary gland has a low rate of oxidation of MCFA, and the major fate is conversion to lipid (Heesom *et al.*, 1992). Clearly, information is required on the steady-state concentrations of MCFA and their acyl-CoA derivatives, and on the relative activities and cellular location of the MCFA acyl-CoA synthetases in the two tissues; with regard to the latter it is likely that activation occurs primarily in mitochondria in liver and in cytosol in mammary gland.

Little information is available on the MCFA content of milk. Expressed rat milk contains a measurable amount of non-esterified fatty acids (NEFA; 2.06 mg/ml of milk) of which 24% or 2.5 mM are MCFA (C_6 to C_{14}) (Miller *et al.*, 1970). It is possible that NEFA content is influenced during stasis or storage by the activity of lipases. If so, the concentration of inhibitory MCFA would increase with time. An advantage of MCFA as feed-back inhibitors is that they appear to be tissue-specific, and any excess escaping in to the circulation can be rapidly metabolized by the liver.

It would make good physiological sense that if, for any reason, MCFA accumulated in mammary tissue lipogenesis should be inhibited to prevent their continued formation and to spare carbohydrate. Whether this occurs physiologically under dietary conditions of the natural environment is another question. The milk lipids of some mammals, as different as guinea pig and seal, contain essentially no MCFA (Smith and Abraham, 1975). It must be stressed that much of the above discussion of the inhibitory effects of MCFA equally applies to long-chain fatty acids (Table 4; Robinson and Williamson, 1978) and in this way the exogenous pathway for the provision of NEFA for milk lipid synthesis exerts a regulatory influence on the *de novo* synthesis pathway and spares carbohydrate.

ACKNOWLEDGEMENTS

We wish to thank Mrs M. Barber for preparation of the typescript. D.H.W. is a member of the Medical Research Council (U.K.) External Scientific Staff.

REFERENCES

Agius, L.A. and Williamson, D.H., 1980, Rapid inhibition of lipogenesis *in vivo* in lactating mammary gland by medium or long-chain triacylglycerols and partial reversal by insulin, *Biochem. J.* 192:361.

Bach, A.C. and Babayan, V.K., 1982, Medium chain triglycerides: an update, *Amer. J. Clin. Nut.* 36:950.

Bauman, D.E. and Davis, C.L., 1974, Biosynthesis of milk fat, *in*: "Lactation: A Comprehensive Treatise", Vol. 2., Larson, B.L. and Smith, V.R.,eds., Academic Press, New York, pp. 31.

Baxter, M.A. and Coore, H.G., 1978, The mode of regulation of pyruvate dehydrogenase of lactating rat mammary gland. Effects of starvation and insulin, *Biochem. J.* 174:553.

Burnol, A.-F., Ebner, S., Ferri, P. and Girard, J., 1988, Regulation by insulin of glucose metabolism in mammary gland of anaesthetized lactating rats. Stimulation of phosphofructokinase-1 by fructose 2,6-bis phosphate and activation of acetyl-CoA-carboxylase, *Biochem. J.* 254:11.

Da Costa, T.H.M. and Williamson, D.H., 1993, Effects of exogenous insulin or vanadate on disposal of dietary triacyl glycerols between mammary gland and adipose tissue in the lactating rat: insulin resistance in white adipose tissue, *Biochem. J.* 290:557.

Da Costa, T.H.M. and Williamson, D.H., 1994, Regulation of rat mammary-gland uptake of orally administered [1-^{14}C]triolein by insulin and prolactin: evidence for bihormonal control of lipoprotein lipase activity, *Biochem. J.* 300:257.

Dils, R.R., 1983, Milk fat synthesis, *in*: "Biochemistry of Lactation", Mepham, T.B., ed., Elsevier, Amsterdam, New York, pp. 141.

Gibbons, G.F., 1990, Assembly and secretion of hepatic very-low-density lipoprotein, *Biochem. J.* 268:1.

Greenberger, N.J. and Silkman, T.G., 1969, Medium-chain triglycerides - physiological considerations and clinical implications, *N. Engl. J. Med.* 280:1045.

Grigor, M.R. and Warren, S.M., 1980, Dietary regulation of mammary lipogenesis in lactating rats, *Biochem. J.* 188:61.

Grigor, M.R., Sneyd, M.J., Geursen, A. and Gain, K.R., 1984, Effects of changes in litter size at mid-lactation on lactation in rats, *J. Endocrinol.* 101:69.

Grigor, M.R., Poczwa, Z. and Arthur, P.G., 1986, Milk lipid synthesis and secretion during milk stasis in the rat, *J. Nutr.* 116:1789.

Hanwell, A. and Linzell, J.L., 1973, The effects of engorgement with milk and of suckling on mammary blood flow in the rat, *J. Physiol.* 233:111.

Hawkins, R.A. and Williamson, D.H., 1972, Measurements of substrate uptake by mammary gland of the rat, *Biochem. J.* 129:1171.

Heesom, K.J., Souza, P.F.A., Ilic, V. and Williamson, D.H., 1992, Chain-length dependency of interactions of medium-chain fatty acids with glucose metabolism in acini isolated from lactating rat mammary glands. A putative feed-back to control milk lipid synthesis from glucose, *Biochem. J.* 281:273.

Henderson, A.J. and Peaker, M., 1984, Feed-back control of milk secretion in the goat by a chemical in milk, *J. Physiol. Lond.* 351:39.

Henderson, A.J. and Peaker, M., 1987, Effects of removing milk from the mammary ducts and alveoli, or of diluting stored milk, on the rate of milk secretion in the goat, *Q. J. Exp. Physiol.* 72:13.

Jones, E.A., 1967, Changes in the enzyme pattern of the mammary gland of the lactating rat after hypophysectomy and weaning, *Biochem. J.* 103:420.

Jones, E.A., 1968, The relationship between milk accumulation and enzyme activities in the involuting rat mammary gland, *Biochim. Biophys. Acta* 177:158.

Jones, R.G., Ilic, V. and Williamson, D.H., 1984, Regulation of lactating-rat mammary-gland lipogenesis by insulin and glucagon *in vivo*. The role and site of action of insulin in the transition to the starved state, *Biochem. J.* 223:345.

Kankel, K.F. and Reinauer, H., 1976, Activity of pyruvate dehydrogenase complex in mammary gland of normal and diabetic rats, *Diabetologia* 12:149.

Knudsen, J., Clark, S. and Dils, R., 1976, Purification and some properties of a medium-chain acyl-thioester hydrolase from lactating rabbit mammary gland which terminates chain elongation in fatty acid synthesis, *Biochem. J.* 160:683.

Kuhn, N.J., 1989, Insulin action on the lactating-rat mammary gland: intracellular glucose and its phosphorylation, *Biochem. J.* 257:933.

Levy, H.R., 1964, The effects of weaning and milk on mammary gland fatty acid synthesis, *Biochim. Biophys. Acta* 84:229.

Libertini, L.J. and Smith, S., 1978, Purification and properties of a thioesterase from lactating rat mammary gland which modifies the product specificity of fatty synthesis, *J. Biol. Chem.* 253:1393.

Lossow, W.J. and Chaikoff, I.L., 1958, Secretion of intravenously administered tripalmitin-1-C^{14} and octanoate-1-C^{14} into milk by the lactating rat, *J. Biol. Chem.* 230:149.

McLean, P., 1964, Interrelationship of carbohydrate and fat metabolism in the involuting mammary gland, *Biochem. J.* 90:271.

Mepham, B., 1976, *in*: "The Secretion of Milk", Edward Arnold, London, p. 10.

Mercer, S.W. and Williamson, D.H., 1987, The regulation of lipogenesis *in vivo* in the lactating mammary gland of the rat during the starved-refed transition. Studies with acarbose, a glucosidase inhibitor, *Biochem. J.* 242:235.

Miller, A.L., Geroch, M.E. and Levy, H.R., 1970, Rat mammary-gland acetyl-coenzyme A carboxylase. Interaction with milk fatty acids, *Biochem. J.* 118:645.

Munday, M.R. and Williamson, D.H., 1981, Role of pyruvate dehydrogenase and insulin in the regulation of lipogenesis in the lactating mammary gland of the rat during the starved-refed transition, *Biochem. J.* 196:831.

Munday, M.R. and Williamson, D.H., 1983, Diurnal variations in food intake and in lipogenesis in mammary gland and liver of lactating rats, *Biochem. J.* 214:183.

Oller do Nascimento, C.M. and Williamson, D.H., 1986, Evidence for conservation of dietary lipid in the rat during lactation and the immediate period after removal of the litter. Decreased oxidation of oral [1-^{14}C] triolein, *Biochem. J.* 239:233.

Oller do Nascimento, C.M. and Williamson, D.H., 1988, Tissue-specific effects of starvation and refeeding on the disposal of oral [1-14 C]triolein in the rat during lactation and on removal of litter, *Biochem. J.* 254:539.

Oller do Nascimento, C.M., Ilic, V. and Williamson, D.H., 1989, Re-examination of the putative roles of insulin and prolactin in the regulation of lipid deposition *in vivo* in mammary gland and white and brown adipose tissue of lactating rats and litter-removed rats, *Biochem. J.* 258:273.

Roberts, A.F.C., Viqa, J.R., Munday, M.R., Farnell, R. and Williamson, D.H., 1982, Effects of inhibition of protein synthesis by cycloheximide on lipogenesis in mammary gland and liver of lactating rats, *Biochem. J.* 204:417.

Robinson, A.M. and Williamson, D.H., 1977a, Effects of acetoacetate administration on glucose metabolism in mammary gland of fed lactating rats, *Biochem. J.* 164:749.

Robinson, A.M. and Williamson, D.H., 1977b, Control of glucose metabolism in isolated acini of the lactating mammary gland of the rat. The ability of glycerol to mimic some of the effects of insulin, *Biochem. J.* 168:465.

Robinson, A.M. and Williamson, D.H., 1978, Control of glucose metabolism in isolated acini of the lactating mammary gland of the rat. Effects of oleate on glucose utilization and lipogenesis, *Biochem. J.* 170:609.

Robinson, A.M., Girard, J.R. and Williamson, D.H., 1978, Evidence for a role of insulin in the regulation of lipogenesis in lactating rat mammary gland. Measurements of lipogenesis *in vivo* and plasma hormone concentrations in response to starvation and refeeding, *Biochem. J.* 176:343.

Scow, R.O. and Chernick, S.S., 1987, Role of lipoprotein lipase during lactation, *in*: "Lipoprotein Lipase", Borensztajn, J., ed., Evener, Chicago, p. 149.

Scow, R.A., Chernick, S.S. and Fleck, T.R., 1977, Lipoprotein lipase and uptake of triacylglycerol, cholesterol and phosphatidylcholine from chylomicrons by mammary and adipose tissue of lactating rats *in vivo*, *Biochim. Biophys. Acta* 487:297.

Smith, S. and Abraham, S., 1975, The composition and biosynthesis of milk fat, *Adv. Lipid Res.* 13:195.

Souza, P.F.A. and Williamson, D.H., 1993, Effects of feeding medium-chain triacyl glycerols on maternal lipid metabolism and pup growth in lactating rats, *Brit. J. Nutr.* 69:779.

Tso, P. and Balint, J.A., 1986, Formation and transport of chylomicrons by enterocytes to the lymphatics, *Amer. J. Physiol.* 250:G715.

Whitehouse, S. and Randle, P.J., 1973, Activation of pyruvate dehydrogenase in perfused rat heart by dichloroacetate, *Biochem. J.* 134:651.

Wilde, C.J. and Peaker, M., 1990, Autocrine control in milk secretion, *J. Agric. Sci.* 114:235.

Wilde, C.J., Calvert, D.T., Daly, A. and Peaker, M., 1987, The effects of goat milk fractions on synthesis of milk constituents by rabbit mammary explants and on milk yield *in vivo*: evidence for autocrine control of milk secretion, *Biochem. J.* 242:285.

Williamson, D.H., 1990, The lactating mammary gland of the rat and the starved-refed transition: a model system for the study of the temporal regulation of substrate utilization, *Biochem. Soc. Trans.* 18:853.

Williamson, D.H. and Da Costa, T.H.M., 1993, The regulation of lipid metabolism in lactation, *in*: "Physiologic Basis of Perinatal Care", Medina, J.M. and Quero, J., eds., Ergon, Salamanca, p. 63.

Williamson, D.H., McKeown, S.R. and Ilic, V., 1975, Metabolic interactions of glucose, acetoacetate and insulin in mammary-gland slices of lactating rats, *Biochem. J.* 150:145.

THE SECRETORY PATHWAY FOR MILK PROTEIN SECRETION AND ITS REGULATION

Robert D. Burgoyne[1], Susan E. Handel[1], Allan W. Sudlow[1],
Mark D. Turner[1], Satish Kumar[2,3], J. Paul Simons[2,4],
David R. Blatchford[5] and Colin J. Wilde[5]

[1]The Physiological Laboratory, University of Liverpool,
PO Box 147, Liverpool, L69 3BX, U.K.
[2]AFRC Roslin Institute
Roslin, Midlothian, EH25 9PS, U.K.
[3]National Institute of Animal Genetics
Karnal - 132001, Haryana, India
[4]Department of Anatomy and Developmental Biology
Royal Free Hospital School of Medicine
Rowland Hill Street, London, NW3 2PF, U.K.
[5]Hannah Research Institute
Ayr KA6 5HL, U.K.

INTRODUCTION

Following their differentiation in response to lactogenic hormones during pregnancy and after parturition, mammary epithelial cells synthesise and secrete high levels of milk proteins including the caseins, α-lactalbumin, whey acidic protein, lactoferrin and transferrin. During lactation in the mouse, the caseins are by far the major newly synthesised proteins detectable within mammary epithelial cells. The characteristics of protein secretory pathways have been studied intensively in recent years but relatively little work has been carried out on the cell biology of milk protein secretion or its regulation. Milk protein secretion is of interest due to its physiological importance, from the point of view of the study of a high-efficiency secretory pathway and because of increasing interest in the use of transgenic manipulation of mammary gland gene expression to allow modification of milk composition, including the secretion of pharmaceutical proteins. In this chapter we will review what is known about the protein secretory pathway in lactating mammary cells and focus on possible physiological sites of regulation within the secretory pathway based on our studies on isolated lactating mouse mammary acini.

Intercellular Signalling in the Mammary Gland
Edited by C.J. Wilde *et al.*, Plenum Press, New York, 1995

CASEIN SYNTHESIS AND PROCESSING

The caseins are synthesised within the ER and after transport to the Golgi are phosphorylated. It has been known for some time that casein phosphorylation occurs at some site within the Golgi complex (West and Clegg, 1984) and more recent work has investigated the sites of casein phosphorylation in more detail. The drug brefeldin A blocks secretion in many cell types including mammary cells (Turner *et al.*, 1993) by preventing transport from the endoplasmic reticulum (ER) to Golgi. In brefeldin A-treated cells the Golgi cisternae collapse into the ER separating them from the *trans*-Golgi network. Brefeldin A-treatment was found to inhibit phosphorylation of β- and γ-casein but not α-casein in isolated lactating mouse acini (Turner *et al.*, 1993). These data suggest that at least two casein kinases exist in mammary cells and that the kinase responsible for α-casein phosphorylation resides in the Golgi cisternae. In contrast, the kinase acting on β- and γ-casein probably resides in the *trans*-Golgi network and so would not have access to newly synthesised caseins in brefeldin A-treated cells. Previous biochemical studies have suggested the existence of two casein kinases with different properties and substrate specificities in bovine mammary cells (Brooks, 1989).

Casein phosphorylation on multiple sites is required to allow Ca^{2+} binding and aggregation of the casein to form micelles. Micelle formation occurs in the trans-Golgi network or in presecretory granules after phosphorylation will have occurred (Clermont *et al.*, 1993). The micelles are packaged into secretory vesicles which contain variable numbers of micelles. Casein secretion then occurs by exocytotic fusion of the vesicles with the apical plasma membrane to release the micelles into the acinar lumen (Wooding, 1977; Franke *et al.*, 1976). The kinetics of casein secretion measured *in vivo* are consistent with at least some of the newly synthesised casein moving through the secretory pathway to be secreted constitutively as soon as the casein vesicles reach the plasma membrane (Saacke and Heald, 1974). The kinetics of casein secretion can, however, be monitored in more detail in isolated acini *in vitro* and data from such experiments suggest that not all of the caseins are secreted constitutively.

CASEINS ARE SECRETED BY BOTH CONSTITUTIVE AND REGULATED SECRETORY PATHWAYS

Labelling of newly synthesised proteins with [^{35}S]-methionine was used to follow the kinetics of casein synthesis and secretion in isolated acini from the lactating mouse mammary gland (Turner *et al.*, 1992a,b; Rennison *et al.*, 1992). From these studies it was clear that α- and β-caseins were major newly synthesised proteins within the mammary cells, accounting for greater than 50% of total protein synthesised (Turner *et al.*, 1992a), and that newly synthesised caseins were secreted by both constitutive and regulated secretory pathways. A portion (around 30%) of newly synthesised caseins were secreted after a lag period of around 60 min, and in pulse chase experiments this burst of secretion was complete within a further 60 min (Turner *et al.*, 1992a). These kinetics are consistent with constitutive secretion occurring immediately after synthesis, phosphorylation, packaging and movement of the casein vesicles to the plasma membrane, with the transit time through the secretory pathway being around 60 min. It was surprising that a larger proportion of newly synthesised casein was not secreted in these experiments. In addition, electron microscopical images of mammary cells usually show a high number of casein vesicles in the cells (Wooding, 1977; Franke *et al.*, 1976), more consistent with what is expected for cells that show regulated secretion. Constitutive secretory vesicles have such a short half-life in cells that they are usually rarely seen by electron microscopy but regulated vesicles accumulate until the cell receives the necessary stimulus to trigger exocytosis. Therefore, it seemed possible that mammary cells possess a regulated secretory pathway for casein secretion. This was shown to be the case since, at times after constitutive secretion had terminated, a further burst of secretion of caseins could be elicited from isolated lactating mouse acini (Turner et al., 1992a) or mammary cells in EHS cultures (Wilde et al., unpublished observations) by raising cytosolic Ca^{2+} concentration using the calcium ionophore ionomycin. The constitutive and regulated pathways showed distinct Ca^{2+} - dependencies with the constitutive secretion being totally independent of cytosolic Ca^{2+} concentration (Turner et al., 1992a). The presence of two distinct pathways was confirmed using digitonin-permeabilised mammary acini, in which the Ca^{2+} concentration could be directly manipulated, to reveal both Ca^{2+}-independent and Ca^{2+}-dependent secretion (Turner et al., 1992b).

Ca^{2+}-induced secretion from mammary acini appeared to occur due to activation of a late stage in the secretory pathway (ie. exocytosis) as in conventional regulated secretory cells. Microtubule disassembly using the drug nocodazole completely blocked casein traffic through early stages of the secretory pathway due to an essential requirement for microtubules, probably due to their involvement in ER to Golgi, intra-Golgi transport and the maintenance of Golgi structure, but had no effect on ionomycin-induced secretion (Rennison et al., 1992), indicating that the Ca^{2+} rise due to the ionophore stimulated a step beyond those requiring intact microtubules.

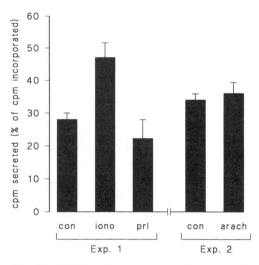

Figure 1. Prolactin and arachidonic acid fail to stimulate acute casein secretion from lactating mammary acini. Mammary acini were isolated from the lactating mouse gland as described previously (Saacke and Heald, 1974) and proteins labelled by a 60 min pulse of [^{35}S]-methionine in the absence of prolactin. Secretion was followed over a subsequent 60 min chase period to measure secretion of prelabelled protein in control incubations (con) or in the presence of ionomycin (iono, 10μM), prolactin (prl, 10μg/ml) or arachidonic acid (arach, 25μM). The data are derived from experiments on two cell preparations from different mice.

Lactating mammary epithelial cells possess Ins(1,4,5)P$_3$-sensitive intracellular Ca^{2+} stores (Yoshimoto et al., 1990; Enomoto et al., 1993) suggesting that cytosolic Ca^{2+} concentration is likely to be under hormonal control in these cells. In addition, cytosolic

Ca^{2+} concentration in mammary cells in culture is elevated by ATP acting on P$_2$-purinergic receptors and by mechanical stimulation (Furuya *et al.*, 1993). The physiological stimulus required to elevate cytosolic Ca^{2+} concentration and initiate regulated exocytosis in the intact gland is unknown. It has been suggested that prolactin can acutely stimulate casein secretion from rabbit mammary gland fragments (Ollivier-Bousquet, 1978) and that this is mediated by arachidonic acid (Ollivier-Bousquet *et al.*, 1991) but we have been unable to detect any stimulatory effects of either prolactin or arachidonic acid on secretion from isolated lactating mouse mammary acini (Figure 1). It should be noted that the freshly isolated lactating mouse acini (Turner *et al.*, 1992a) showed a considerably higher secretory activity, in terms of the percentage of newly synthesised proteins secreted, than the rabbit gland fragments used in the earlier study (Ollivier-Bousquet, 1978). The mechanism of action of Ca^{2+} in activating casein exocytosis is not known, but one candidate protein is the Ca^{2+}- and phospholipid-binding protein annexin II (Burgoyne and Geisow, 1989) which is diffusely distributed in mammary cells during pregnancy but becomes specifically expressed on the apical plasma membrane of mammary epithelial cells during lactation (Handel *et al.*, 1991).

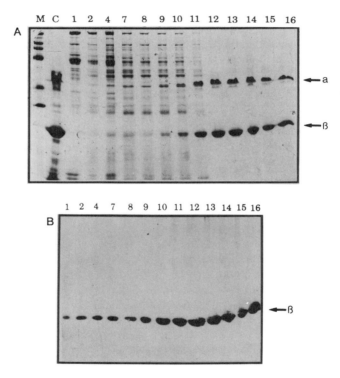

Figure 2. Characterisation of anti-β-casein antiserum. Lactating mouse mammary tissue was fractioned on a 25-65% sucrose gradient and the fractions analysed by SDS-polyacrylamide gel electrophoresis to show the Coomassie blue-stained pattern of polypeptides present (A) and immunoblotting with anti-β-casein antiserum (B). Lanes 1-16 are gradient fractions from the top to the bottom of the gradient with fraction 1 being soluble proteins. Lane M contains molecular weight standards of 205, 116,97,66,45 and 29kDa from the top to the bottom of the gel. Lane C contains casein standards isolated from mouse milk. The position of migration of α- and β-caseins is indicated on the right of the figures.

LOCALIZATION OF CASEINS WITHIN MAMMARY EPITHELIAL CELLS

The finding that caseins can be secreted by both constitutive and regulated pathways (Turner *et al.*, 1992a) raises the issue of whether caseins are packaged into two recognisably distinct secretory vesicles and whether both pathways involve secretion of casein micelles. In addition, it was found that the regulated protein secretion was enriched in α- and γ-caseins over β-casein leading to the possibility of differential distribution of caseins between secretory vesicles (Turner *et al.*, 1993). The majority of caseins in milk are in micellar form and so it seemed likely that casein secretion from both pathways was, in fact, essentially all due to release of micelles (Turner *et al.*, 1992a). Data from immunogold labelling of isolated lactating mouse acini with anti-casein antisera suggested that this was the case since little casein was detected in secretory vesicles in a non-micellar form. The immunogold technique was used on epon-embedded sections using a general anti-mouse casein antiserum or a specific antiserum directed against β-casein (Figure 2). With both antisera, labelling was detected over Golgi structures but was most concentrated over micelles in secretory vesicles (Figure 3). In some secretory vesicles, labelling was

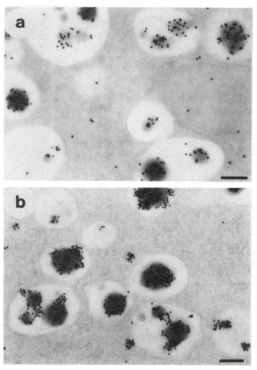

Figure 3. Immunogold labelling with anti-casein antisera. Isolated lactating mammary acini were processed for electron microscopy (Turner *et al.*, 1992a) and sections incubated with anti-β-casein antiserum (a) or a general anti-casein antiserum (b) followed by collodial gold - anti-rabbit-IgG antiserum. A high density of immunogold staining was found over casein micelles with gold particles distributed across the entire diameter of the micelle. Occasionally labelling of presumed sub-micellar structures within the secretory vesicles was also seen. Scale bar represents 200 nm.

also detected over what may have been sub-micellar structures but no extensive non-micelle labelling was detected in vesicles near the apical membrane. Therefore, it appears likely that both constitutive and regulated secretion involves exocytosis of micelle-containing secretory vesicles. The difference between the vesicles in each pathway is unknown. Around 18% of the micelles were unlabelled by the general casein antiserum and 35% were unlabelled by the β-casein antiserum. The majority of the micelles are likely to contain a mixture of caseins but we can not rule out the presence of subpopulations enriched in one or more caseins. No evidence was seen for any specific distribution of β-casein within micelles; immunogold labelling was seen across the entire micelle (Figure 3).

EFFECT OF β-CASEIN KNOCK-OUT ON CASEIN SECRETION

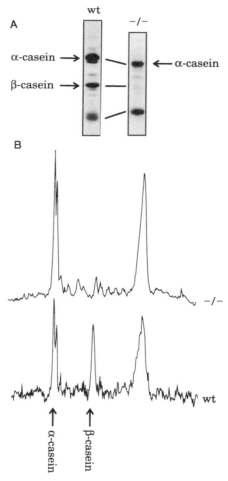

Figure 4. Proteins synthesised by lactating acini from wild type and β-casein-deficient mice. Acini were isolated from control wild type (wt) and heterozygous (-/-) β-casein-deficient mice, incubated with [^{35}S]-methionine for 60 min and newly synthesised proteins within the cells detected by polyacrylamide gel electrophoresis and fluorography (A). The major polypeptides synthesised in control mice are α-casein, β-casein and an unidentified low molecular weight polypeptide. [^{35}S]-methionine incorporation was quantified by using a Molecular Dynamics phosphorimager and line plots representing incorporation in representative tracks for wild type and -/- acini are shown (B). In wild type acini 22% of total incorporation was into β-casein whereas no detectable incorporation into β-casein was detected in -/- acini.

258

A β-casein-deficient mouse line has been generated by targeted gene knock-out (Kumar *et al.*, 1994). Since β-casein is present as a major component of the majority of intracellular micelles, and micelle formation is dependent on co-aggregation of phosphorylated caseins, it is of interest to know whether the absence of β-casein, or indeed any other casein, would affect micelle formation and secretion. The β-casein-deficient animals produce milk and the milk caseins are in a micellar form (Kumar *et al.*, 1994). We have now examined the extent of intracellular micelle formation and secretion. Following [^{35}S]-methionine labelling of proteins synthesised by acini isolated from the gland of lactating homozygous mutant mice, it was clear that β casein was absent with α-casein being the major casein synthesised (Figure 4) and secreted. We examined the mammary glands of lactating homozygous mutant mice by electron microscopy (Figure 5) and found that the

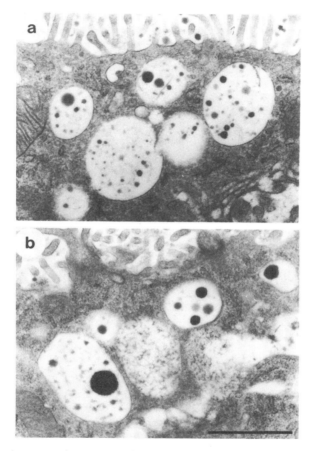

Figure 5. Electron microscopy of mammary cells in wild type and β-casein-deficient mice. Mammary tissue was taken from wild type and -/- mice at 10 days of lactation and fixed and processed for electron microscopy (Turner *et al.*, 1992a). No differences were seen in cell morphology between the wild type (a) and the -/- cells (b). The -/- cells had many secretory vesicles containing apparently normal casein micellar structures. Scale bar, 1μM.

general morphology of the secretory cells were not noticeably different from those in control (wild-type) mice. Intracellular casein micelle formation appeared to be normal. In wild-type mice β-casein appears to be secreted predominantly by the constitutive pathway (Brooks, 1989) and therefore, we have paid particular attention to this pathway in β-casein deficient mice. Constitutive casein secretion normally occurs after a lag period of at least 60 min (Turner et al., 1993). The time course of constitutive secretion from β-casein-deficient mice (Figure 6) did not appear to be markedly different from that seen in control animals although the lag period before secretion began was apparently longer with -/- acini. This latter effect could simply reflect variability between animals. The extent of secretion was not significantly different between -/- and wild type acini (around 20% of total synthesised protein in each case). We can conclude therefore that β-casein-deficient mice are still able to form micelles intracellularly and secrete caseins constitutively with a high degree of efficiency. There is, however, some effect of the mutation on micelle structure since the micelles in milk of -/- mice have a reduced diameter (Kumar et al., 1994).

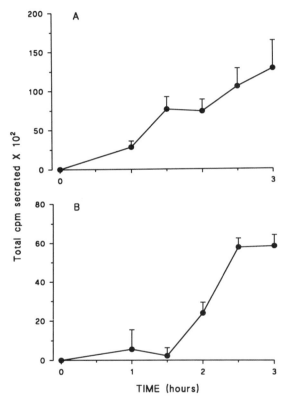

Figure 6. Time course of constitutive secretion from lactating acini of wild-type and β-casein deficient mice. Acini were isolated from a wild type (A) and a -/- mouse (B) at 10 days of lactation. Proteins were labelled by incubation for various times with [^{35}S]-methionine and cell-associated label and labelled protein in the medium assayed (Turner et al., 1992a) to provide a measure of constitutive secretion (B).

INHIBITORY CONTROL OF CONSTITUTIVE SECRETION FROM MAMMARY CELLS

A variety of physiological evidence has suggested that a feedback inhibitor switches off casein secretion as milk accumulates in the alveolar lumen (Blatchford and Peaker, 1982;

Wilde *et al.*, 1990). One possible inhibitor of secretory function is TGFβ which was suggested to function in this way on the basis of its high concentration in the mammary gland during pregnancy but not lactation and its ability to inhibit casein secretion from mammary cells in culture over a 4-day incubation (Robinson *et al.*, 1993). These data suggest that TGFβ is not an acute regulator of casein secretion but may be involved in the suppression of secretion during pregnancy. An additional peptide regulator termed the feedback inhibitor of lactation (FIL) has been identified (Wilde *et al.*, 1987; Wilde *et al.*, 1995) that acts as an acute inhibitor. FIL was purified on the basis of its ability to inhibit casein synthesis in mammary gland explant cultures. In studies of the effect of FIL on isolated mammary acini this effect on casein synthesis was confirmed and it was found that FIL also partially inhibited the constitutive secretion of already synthesised protein (Rennison *et al.*, 1993). Ionomycin-induced secretion was unaffected by FIL suggesting that its inhibitory action was not at the level of exocytosis but at an earlier stage of the secretory pathway. Examination of FIL-treated cells by electron microscopy showed that the ER was swollen and dispersed (Figure 7). The effect of FIL on synthesis, secretion and ER morphology was completely reversible and mimicked by brefeldin A, a well characterised drug that acts as an inhibitor of ER to Golgi transport (Rennison *et al.*, 1993). An explanation for the effects of FIL is that its primary action results in a block of transport from the ER to the Golgi. This would lead to a reduction in the extent of constitutive secretion and the accumulation of casein in the ER would lead to swelling of the ER and to an inhibition of further protein synthesis. The signalling pathway by which FIL, presumably acting at the apical plasma membrane, is able to disrupt vesicular traffic is not known but we have been able to rule out most of the common second messenger pathways (Rennison *et al.*, 1993).

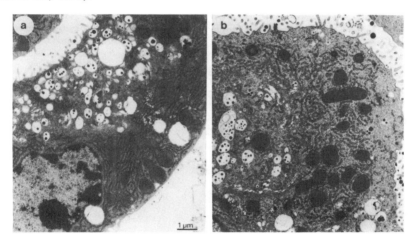

Figure 7. Electron microscopy of isolated lactating acini showing the effects of treatment with FIL. Isolated mammary acini were incubated with or without 8μg/ml of FIL for 3h and then fixed and processed for electron microscopy (Rennison *et al.*, 1993). FIL treated cells (b) contained ER that was swollen and dispersed compared to that in control cells (a). The effects demonstrated here were not the most severe and in some cells the ER was difficult to discern within the cytoplasm. Scale bar 1μM.

CONCLUSIONS

Casein secretion from lactating mammary cells can no longer be seen as an unregulated process. Autocrine regulation exerts a rapid inhibitory control on casein synthesis and on secretion of pre-formed casein through effects on early stages of the secretory pathway. The detailed cellular mechanisms involved remain to be identified. In addition, casein secretion can be enhanced by a Ca^{2+}-regulated pathway for exocytosis. The physiological stimulus or stimuli that activate regulated exocytosis is an important area for further study, in particular, to determine whether the major signal is purely mechanical or whether positive control of milk protein exocytosis is exerted by hormones or neurotransmitters.

ACKNOWLEDGEMENTS

This work was supported by the AFRC and in part by the Scottish Office Agriculture and Fisheries Department.

REFERENCES

Blatchford, D.R. and Peaker, M., 1982, Effects of frequent milking on milk secretion during lactation in the goat: relation to factors which limit the rate of secretion, *Quart. J. Exp. Physiol.* 67:303.

Brooks, C.L., 1989, Two physiological substrate-specific casein kinases are present in the bovine mammary gland, *FEBS Lett.* 243:385.

Burgoyne, R.D. and Geisow, M.J., 1989, The annexin family of calcium-binding proteins, *Cell Calcium* 10:1.

Clermont, Y, Xia, L., Rambourg, A., Turner, J.D. and Hermo, L., 1993, Transport of casein submicelles and formation of secretion granules in the Golgi apparatus of epithelial cells of the lactating mammary gland of the rat, *Anat. Rec.* 235:363.

Enomoto, K.-I., Furuya, K., Yamagishi S. and Maeno, T., 1993, Proliferation-associated increase in the sensitivity of mammary epithelial cells to inositol-1,4,5-trisphosphate, *Cell. Biochem. Funct.* 11:55.

Franke, W.W., Luder, M.R., Kartenbeck, J., Zerban H. and Keenan, T.W., 1976, Involvement of vesicle coat material in casein secretion and surface regeneration, *J. Cell Biol.* 69:173.

Furuya, K., Enomoto K. and Yamagishi, S., 1993, Spontaneous calcium oscillations and mechanically and chemically induced calcium reponses in mammary epithelial cells, *Pflugers Arch.* 422:295.

Handel, S.E., Rennison, M.E., Wilde C.J. and Burgoyne, R.D., 1991, Annexin II (Calpactin I) in the mouse mammary gland: immunolocalisation by light and electron microscopy, *Cell Tiss. Res.* 264:549.

Kumar, S., Clarke, A.L., Hooper, M.L., Horne, D.S., Law, A.J.R., Leaver, J., Springbett, A., Stevenson E. and Simons, J.P., 1994, Milk composition and lactation of β-casein-deficient mice, *Proc. Natl. Acad. Sci. USA.* 91:6138.

Ollivier-Bousquet, M., 1978, Early effects of prolactin on lactating rabbit mammary gland, *Cell Tiss. Res.* 187:25.

Ollivier-Bousquet, M., Radvanyi F. and Bon, C., 1991, Crotoxin, a phospholipase A2 neurotoxin from snake venom, interacts with epithelial mammary cells, is internalized and induces secretion, *Mol. Cell. Endocrinol.* 82:41.

Rennison, M.E., Handel, S.E., Wilde C.J. and Burgoyne, R.D., 1992, Investigation of the role of microtubules in protein secretion from lactating mouse mammary epithelial cells, *J. Cell Sci.* 102:239.

Rennison, M.E., Kerr, M., Addey, C.V.P., Handel, S.E., Turner, M.D., Wilde, C.J. and Burgoyne, R.D., 1993, Inhibition of constitutive protein secretion from lactating mouse mammary epithelial cells by FIL (feedback inhibitor of lactation) a secreted milk protein, *J. Cell Sci.* 106:641.

Robinson, S.D., Roberts A.B. and Daniel, C.W., 1993, TGFβ suppresses casein synthesis in mouse mammary explants and may play a role in controlling milk levels during pregnancy, *J. Cell Biol.* 120:245.

Saacke, R.G. and Heald, C.W., 1974, Cytological aspects of milk formation and secretion, *in*: "Lactation: A Comprehensive Treatise", volume 2, Larson, B.L. and Smith, V.R. eds., Academic Press, New York 147.

Turner, M.D., Handel, S.E., Wilde, C.J. and Burgoyne, R.D., 1993, Differential effect of brefeldin A on phosphorylation of the caseins in lactating mouse mammary epithelial cells, *J. Cell Sci.* 106:1221.

Turner, M.D., Rennison, M.E., Handel, S.E., Wilde, C.J. and Burgoyne, R.D., 1992a, Proteins are secreted by both constitutive and regulated secretory pathways in lactating mouse mammary epithelial cells, *J. Cell Biol.* 117:269.

Turner, M.D., Wilde, C.J. and Burgoyne, R.D., 1992b, Exocytosis from permeabilized lactating mouse mammary epithelial cells: stimulation by Ca^{2+} and phorbol ester, but inhibition of regulated exocytosis by GTP$_\gamma$S, *Biochem. J.* 286:13.

West, D.W. and Clegg, R.A., 1984, Casein kinase activity in rat mammary gland Golgi vesicles. Demonstration of latency and requirement for a transmembrane ATP carrier, *Biochem. J.* 219:181.

Wilde, C.J., Addey, C.V.P., Boddy-Finch, L.M. and Peaker, M., 1995, Autocrine regulation of milk secretion by a protein in milk, *Biochem. J.* in press.

Wilde, C.J., Calvert, D.T., Daly A. and Peaker, M., 1987, The effect of goat milk fractions on synthesis of milk constituents by rabbit mammary explants and on milk yield *in vivo*, *Biochem. J.* 242:285.

Wilde, C.J., Knight, C.H., Addey, C.V.P., Blatchford, D.R., Travers, M., Bennett C.N. and Peaker, M., 1990, Autocrine regulation of mammary cell differentiation, *Protoplasma* 159:112.

Wooding, F.B.P., 1977, Comparative mammary fine structure, *Symp. zool. Soc. Lond.* 41:1.

Yoshimoto, A., Nakanishi, K., Anzai, T. and Komine, S., 1990, Effects of inositol 1,4,5-trisphosphate on calcium release from the endoplasmic reticulum and golgi apparatus in mouse mammary epithelial cells: a comparison during pregnancy and lactation, *Cell. Biochem. Funct.* 8:191.

IN VITRO EFFECTS OF OXYTOCIN AND IONOMYCIN ON LIPID SECRETION BY RAT MAMMARY GLAND. ROLE OF THE MYOEPITHELIAL CELLS

Teresa H.M. Da Costa, Vera Ilic and Dermot H. Williamson

Metabolic Research Laboratory
Nuffield Department of Clinical Medicine
Radcliffe Infirmary, Oxford, OX2 6HE U.K.

Milk is expelled from the alveolar lumen by contraction of the myoepithelial cells in response to the suckling stimuli and subsequent release of oxytocin (Robinson, 1986). The action of oxytocin is dependent on availability of external calcium (Olins and Bremel, 1984). Ionomycin is a calcium ionophore that increases protein secretion from mammary gland acini and facilitates lipid secretion by apparent leakage from lipid droplets (Turner *et al.*, 1992). Oxytocin and ionomycin were used in an *in vitro* system developed to study the process of milk lipid secretion. We describe the stimulatory effects of oxytocin and ionomycin on secretion of milk constituents by their action on myoepithelial cells in the presence or absence of external calcium (Ca^{2+}).

Mammary gland slices were obtained from rats at peak lactation (10-12d).The slices were washed in fresh Krebs-Hesenleit (KH) saline containing 5mM glucose and 2.4mM calcium chloride (37°C gassed O_2/CO_2, 19:1). To pre-label the triacylglycerol pool with ³H-oleate about 3.0g of slices were transferred to an Erlenmeyer flask containing the pre-labelling medium and incubated after gassing at 37°C for 30 min in a Dubnoff shaker. The medium consisted of 48.0 ml saline as above, with 2% bovine serum albumin and 2.0 ml ³H-oleate (0.62μmol/ml; 0.84μCi/ml). After pre-labelling the slices were washed 3 times with 10ml saline to remove excess radioactivity. Where appropriate, Ca^{2+}-free saline was used. For the time course ³H-labelled slices (about 1.5g) were incubated with 50 ml KH saline (as above) with additions of oxytocin (50 ng/ml) or ionomycin (2.5 μg/ml). The flasks were incubated for 90 min in a water-bath at 37°C with gentle agitation. Samples of medium (1 ml) were obtained at intervals and assayed for turbidity (Patton *et al.*, 1980), triacylglycerol (Triacylglycerol GPO-PAP kit, Boehringer Mannheim), radioactivity, and glucose utilization (Slein, 1963). Results are expressed per 100 mg of tissue.

Secretion of radioactivity (dpm) increase 2-fold after an oxytocin challenge (Figure 1). A second challenge with oxytocin did not produce a further "burst" of dpm secretion from the slices. Turbidity and triacylglycerol secretion showed a similar pattern. The lack of

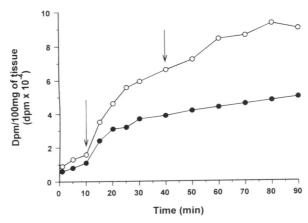

Figure 1. Effect of oxytocin on secretion of radioactivity from mammary gland slices in a Ca^{2+}-containing ($\circ$) or a Ca^{2+}-free medium ($\bullet$). Arrows indicate oxytocin addition.

oxytocin effect on a second challenge implies that either the myoepithelial cells are desensitised or that the first challenge expels the milk already secreted to the alveolar lumen. Ionomycin increased secretion of radioactivity to higher levels (13-fold) when compared with oxytocin, possibly reflecting preferential secretion of newly synthesized lipid. Turbidity and triacylglycerol secretion also increased after ionomycin addition but the results were variable. Ionomycin decreased glucose utilization (41%); this was not observed with oxytocin. With Ca^{2+}-free medium oxytocin caused a diminished initial increase in secretion of radioactivity which then tended to a plateau (Figure 1). Ionomycin was ineffective in Ca^{2+}-free medium, confirming that Ca^{2+} influx is required for its effect (Turner *et al.*, 1992).

The action of oxytocin in increasing secretion of ^{3}H-lipid and other milk components was evident in a Ca^{2+}-free medium. The response to oxytocin may involve mobilization of intracellular Ca^{2+} stores (Yoshimoto *et al.*, 1990). It is likely that the observed effect of oxytocin is mainly on the milk components already present within the alveoli lumen and a slower rate of secretion does not supply enough milk for a second "burst" of secretion from the epithelial cells. We postulate therefore that Ca^{2+} is required for maximal basal secretion from epithelial cells and that oxytocin acts on myoepithelial cells to release milk products already secreted. Ionomycin has more marked effects, but may well alter the metabolic integrity of the cells because it decreases glucose utilization.

REFERENCES

Olins, G.M. and Bremel, R.D., 1984, Oxytocin-stimulated myosin phosphorylation in mammary myoepithelial cells: role of calcium ions and cyclic nucleotides, *Endocrinology*, 114:1617.

Patton, S., Stemberger, B.H., Horton, A. and McCarl, R.L., 1980, Suppression of milk secretion (exocytosis) by concanavalin A *in vitro*, *Biochim. Biophys. Acta* 630:530.

Robinson, I.C.A.F., 1986, *in* "Current Topics in Neuro-endocrinology", Gaten, D. and Pfaff, D., eds., Springer-Verlag, Berlin.

Slein, M.W., 1963, *in* "Methods of Enzymatic Analysis", Bergmeyer, H.U., ed., Verlag, New York.

Turner, M.D., Rennison, M.E., Handel, S.E., Wilde, C.J. and Burgoyne, R.D., 1992, Proteins are secreted by both constitutive and regulated secretory pathways in lactating mouse mammary epithelial cells, *J. Cell Biol.* 117:269.

Yoshimoto, A., Nakanishi, K., Anzai, T. and Komine, S, 1990, Effects of inositol, 1,4,5-trisphosphate on calcium release from the endoplasmic reticulum and Golgi apparatus in mouse mammary epithelial cells: a comparison during pregnancy and lactation, *Cell Biochem. Funct.* 8:191.

DOSE-DEPENDENT EFFECTS OF OXYTOCIN ON THE MICROCIRCULATION IN THE MAMMARY GLAND OF THE LACTATING RAT

Stephen R. Davis, Vicki C. Farr and Colin G. Prosser

Dairying Research Corporation
Ruakura Agricultural Research Centre
Private Bag 3123
Hamilton, New Zealand

INTRODUCTION

A number of studies have described the microcirculation of the rodent mammary gland usually through histological or vascular corrosion casting methodology (e.g. Yasugi *et al.*, 1989). The objective of the study described below was to develop a method for *in vivo* microscopy of the mammary gland of the lactating rat and to observe the behaviour of the microcirculation during milk ejection following injection of oxytocin.

Blood flowing through capillaries supplying superficial alveoli in the mammary glands of lactating rats (n = 10) was visualised by fluorescence microscopy for several hours. Rats around peak lactation (days 10-15) were removed from pups and anaesthetised with sodium pentobarbitone (50mg/kg) and a catheter placed in a jugular vein. A teat of an inguinal or abdominal gland was cannulated and 0.2ml of FITC-dextran (150 kDa; 20mg/ml; Sigma Chemical Co, St Louis, USA) injected into the gland. The skin above the alveolar surface was reflected to expose approximately 1cm^2 of secretory tissue. This area was covered with a coverslip mounted on a perforated glass slide and the rat placed on a heated (37°C) platform on an inverted microscope (Nikon Diaphot-TMD). Glands were visualised by fluorescence (Nikon filter cassette B-IA) using objectives up to x40 and the image recorded (Panasonic, AG7450) via a video camera (JVC, TK128OE) onto S-VHS tape.

Using a x20 objective a field of view approximating 0.1mm^2 was obtained, containing 10-20 alveoli. Oxytocin injections were made via a jugular vein at physiological (0.5-1.0mIU) and supraphysiological doses (10 and 25mIU).

Diameters of the superficial alveoli were highly variable (30-150μM). Each alveolus was surrounded by a ring of capillaries and the larger alveoli, in particular, had capillaries traversing the alveolar surface. Flow in these capillaries was intermittent, movement of red blood cells often ceasing for periods of up to a minute, although longer periods of stasis were not unusual.

Following injection of 1mIU oxytocin, 1-4 asynchronous contractions of alveoli were observed over a period of 20-40 sec. Contractions were brief (1-3 sec) but of relatively high amplitude (as judged by the movement of the fluorescence), in some cases leading to a complete expulsion of fluorescence from the alveolus. During the relaxation phase, some alveoli showed visible distension as other alveoli were contracting. Capillary flow was disturbed only briefly during alveolar contraction but otherwise continued unabated.

Injection of 10 and 25mIU oxytocin caused a low amplitude, sustained alveolar contraction which lasted 2-3 min (10 mIU) or 3-4 min (25mIU). Flow in blood capillaries ceased during this period. Following this phase, asynchronous alveolar contractions began, in association with a resumption of flow in the capillaries. These contractions ceased after a further 1-3 min.

The results confirm, by direct observation, the supposition of Cross and Silver (1962), that flow in the mammary capillaries is reduced following oxytocin injection. However, the subjective assessment made in the present study indicates that, at physiological doses, oxytocin only causes a brief disruption to flow in the microcirculation, suggesting that closure of the microcirculation by oxytocin is a pharmacological rather than a physiological phenomenon.

Our observations also suggest that there may be considerable variation in sensitivity to oxytocin between alveoli within the same gland. In some cases we observed adjacent alveoli being responsive and unresponsive to oxytocin, despite apparently normal capillary flow in both areas.

The increasing synchrony of contraction of alveoli with increasing dose of oxytocin explains the dose-response relationship of oxytocin and intra-mammary pressure (Bisset *et al.*, 1967). At physiological doses, milk ejection is a dynamic phenomenon with the rise in intramammary pressure being the net result of alveolar contraction and relaxation. At supraphysiological doses increased synchrony and a lower amplitude of contraction occurs, coupled with cessation of flow in the microcirculation. The latter may be a direct effect on mammary blood vessels, perhaps through a pharmacological action of oxytocin on vasopressin receptors. In ruminants, injection of oxytocin leads to increased mammary blood flow during milk ejection. The present data suggest that this increase in flow may be largely due to increased flow through 'thoroughfare' channels rather than through increased flow in capillaries.

REFERENCES

Bissett, G.W., Clark, B.J., Haldar, J., Harris, M.C., Lewis, G.P., Rocha e Silva, M., 1967, The assay of milk ejecting activity in the lactating rat, *Br. J. Pharmac. Chemother.* 31:537.

Cross, B.A. and Silver, I.A., 1962, Mammary oxygen tension and the milk ejection mechanism, *J. Endocrinology* 23:375.

Yasugi, T., Kaido, T. and Uehara, Y., 1989, Changes in density and architecture of microvessels of the mammary gland during pregnancy and lactation, *Arch. Histol. Cytol.* 52:115.

EFFECTS OF TRIIODOTHYRONINE ADMINISTRATION ON THE DISPOSAL OF ORAL [1-¹⁴C]-TRIOLEIN, LIPOPROTEIN LIPASE ACTIVITY AND LIPOGENESIS IN THE RAT DURING LACTATION AND ON REMOVAL OF THE LITTER

Martha Del Prado*, Teresa M Da Costa and
Dermot H Williamson**

*Nutrition Research Unit, IMSS,
Mexico City, Mexico
**Metabolic Research Laboratory
Nuffield Department of Clinical Medicine
Radcliffe Infirmary, Oxford, U.K.

INTRODUCTION

Lactation induces a state of relative hypothyroidism. Nevertheless, thyroid hormones have an important role in mammary gland development and function. Chronic administration of thyroid hormones to pregnant rats has adverse effects on subsequent maternal behaviour and on lactation performance. The present work was undertaken to study the effect of triiodothyronine (T_3) administration on the utilization of dietary [¹⁴C] lipid and lipoprotein lipase activity (LPL) in the mammary gland (MG), white adipose tissue (WAT) and muscle (SM) of lactating and litter-removed rats.

After an oral load of [1-¹⁴C] triolein, the lactating rats treated with T_3 (50 μg / 100g body wt) over 24 h showed an increase in $^{14}CO_2$ production and a decrease in the total [¹⁴C] lipid transferred through the mammary gland that was paralleled by a decrease in mammary gland lipoprotein lipase activity (controls: 11.75 (SD 3.57) μmol/min/total tissue; T_3-treated: 5.95 (SD 1.60); $P < 0.01$).

Triiodothyronine administration decreased plasma prolactin in the lactating rats (controls: 114.4 (SD 33.4) ng/ml; T_3-treated: 30.4 (SD 16.0); $P < 0.01$). Prolactin replacement in T_3-treated rats restored LPL activity in the mammary gland but did not increase the amount of dietary [¹⁴C] lipid transferred to the milk. Chronic T_3 administration (4 days) to lactating rats did not affect pup growth or the lipogenic rate in the mammary gland (controls: 124.6 (17.6) μmol ^3H_2O/h/g tissue; T_3-treated: 115.9 (34.2); not significant). The administration of T_3 to litter-removed rats inhibited the increase of LPL activity and the [^{14}C] lipid accumulation in white adipose tissue and was accompanied by increased $^{14}CO_2$ production and [^{14}C] lipid accumulation in skeletal muscle and heart.

Table 1. Effects of T_3 administration on $^{14}CO_2$ production and [^{14}C] lipid accumulation in mammary gland, adipose tissue and muscle

| Experimental group | CO$_2$ production (%absorbed dose) | Tissue [^{14}C] lipid accumulation (% absorbed dose/ 5h) | | |
		MG+pups + milk clot	WAT (per g)	SM (per g)
Lactating	7.9 (2.8)	60.6 (10.5)	0.04 (0.01)	1.4 (0.2)
+ T$_3$	12.0 (2.1)*	43.4 (7.0)**	0.07 (0.02)	3.1 (0.8)**
+ T$_3$ + PRL		44.4 (10.4)**		
Pups-removed	12.1 (3.4)		2.53 (0.70)	5.4 (0.4)
+ T$_3$	20.1 (4.9)*		1.00 (0.55)**	8.2 (1.9)**

Values are mean (SD). Significantly different from no treatment group *$P<0.04$, **$P<0.01$.

It is concluded that hyperthyroidism depresses LPL activity in mammary gland and white adipose tissue, but not in muscle. The increased accumulation of [^{14}C] lipid in muscle and the increased production of $^{14}CO_2$ in lactating and litter-removed rats treated with T_3 is in part due to the decreased total LPL activity in mammary gland and adipose tissue respectively, which are therefore less able to compete with muscle for the available plasma triacylglycerols.

INDUCTION OF IMMUNE RESPONSES IN THE BOVINE MAMMARY GLAND

Julie L. Fitzpatrick

Department of Veterinary Medicine
University of Glasgow Veterinary School
Bearsden, Glasgow G61 1QH, U.K.

Mammary gland infections of cattle adversely influence animal welfare, food quality and the profitability of dairy farming. Improved understanding of local immune responses in the udder may lead to better prophylaxis against mastitis. Normal milk from dairy cows contains approximately 2×10^5 cells/ml and cells of the monocyte/macrophage series are the major type in mammary secretions for most of the lactational cycle. In addition to their phagocytic role in the mammary gland, mammary macrophages express major histocompatability complex (MHC) class II antigens and may, therefore, be one cell type involved in local presentation of antigen to T lymphocytes. The epithelium of the bovine mammary gland is the first tissue to become exposed to bacteria following invasion of the teat duct in the early stages of mammary gland infections, however the potential role of mammary epithelium in regulating immune responses in the udder is unknown.

Proliferation of unprimed bovine T lymphocytes to soluble and bacterial antigens *in vitro* was measured and the accessory cell function of peripheral blood mononuclear cells (PBM) and milk mononuclear cells compared. Responder T lymphocytes were isolated from defibrinated peripheral blood by density centrifugation and plastic and nylon wool adherence techniques. Milk mononuclear cells were isolated from mammary secretions, and PBM from blood, by density centrifugation. The addition of PBM APC to T lymphocytes resulted in statistically significant proliferation for a range of antigens: ovalbumin, KLH, HSA and *Streptococcus uberis*. Similarly, significant proliferation was observed in the presence of milk APC and the antigens ovalbumin and *S. uberis* (Fitzpatrick *et al.*, 1994). Differences in the proliferative response were seen in the presence of PBM APC or milk APC with peak proliferation occurring two to three days earlier in the presence of PBM APC than in the presence of milk APC. A higher concentration of milk cells, relative to PBM, therefore may be required to provide sufficient APC to initiate T cell proliferation (Fitzpatrick *et al.*, 1994).

MHC class II expression on epithelial cells *in vivo* was investigated by infusing formalin-killed *S. uberis* bacteria into one udder quarter, the soluble protein ovalbumin into another udder quarter, and phosphate buffered saline into the remaining two control quarters in non-lactating cows. Tissue sections, removed at slaughter, were fixed and MHC class II antigens identified using an indirect immunoperoxidase technique with and anti-bovine class II monoclonal antibody (IL-A21). Infusion of formalin-killed *S. uberis* into the mammary gland of dry cows *in vivo* significantly increased MHC class II expression on cells in the connective tissue and in the epithelium of the gland cistern compared to either control quarters or ovalbumin-infused quarters (Fitzpatrick *et al.*, 1992).

MHC class II expression on epithelial cells *in vitro* was investigated by culturing isolated cells as monolayers. Bovine mammary gland epithelial cells were isolated by brushing the surface of the gland cistern and large mammary ducts of the udder following slaughter. Isolated cells were cultured as monolayers on collagen gels *in vitro* and shown to be of epithelial origin by immunofluorescent staining with anti-cytokeratin antibodies. Recombinant bovine interferon-γ or mitogen-induced cytokine supernatants were incubated with the epithelial monolayers and were shown to induce MHC class II antigens.

Our results show that bovine mammary gland epithelial cells express MHC class II both *in vivo* and *in vitro* and that milk cells capable of presenting antigen to unprimed T cells exist within the local mammary gland environment. Further investigation of the role of macrophages, epithelial cells and T lymphocytes in local mammary gland immunity may prove to be beneficial in providing protection against mastitis pathogens in the future.

ACKNOWLEDGEMENTS

The author would like to thank the Agricultural and Food Research Council for their financial support; J.L. Fitzpatrick being in receipt of an AFRC Veterinary Fellowship. I am grateful to Dr W. Ivan Morrison and Dr Robert Collins of The Institute for Animal Health, Compton for the gifts of the anti-MHC class II monoclonal antibody and the recombinant bovine interferon-γ.

REFERENCES

Fitzpatrick, J.L., Cripps, P.J., Hill, W.W., Bland P.W. and Stokes, C.R., 1992, MHC class II expression in the bovine mammary gland. *Vet. Immunol. Immunopathol.* 32:13.
Fitzpatrick, J.L., Mayer, S.J., Vilela, C., Bland P.W. and Stokes, C.R., 1994, Cytokine induced major histocompatibility complex class II expression in the bovine mammary gland. *J. Dairy Sci.*, in press.
Fitzpatrick, J.L., Williams, N.A., Bailey, M., Bland P.W. and Stokes C.R., 1994, Presentation of soluble and bacterial antigens by milk-derived cells to unprimed bovine T cells *in vitro*. *Vet. Immunol. Immunopathol.*, in press.

CULTURED MAMMARY EPITHELIAL CELLS SECRETE A FACTOR WHICH SUPPRESSES LIPOPROTEIN LIPASE PRODUCTION BY 3T3-L1 ADIPOCYTES

Sean J.P. Gavigan and Margaret C. Neville

University of Colorado
Health Sciences Center, Denver, CO, U.S.A.

The enzyme lipoprotein lipase (LPL) is made in tissues which use plasma lipoprotein triglycerides either as a source of energy, e.g. muscle, or for storage, adipose tissue (Eckel, 1989). In rodents, LPL activity is low in the mammary gland and high in adipose tissue during pregnancy. At birth there are reciprocal changes with LPL activity in the mammary gland rising to very high levels while it is barely detectable in adipose tissues. This allows triglycerides to be diverted from storage in adipose tissue towards the production of milk fat (Hamosh *et al.*, 1970; Jensen *et al.*, 1994). However, LPL is synthesized by the adipocytes in the mammary gland (Jensen *et al.*, 1991), raising the question: how are the same cell types in two different regions of the body controlled independently?

We examined short-range interactions between mammary epithelial cells and adipocytes using co-cultured cells. NMMG mammary epithelial cells have been shown to secrete a factor which blocks triglyceride accumulation in 3T3-L1 adipocytes, although this factor has not been purified (Calvo *et al.*, 1992). We used CIT_3 cells, a sub-population of the Comma 1D cell line, which are mammary epithelial cells obtained from a mid-pregnant mouse (Danielson *et al.*, 1984). 3T3-L1 cells are fibroblast-like cells which can be induced to differentiate into adipocytes in culture (Green and Meuth, 1974). Adipocytes synthesize and secrete LPL which then attaches to heparan sulphate proteoglycans on the extracellular surface. The enzyme is released into the medium by incubating the cells with heparin. LPL is then assayed by measuring the release of [^{14}C]-oleate from [^{14}C]-triolein (Jensen *et al.*, 1991). As expected 3T3-L1 adipocytes had high levels of LPL activity (1525.9 ± 363.1 nmoles triglyceride hydrolised/min/dish) while CIT_3 cells had no detectable activity (25.9 ± 77.6). There was also no detectable LPL activity in the co-cultures (155.2 ± 51.7).

Conditioned medium from the CIT_3 cells was sufficient to suppress LPL activity by 93% in the adipocytes. Conditioned medium from bovine (86% inhibition) and rat (70%) mammary epithelial cell lines was also able to suppress LPL activity in adipocytes as was medium from some non-mammary epithelial cells such as CHO-K1 (86% inhibition) and human hepatocytes (68%). However, conditioned medium from neither human kidney epithelial cells (< 10% inhibition) nor rat myoblast L6 cells (no inhibition) was able to elicit this effect. Although CIT_3 cells can be induced to display some differentiated functions in the presence of lactogenic hormones, these hormones had no effect on the ability to produce the factor. Conditioned medium added directly to the LPL assay had no effect on activity, demonstrating that the mammary factor was acting on the adipocytes rather than directly on the enzyme. This was supported by the time course since the factor required at least 6 h to lower LPL activity in the adipocytes. The effect was reversible by replacing the conditioned medium with normal growth medium for 24 h.

To further characterize the factor, we showed that it was precipitated by 35% (w/v) ammonium sulphate (97% inhibition), heat-stable (94% inhibition after boiling conditioned medium for 10 min) and was retained by a 10 kD molecular weight cut-off filter (87% inhibition). This last property later allowed us to dialyse column fractions against tissue culture medium. The factor was also secreted by CIT_3 cells grown in serum-free medium giving a relatively pure starting material. The ammonium sulphate (350 mg/ml) precipitate from conditioned medium was redissolved and run through a Sephacryl S300 column. LPL suppressing activity eluted with an apparent molecular weight (mw) of 70 kD. Although several cytokines have the ability to suppress LPL activity in adipocytes, these all have lower molecular weights. Serum-free medium was adjusted to 4M NaCl and loaded onto a phenyl sepharose column. Fractions were collected from a continuous, descending NaCl gradient (4M to OM in the presence of 10nM HEPES, pH 7.4). The LPL suppressing activity smeared across the entire column but 10-fold dilutions of each fraction revealed that maximum activity was eluting in the early fractions where there was very little other protein. When these early fractions were concentrated and run on SDS-polyacrylamide gels, a major band of approximately 60 kD was seen, consistent with the molecular weight of approximately 70 kD found with the Sephacryl S300 column. In the absence of ß-mercaptoethanol, the major band has an apparent mw of ~55 kD. This suggests that this band is a protein stabilized by disulphide bonds, which is consistent with the heat stability of the suppressor.

In conclusion, we have partially purified a factor secreted by CIT_3 cells which blocks LPL secretion by adipocytes. It is probably a protein with an approximate mw of 60 kD.

REFERENCES

Calvo, J.C., Chernick, S. and Rodbard, D., 1992, Mouse mammary epithelium produces a heat-sensitive macromolecule that inhibits differentiation of 3T3-L1 preadipocytes, *Proc. Soc. Exp. Biol. Med.* 201:174.

Danielson, K.G., Oborn, C.J., Durban, E.M., Butel, J.S. and Medina, D., 1984, Epithelial mouse mammary cell line exhibiting normal morphogenesis *in vivo* and functional differentiation *in vitro*, *Proc. Natl. Acad. Sci USA* 81:3756.

Eckel, R.H., 1989, Lipoprotein lipase. A multifunctional enzyme relevant to common metabolic diseases, *N. Engl. J. Med.* 320:1060.

Green, H. and Meuth, M., 1974, An established pre-adipose cell line and its differentiation in culture, *Cell* 3:127.

Hamosh, M., Clary, T.R., Chernick, S.S. and Scow, R.O., 1970, Lipoprotein lipase activity of adipose and mammary tissue and plasma triglyceride in pregnant and lactating rats, *Biochim. Biophys. Acta* 210:473.

Jensen, D.R., Bessesen, D.H., Etienne, J., Eckel, R.H. and Neville, M.C., 1991, Distribution and source of lipoprotein lipase in mouse mammary gland, *J. Lipid Res.* 32:733.

Jensen, D.R., Gavigan, S., Sawicki, V., Witsell, D.L., Eckel, R.H. and Neville, M.C., 1994, Regulation of lipoprotein lipase activity and mRNA in the mammary gland of the lactating mouse, *Biochem. J.* 298:321.

LOCALISATION OF ANNEXIN V AND ANNEXIN VI IN LACTATING MAMMARY EPITHELIAL CELLS

Françoise Lavialle, Georges Durand* and
Michèle Ollivier-Bousquet

Laboratoire de Biologie Cellulaire et Moléculaire
*Laboratoire de Nutrition et Sécurité Alimentaire
Institut National de la Recherche Agronomique
78352 Jouy-en-Josas Cédex, France

Casein secretion which results from "teamwork in the community of mammary cells" (Pitelka, 1985) is controlled by lactogenic hormones including prolactin (PRL). The vesicular transport and calcium-dependent secretagogue effects of PRL are modified in epithelial cells (MEC) of lactating rats fed with lipid-deficient diets (Ollivier-Bousquet *et al.*, 1993). Annexins, which reversibly bind acidic phospholipids in a calcium dependent manner, but also behave as integral membrane proteins (Moufqia *et al.*, 1993), display differing expression during epithelial differentiation and during lactation (Schwartz-Albiez *et al.*, 1993). To determine if annexins are involved in the signal transmission of PRL, we first studied the cellular localisation of annexin V, which shares features with calcium-channel forming proteins, and that of annexin VI, which seems to be associated with endocytotic events (Lin *et al.*, 1992; Jäckle *et al.*, 1994).

Mammary tissue was collected from lactating rats fed with control (C) and lipid-deficient (L) diets immediately after decapitation and analysed for PUFA composition and casein secretion as previously described (Ollivier-Bousquet *et al.*, 1993). Sections of mammary gland fragments were treated for immunofluorescence. The content of annexin VI was studied by immunoblot after SDS-PAGE on three fractions named A, B and C. A and B corresponded to 105,000 g supernatants obtained from mammary tissue lysates incubated for 5 h and 17 h respectively in 10 mM Hepes pH 7.5 containing 10 mM EDTA and 0.5 M NaCl. Fraction C contained the membraneous material.

In MEC membranes of L rats, the (n-6) + (n-3) PUFA content was significantly lower than in C rats (25.4 $\pm$ 3.4 versus 44.0 $\pm$ 1.8). This change in membrane composition was associated with a decrease in casein secretion and abolition of PRL's secretagogue effect, suggesting the importance of membrane organisation in cell signalling events. Immunofluorescent detection of annexin V showed that irrespective of the diet, the protein is cytoplasmic. In MEC of C rats, annexin VI was mainly located close to the apical and basolateral membrane domains, but a faint cytoplasmic signal was also observed (Figure 1a). Individual acini varied in degree of labelling. In contrast, in L rats, strong labelling surrounded the cells of all acini (Figure 1b).

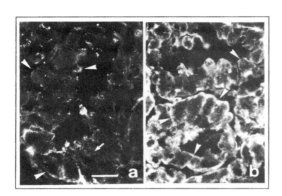

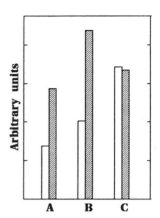

Figure 1: Immunofluorescent localisation of annexin VI in MEC of a) C rats, b) L rats. Anti-annexin VI antibodies were from D. Rainteau. Second antibody was anti-rabbit IgG-FITC. Bar = 10 μm.

Figure 2: Immunoblot detection of annexin VI in the polar (A), interfacial (B) and hydrophobic (C) membrane regions of MEC. Open and hatched bars correspond to C and L rats respectively.

Figure 2 indicates that annexin VI interacted with the polar, interfacial and hydrophobic regions of membranes. In agreement with the above data, L rat membranes contained greater amounts of annexin VI as compared to the C rat membranes. This difference was in the pool of annexin VI interacting with the polar and the interfacial regions of the lipid bilayer. Together, these data suggest that annexin VI interferes with signal transmission during lactation, and may account for CEM function of rats.

REFERENCES

Jäckle, S., Beisiegl, U., Rinninger, F., Buck, F., Grigoleit, A., Block, A., Gröger, I., Greten, H. and Windler, E., 1994, Annexin VI, a marker protein of hepatocytic endosomes, *J. Biol. Chem.* 2:1026.

Lin, H.C., Südhof, T.C. and Anderson, R.G.W., 1992, Annexin VI is required for budding of clathrin-coated pits, *Cell* 70:283.

Moufqia, J., Rothhut, B., Comera, C., Alfsen, A., Russo-Marie, F. and Lavialle, F., 1993, Annexins from bovine adrenal cortex exhibit specific cytosol/membrane solvation, *Biochem. Biophys. Res. Commun.* 1:132.

Ollivier-Bousquet, M., Kann, G. and Durand, G., 1993, Prolactin transit through mammary epithelial cells and appearance in milk, *Endocr. Regul.* 27:103.

Pitelka, D.R., 1985, Teamwork in the community of mammary cells, *Hannah Institute Yearbook* 63.

Schwartz-Albiez, R., Koretz, K., Möller, P. and Wirl, G., 1993, Differential expression of annexins I and II in normal and malignant human mammary epithelial cells, *Differentiation* 52:229.

SCIENTIFIC PAPERS PUBLISHED SINCE 1984 BY SPEAKERS AT THIS SYMPOSIUM IN THE CAB ABSTRACTS DATABASE

Paul D. Wilson

Department of Dairy Science and Technology
CAB International
Wallingford, Oxfordshire OX10 8DE, U.K.

INTRODUCTION

A search was made on the CAB ABSTRACTS database for all scientific papers by the 16 speakers at this Symposium (*Intercellular Signalling in the Mammary Gland*, Hannah Research Institute, Ayr, UK, 13-15 April 1994) published since 1984. A total of 475 records was retrieved from the database.

RESULTS

The number of papers found for each speaker varied from 3 to 84 (mean of 34), to give a total of 538 papers for the 16 speakers. Because 56 of these were by more than one speaker, the total number of papers retrieved was 475. The number of papers found for each speaker is shown in Table 1 (joint authorship papers in parenthesis).

As the period reviewed covers all papers added to the CAB ABSTRACTS database since 1984 and up to March 1994 (i.e. literature going back more than 10 years), some authors have changed their research interests during this period. In an attempt to limit the selection of papers to those relevant to the subject of this Symposium, a search was made for those papers containing the indexing terms 'mammary glands' and 'mammary development' by the Symposium speakers. This gave a reduced number of papers (203): for each speaker, the number varied from 1 to 29, this representing a proportion of the total number of papers varying between 11 and 100% (44% overall). The total of 239 papers for the 16 speakers included 34 by more than one speaker, so only 203 were retrieved. The number of 'mammary' papers for each speaker is shown in Table 2 (percentage of total in parenthesis).

The 'mammary' papers appeared in a range of CAB International printed products, including *AgBiotech News and Information, Animal Breeding Abstracts, Dairy Science Abstracts, Index Veterinarius, Nutrition Abstracts and Reviews, Series A-Human and Experimental* and *Series B-Livestock Feeds and Feeding, Pig News and Information*, and *Veterinary Bulletin*, all produced by the Division of Animal Production and Health of CAB International, as well as in the BEAST-CD (compact disc read-only memory) electronic product. Papers may appear in more than one printed product if their subject matter is appropriate. The breakdown between products for each speaker varied between 32 in *Dairy Science Abstracts* (238 overall) and 1 or none in all the other products (overall totals were between 3 in *Nutrition Abstracts and Reviews, Series B* and *Veterinary Bulletin*, and 14 in *Nutrition Abstracts and Reviews, Series A*). All but one of the 239 'mammary' papers appeared in *Dairy Science Abstracts*.

CONCLUSION

Because of the difficulty of finding a combination of suitable indexing terms to search the CAB ABSTRACTS database for papers relating to the subject of this Symposium, a broader approach was used to retrieve records published by the 16 Symposium speakers. To reduce the number of records unrelated to the subject of the Symposium, a further set of records containing only the indexing terms 'mammary glands' and 'mammary development' was retrieved. All but one of these has appeared in *Dairy Science Abstracts*. A small number of the records was also published in one or more of 7 other printed products produced by the Division of Animal Production and Health of CAB International. The 4 members of staff of the Hannah Research Institute who are Symposium speakers were responsible for 68 (33%) of the 203 'mammary' papers.

Table 1. Total papers (joint authors)

Bissell, M.J.	31	(3)
Bremel, R.D.	43	
Burgoyne, R.D.	5	(5)
Collier, R.J.	84	
Daniel, C.W.	17	
Edwards, P.A.W.	8	
Flint, D.J.	53	(7)
Friis, R.R.	3	
Grosse, R.	17	
Hartmann, P.E.	61	(1)
Knight, C.H.	55	(35)
Nicholas, K.R.	18	(1)
Peaker, M.	50	(20)
Streuli, C.H.	3	(3)
Wilde, C.J.	48	(40)
Williamson, D.H.	42	
Total	538	

Table 2. 'Mammary' papers (% of total)

Bissell, M.J.	20	(65)
Bremel, R.D.	14	(33)
Burgoyne, R.D.	5	(100)
Collier, R.J.	15	(18)
Daniel, C.W.	17	(100)
Edwards, P.A.W.	6	(75)
Flint, D.J.	12	(23)
Friis, R.R.	1	(33)
Grosse, R.	17	(100)
Hartmann, P.E.	7	(11)
Knight, C.H.	29	(53)
Nicholas, K.R.	13	(72)
Peaker, M.	23	(46)
Streuli, C.H.	3	(100)
Wilde, C.J.	32	(67)
Williamson, D.H.	25	(60)
Total	239	(44)

CONTRIBUTORS

Clare L. Abram
   Division of Cell and Genetic Pathology,
   Department of Pathology, University of Cambridge,
   Tennis Court Road, Cambridge, CB2 1QP, U.K.

Aldrin A. Adamos
   Pathology Research, Veterans Affairs Medical Center,
   3801 Miranda Avenue, Palo Alto, CA 94304, U.S.A.

Caroline V.P. Addey
   Hannah Research Institute, Ayr, KA6 5HL, U.K.

Ton van Agthoven
   Department of Molecular Biology,
   Dr Daniel den Hoed Cancer Center, PO Box 5201,
   3008 AE Rotterdam, Netherlands.

Werner Amselgruber
   Institute of Veterinary Anatomy II, Histology and Embryology,
   University of Munich, Germany.

Elizabeth Anderson
   Department of Clinical Research, Christie Hospital NHS Trust,
   Manchester, M20 9BX, U.K.

Craig S. Atwood
   Department of Biochemistry, University of Western Australia,
   Nedlands, W. Australia 6009, Australia.

Itamar Barash
   Institute of Animal Science, ARO,
   The Volcani Center, Bet-Dagan 50250, Israel.

Michael C. Barber
   Hannah Research Institute, Ayr, KA6 5HL, U.K.

Roger Barraclough
   Cancer and Polio Research Fund Laboratories,
   Department of Biochemistry, University of Liverpool,
   PO Box 147, Liverpool, L69 3BX, U.K.

James Beattie
Hannah Research Institute, Ayr, KA6 5HL, U.K.

Alan W. Bell
Department of Animal Sciences, Cornell University,
Ithaca, NY, 14853, U.S.A.

Wolfgang Bielke
Laboratory for Clinical and Experimental Research,
University of Bern, Tiefenaustrasse 120, CH-3004,
Bern, Switzerland.

Bert Binas
BBSRC Roslin Institute, Roslin,
Midlothian, EH25 9PS, U.K.

Peter H. Bird
CSIRO, Division of Wildlife and Ecology, PO Box 84,
Lyneham, ACT 2602, Australia *and*
Hannah Research Institute, Ayr, KA6 5HL, U.K.

Mina J. Bissell
Cell and Molecular Biology Department, Life Science Division,
Lawrence Berkeley Laboratories, Berkeley, CA, U.S.A.

David R. Blatchford
Hannah Research Institute, Ayr, KA6 5HL, U.K.

Gunnar Bjursell
Department of Molecular Biology, University of Göteborg,
Medicinaregatan 9c, S-413 90, Göteborg, Sweden.

Lynn M. Boddy-Finch
Hannah Research Institute, Ayr, KA6 5HL, U.K.

Christine E. Bowden
Department of Animal and Food Science,
University of Vermont, Burlington, VT 05405, U.S.A.

Jane M. Bradbury
Division of Cell and Genetic Pathology,
Department of Pathology, University of Cambridge,
Tennis Court Road, Cambridge, CB2 1QP, U.K.

Robert D. Bremel
Department of Dairy Science,
Endocrinology-Reproductive Physiology Program,
University of Wisconsin-Madison,
266 Animal Sciences Building, 1675 Observatory Drive,
Madison, WI, 53706-1284, U.S.A.

Susan A. Brooks
Breast Cancer Research Group, Department of Surgery,
University College London Medical School, London, W1P 7PN, U.K.

Gillian D. Bryant-Greenwood
Department of Anatomy and Reproductive Biology,
University of Hawaii at Manoa, Honolulu, Hawaii, 96822, U.S.A.

John C. Byatt
  Monsanto Company, The Agricultural Group,
  Animal Sciences Division, 700 Chesterfield Parkway North,
  St. Louis, MO, U.S.A.

Robert D. Burgoyne
  The Physiological Laboratory, University of Liverpool,
  PO Box 147, Liverpool, L69 3BX, U.K.

Anthony V. Capuco
  Milk Secretion and Mastitis Laboratory,
  USDA-ARS, Beltsville, MD, U.S.A.

Gerald F. Casperon
  Molecular and Cell Biology, Searle, St. Louis, MO, U.S.A.

Hai-Lan Chen
  Cancer and Polio Research Fund Laboratories,
  Department of Biochemistry, University of Liverpool,
  PO Box 147, Liverpool, L69 3BX, U.K.

Michael S.C. Chen
  Pathology Research, Veterans Affairs Medical Center,
  3801 Miranda Avenue, Palo Alto, CA, 94304, U.S.A.

A. John Clark
  BBSRC Roslin Institute, Roslin,
  Midlothian, EH25 9PS, U.K.

Catherine Clarke
  Institute of Cancer Research, Haddow Laboratories,
  15 Cotswold Road, Sutton, Surrey SM2 5NG, U.K.

Robert B. Clarke
  Department of Clinical Research, Christie Hospital NHS Trust,
  Manchester, M20 9BX, U.K.

Robert J. Collier
  Monsanto Company, The Agricultural Group,
  Animal Sciences Division, 700 Chesterfield Parkway North,
  St. Louis, MO, 63198, U.S.A.

David B. Cox
  Department of Biochemistry, University of Western Australia,
  Nedlands, W. Australia 6009, Australia.

Teresa H.M. Da Costa
  Metabolic Research Laboratory, Nuffield Department of Clinical Medicine,
  Radcliffe Infirmary, Woodstock Road, Oxford, OX2 6HE, U.K.

Trevor C. Dale
  Institute of Cancer Research, Haddow Laboratories,
  15 Cotswold Road, Sutton, Surrey SM2 5NG, U.K.

Steven E.J. Daly
  Department of Biochemistry, University of Western Australia,
  Nedlands, W. Australia 6009, Australia.

Charles W. Daniel
  Department of Biology, Sinsheimer Laboratories,
  University of California, Santa Cruz, CA, 95064, U.S.A.

Philippa D. Darbre
  Department of Biochemistry and Physiology,
  University of Reading, Whiteknights, Reading, RG6 2AJ, U.K.

Helen W. Davey
  AgResearch, Ruakura Agricultural Centre, Hamilton, New Zealand.

Stephen R. Davis
  Dairying Research Corporation, Ruakura Agricultural Centre,
  Private Bag 3123, Hamilton, New Zealand.

Martha Del Prado
  Nutrition Research Unit, IMSS, Mexico City, Mexico *and*
  Metabolic Research Laboratory, Nuffield Department of Clinical Medicine,
  Radcliffe Infirmary, Oxford, OX2 6HE, U.K.

Eve Devinoy
  National Institute for Research in Agronomy (INRA),
  Laboratory of Cellular and Molecular Biology,
  78352, Jouy-en-Josas, France.

Lambert C.J. Dorssers
  Department of Molecular Biology,
  Dr Daniel den Hoed Cancer Center, PO Box 5201,
  AE Rotterdam, Netherlands.

Georges Durand
  Laboratoire de Nutrition et Sécurité Alimentaire,
  Institut National de la Recherche Agronomique (INRA),
  78352, Jouy-en-Josas Cédex, France.

Miriam V. Dwek
  Breast Cancer Research Group, Department of Surgery,
  University College London Medical School, London, W1P 7PN, U.K.

Paul A.W. Edwards
  Division of Cell and Genetic Pathology,
  Department of Pathology, University of Cambridge,
  Tennis Court Road, Cambridge, CB2 1QP, U.K.

Ursula K. Ehmann
  Pathology Research, Veterans Affairs Medical Center,
  3801 Miranda Avenue, Palo Alto, CA, 94304, U.S.A.

Philip J. Eppard
  Monsanto Company, The Agricultural Group,
  Animal Sciences Division, 700 Chesterfield Parkway North,
  St. Louis, MO, U.S.A.

W. Howard Evans
  Department of Medical Biochemistry,
  University of Wales College of Medicine,
  Cardiff, CF4 4XN, U.K.

Alexander Faerman
  Institute of Animal Science, ARO,
  The Volcani Center, Bet-Dagan 50250, Israel.

Vicki C. Farr
  Dairying Research Corporation, Ruakura Agricultural Centre,
  Private Bag 3123, Hamilton, New Zealand.

David G. Fernig
  Cancer and Polio Research Fund Laboratories,
  Department of Biochemistry, University of Liverpool,
  PO Box 147, Liverpool, L69 3BX, U.K.

Julie L. Fitzpatrick
  Department of Veterinary Medicine,
  University of Glasgow Veterinary School,
  Bearsden, Glasgow, G61 1QH, U.K.

David J. Flint
  Hannah Research Institute, Ayr KA6 5HL, U.K.

Christine A. Ford
  Centre for Research in Animal Biology, University of Waikato,
  Hamilton, New Zealand.

Isabel A. Forsyth
  Cellular Physiology Department, Babraham Institute,
  Babraham, Cambridge, CB2 4AT, U.K.

Yael Friedmann
  Department of Biology, Sinsheimer Laboratories,
  University of California, Santa Cruz, CA, 95064, U.S.A.

Robert R. Friis
  Institute for Experimental Cancer Research,
  Tiefenauspital, University of Bern, Tiefenaustrasse 120,
  CH-3004, Bern, Switzerland.

Sean J.P. Gavigan
  University of Colorado, Health Sciences Center,
  Denver, CO, U.S.A.

R. Stewart Gilmour
  Cellular Physiology Department, Babraham Institute,
  Babraham, Cambridge, CB2 4AT, U.K.

Katrina E. Gordon
  BBSRC Roslin Institute, Roslin,
  Midlothian, EH25 9PS, U.K.

Andrew R. Green
  Department of Medicine, Medical Research Laboratory,
  Wolfson Building, University of Hull, Hull, HU6 7RX, U.K.

Richard Grosse
  Department of Pathology, The Gade Institute,
  Haukeland Hospital, N-5021, Bergen, Norway.

Susan E. Handel
The Physiological Laboratory, University of Liverpool,
PO Box 147, Liverpool, L69 3BX, U.K.

Peter E. Hartmann
Department of Biochemistry, University of Western Australia,
Nedlands, W. Australia 6009, Australia.

Kay A.K. Hendry
Hannah Research Institute, Ayr, KA6 5HL, U.K.

Susan E. Hiby
Division of Cell and Genetic Pathology,
Department of Pathology, University of Cambridge,
Tennis Court Road, Cambridge, CB2 1QP, U.K.

Louis-Marie Houdebine
National Institute for Research in Agronomy (INRA),
Laboratory of Cellular and Molecular Biology,
78352, Jouy-en-Josas, France.

Anthony Howell
Department of Medical Oncology, Christie Hospital NHS Trust,
Manchester, M20 9BX, U.K.

David R. Hurwitz
Rhone-Poulenc Rorer Central Research, 500 Arcola Road,
PO Box 1200, Collegeville, PA 19426, U.S.A.

Vera Ilic
Metabolic Research Laboratory, Nuffield Department of Clinical Medicine,
Radcliffe Infirmary, Woodstock Road, Oxford, OX2 6HE, U.K.

Guo Ke
Laboratory for Clinical & Experimental Research,
University of Bern, Tiefenaustrasse 120,
CH-3004, Bern, Switzerland.

Youqiang Ke
Cancer and Polio Research Fund Laboratories,
Department of Biochemistry, University of Liverpool,
PO Box 147, Liverpool, L69 3BX, U.K.

Danuta Kleczkowska
Institute of Genetics and Animal Breeding,
Polish Academy of Sciences, Jastrzrbiec, 05-551,
Mroków, Poland.

Christopher H. Knight
Hannah Research Institute, Ayr, KA6 5HL, U.K.

Sylvia Kondo
Department of Anatomy and Reproductive Biology,
University of Hawaii at Manoa, Honolulu, Hawaii, 96822, U.S.A.

Satish Kumar
BBSRC Roslin Institute, Roslin, Midlothian, EH25 9PS, U.K. *and*
National Institute of Animal Genetics, Karnal-132001,
Haryana, India.

Ian J. Laidlaw
    Department of Medical Oncology, Christie Hospital NHS Trust,
    Manchester, M20 9BX, U.K.

Françoise Lavialle
    Laboratoire de Biologie Cellulaire et Moléculaire,
    Institut National de la Recherche Agronomique (INRA),
    78352, Jouy-en-Josas Cédex, France.

Anthony J.C. Leathem
    Breast Cancer Research Group, Department of Surgery,
    University College London Medical School, London, W1P 7PN, U.K.

Ulf Lidberg
    Department of Molecular Biology, University of Göteborg,
    Medicinaregatan 9c, S-413 90, Göteborg, Sweden.

Georg Löffler
    Institute of Biochemistry, Genetics and Microbiology,
    University of Regensburg, D-93040, Regensburg, F.R.G.

Kerry Loomes
    Biochemistry and Molecular Biology Group,
    School of Biological Sciences, University of Auckland,
    Auckland, New Zealand.

Patricia Lund
    Metabolic Research Laboratory, Nuffield Department of Clinical Medicine,
    Radcliffe Infirmary, Woodstock Road, Oxford, OX2 6HE, U.K.

Rhonda L. Maple
    Department of Animal and Food Science,
    University of Vermont, Burlington, VT 045405, U.S.A.

Michael F. McGrath
    Monsanto Company, The Agricultural Group,
    Animal Sciences Division, 700 Chesterfield Parkway North,
    St. Louis, MO, U.S.A.

Wanda Sokól-Misiak
    Institute of Genetics and Animal Breeding,
    Polish Academy of Sciences, Jastrzrbiec, 05-551,
    Mroków, Poland.

Paul Monaghan
    Institute of Cancer Research, Haddow Laboratories,
    15 Cotswold Road, Sutton, Surrey SM2 5NG, U.K.

Romina Mossi
    Laboratory for Biochemistry I, ETH - Zentrum,
    Universitätstr. 16, CH-8092, Zurich, Switzerland.

Connie A. Myers
    Cell and Molecular Biology Department, Life Science Division,
    Lawrence Berkeley Laboratories, Berkeley, CA, U.S.A.

Margaret Natan
    Rhone-Poulenc Rorer Central Research, 500 Arcola Road,
    PO Box 1200, Collegeville, PA, 19426, U.S.A.

Margaret C. Neville
  University of Colorado, Health Sciences Center,
  Denver, CO, U.S.A.

Kevin R. Nicholas
  CSIRO, Division of Wildlife and Ecology, PO Box 84,
  Lyneham, ACT 2602, Australia.

Wieslaw Niedbalski
  Institute of Genetics and Animal Breeding,
  Polish Academy of Sciences, Jastrzrbiec, 05-551,
  Mroków, Poland.

Christina Niemeyer
  Institute of Cancer Research, Haddow Laboratories,
  15 Cotswold Road, Sutton, Surrey SM2 5NG, U.K.

Jeanette Nilsson
  Department of Molecular Biology, University of Göteborg,
  Medicinaregatan 9c, S-413 90, Göteborg, Sweden.

Michael J. O'Hare
  Institute of Cancer Research, Haddow Laboratories,
  15 Cotswold Road, Sutton, Surrey SM2 5NG, U.K.

Michèle Ollivier-Bousquet
  Laboratoire de Biologie Cellulaire et Moléculaire,
  Institut National de la Recherche Agronomique (INRA),
  78352, Jouy-en-Josas Cédex, France.

Martin J. Page
  Wellcome Foundation Ltd, Langley Court,
  South Eden Park Road, Beckenham, Kent BR3 3BS, U.K.

Malcolm Peaker
  Hannah Research Institute, Ayr, KA6 5HL, U.K.

Anna Perachiotti
  Department of Biochemistry & Physiology,
  University of Reading, Whiteknights, Reading, RG6 2AJ, U.K.

Nina Perusinghe
  Institute of Cancer Research, Haddow Laboratories,
  15 Cotswold Road, Sutton, Surrey, SM2 5NG, U.K.

Deborah Phippard
  Institute of Cancer Research, Haddow Laboratories,
  15 Cotswold Road, Sutton, Surrey, SM2 5NG, U.K.

Angela Platt-Higgins
  Cancer and Polio Research Fund Laboratories,
  Department of Biochemistry, University of Liverpool,
  PO Box 147, Liverpool, L69 3BX, U.K.

Karen Plaut
  Department of Animal and Food Science,
  University of Vermont, Burlington, VT, 05405, U.S.A.

Colin G. Prosser
    Dairying Research Corporation, Ruakura Agricultural Centre,
    Private Bag 3123, Hamilton, New Zealand.

Stig Purup
    National Institute of Animal Science,
    Department of Research in Cattle and Sheep,
    Research Centre Foulum, DK-8830, Tjele, Denmark.

Raisa Puzis
    Institute of Animal Science, ARO,
    The Volcani Center, Bet-Dagan, 50250, Israel.

Lynda H. Quarrie
    Hannah Research Institute, Ayr, KA6 5HL, U.K.

Philip S. Rudland
    Cancer and Polio Research Fund Laboratories,
    Department of Biochemistry, University of Liverpool,
    PO Box 147, Liverpool, L69 3BX, U.K.

Yael Sandowski
    Department of Biochemistry, Food Science and Nutrition,
    The Hebrew University of Jerusalem, Rehovot, 76100, Israel.

Dieter Schams
    Institute of Physiology, Technical University of Munich,
    Freising Weihenstephan, Germany.

Christian Schmidhauser
    Laboratory for Biochemistry I, ETH - Zentrum,
    Universitätstr. 16, CH-8092, Zurich, Switzerland.

Martin Schmidt
    Institute of Biochemistry, Genetics and Microbiology,
    University of Regensburg, D-93040, Regensburg, F.R.G.

Angelika Schnieke
    Pharmaceutical Proteins Ltd, Roslin,
    Midlothian, EH25 9PP, U.K.

Udo Schumacher
    Department of Human Morphology,
    University of Southampton Medical School, Southampton, U.K.

Kris Sejrsen
    National Institute of Animal Science,
    Department of Research in Cattle and Sheep,
    Research Centre Foulum, DK-8830, Tjele, Denmark.

Moshe Shani
    Institute of Animal Science, ARO,
    The Volcani Center, Bet-Dagan, 50250, Israel.

Paul Sharpe
    United Medical & Dental Schools, Guy's Hospital Campus,
    London Bridge, London, SE1 9RT, U.K.

J. Paul Simons
BBSRC Roslin Institute, Roslin, Midlothian, EH25 9PS, U.K. *and*
Department of Anatomy and Developmental Biology,
Royal Free Hospital School of Medicine,
Rowland Hill Street, London, NW3 2PF, U.K.

Fred Sinowatz
Institute of Veterinary Anatomy II, Histology and Embryology,
University of Munich, Munich, Germany.

John A. Smith
Cancer and Polio Research Fund Laboratories,
Department of Biochemistry, University of Liverpool,
PO Box 147, Liverpool, L69 3BX, U.K.

Matthew Smalley
Institute of Cancer Research, Haddow Laboratories,
15 Cotswold Road, Sutton, Surrey, SM2 5NG, U.K.

Valerie Speirs
Department of Medicine, Medical Research Laboratory,
Wolfson Building, University of Hull, Hull, HU6 7RX, U.K.

Alexander Stacey
Pharmaceutical Proteins Ltd, Roslin,
Midlothian, EH25 9PP, U.K.

Robert Strange
AMC Cancer Research Center, 1600 Pierce Street,
Denver, CO, 80214, U.S.A.

Charles H. Streuli
Department of Cell and Structural Biology,
School of Biological Sciences, University of Manchester,
3.239 Stopford Building, Oxford Road, Manchester, M13 9PT, U.K.

Allan W. Sudlow
The Physiological Laboratory, University of Liverpool,
PO Box 147, Liverpool, L69 3BX, U.K.

Eleanor Taylor
Hannah Research Institute, Ayr, KA6 5HL, U.K.

James A. Taylor
Cellular Physiology Department, Babraham Institute,
Babraham, Cambridge, CB2 4AT, U.K.

Jenny Titley
Institute of Cancer Research, Haddow Laboratories,
15 Cotswold Road, Sutton, Surrey, SM2 5NG, U.K.

Victoria L. Todd
Department of Medicine, Medical Research Laboratory,
Wolfson Building, University of Hull, Hull, HU6 7RX, U.K.

Elizabeth Tonner
Hannah Research Institute, Ayr, KA6 5HL, U.K.

Maureen T. Travers
Hannah Research Institute, Ayr, KA6 5HL, U.K.

288

Karen Tregenza
CSIRO, Division of Wildlife and Ecology, PO Box 84,
Lyneham, ACT 2602, Australia *and*
Division of Biochemistry and Molecular Biology,
Australian National University, Canberra, ACT 2600, Australia.

Mark D. Turner
The Physiological Laboratory, University of Liverpool,
PO Box 147, Liverpool, L69 3BX, U.K.

Amanda J. Vallance
Hannah Research Institute, Ayr, KA6 5HL, U.K.

Richard G. Vernon
Hannah Research Institute, Ayr, KA6 5HL, U.K.

Maggy Villa
Cellular Physiology Department, Babraham Institute,
Babraham, Cambridge, CB2 4AT, U.K.

Beverley Warner
CSIRO, Division of Wildlife and Ecology, PO Box 84,
Lyneham, ACT 2602, Australia *and*
Division of Biochemistry and Molecular Biology,
Australian National University, Canberra, ACT 2600, Australia.

Christine J. Watson
BBSRC Roslin Institute, Roslin,
Midlothian, EH25 9PS, U.K.

Stephen Weber-Hall
Institute of Cancer Research, Haddow Laboratories,
15 Cotswold Road, Sutton, Surrey, SM2 5NG, U.K.

Thomas T. Wheeler
AgResearch, Ruakura Agricultural Centre, Hamilton, New Zealand.

Michael C. White
Department of Medicine, Medical Research Laboratory,
Wolfson Building, University of Hull, Hull, HU6 7RX, U.K.

Colin J. Wilde
Hannah Research Institute, Ayr, KA6 5HL, U.K.

Dermot H. Williamson
Metabolic Research Laboratory, Nuffield Department of Clinical Medicine,
Radcliffe Infirmary, Woodstock Road, Oxford, OX2 6HE, U.K.

Richard J. Wilkins
AgResearch, Ruakura Agricultural Centre, Hamilton, New Zealand *and*
Centre for Research in Animal Biology, University of Waikato,
Hamilton, New Zealand.

Mark C. Wilkinson
Cancer and Polio Research Fund Laboratories,
Department of Biochemistry, University of Liverpool,
PO Box 147, Liverpool, L69 3BX, U.K.

Paul D. Wilson
  Department of Dairy Science and Technology,
  CAB International, Wallingford, Oxfordshire, OX10 8DE, U.K.

John H.R. Winstanley
  Cancer and Polio Research Fund Laboratories,
  Department of Biochemistry, University of Liverpool,
  PO Box 147, Liverpool, L69 3BX, U.K.

Lech Zwierzchowski
  Institute of Genetics and Animal Breeding,
  Polish Academy of Sciences, Jastrzrbiec, 05-551,
  Mroków, Poland.

INDEX

MDGI, see mammary-derived growth inhibitor
Medium-chain fatty acids, 240, 244
    inhibition of lipid synthesis, 198, 218, 239-251
Micelles, 254, 257
    tammar wallaby, 157
Milk
    bioactive factors in, 5-6
    and prolactin binding, 6
    stasis, 95
    storage, 1, 7
        effect on milk secretion, 7-8, 214-216, 233
Milk composition
    effect of EGF, 15-16
    effect of IGF-I, 15-16
    effect of oxytocin, 195
Milk ejection, 211-213
    blocking of, 7, 195
Milk production
    concurrent pregnancy and lactation, 20
    effect of EGF, 15
    effect of IGF-1, 15
    effect of growth hormone on, 4-5, 16, 20
    effect of hourly milking, 195
    effect of placental lactogen on, 18, 20
    methodologies, 213
Mouse
    gene expression
        litter size, 127-128
        α-lactalbumin, 187-188
    mammary tissue
        cell sorting, 100
        involution, 95
        reconstitution, 60-65
        Wnt gene expression, 106

Ovarian steroids
    effect on transforming growth factor-ß, 37
    and *Hox* gene expression, 29, 31
    induction of lactation by, 17, 206
    and insulin growth factor-II, 90
    and mammary growth, 4, 35, 37, 77, 204
    oestrogen receptor, 69, 77
    progesterone receptor, 77-78
Oxytocin
    and lipid secretion, 265-266
    and mammary microcirculation, 267-268
    see also milk ejection

Paracrine regulation, 2
    epithelial cell:adipocyte interaction, 273-274
Parental investment, 193
Pentose phosphate pathway, 246
Placental lactogen
    effect on food intake, 21
    effect on growth hormone, 20
    effect on insulin-like growth factors, 19
    effect on milk production, 18, 20
    and mammary growth, 13, 18
    receptor, 20

Progesterone, see ovarian steroids
Prolactin
    and annexins, 275-276
    bromocriptine
        inhibition of milk yield, 132
        and mammary growth, 17, 18
        induction of involution, 103, 135
    and casein gene expression, 112
    and galactopoiesis, 131-138, 209-210
    and lactogenesis, 121, 160
    and lipid synthesis, 241, 269-270
    and mammary differentiation, 18, 205-207
    and mammary growth, 17, 35
    and protein secretion, 256
    tammar wallaby, 163-164
Prolactin receptor
    effect of FIL, 5
Prostaglandin
    and cell proliferation, 15
    mammary synthesis of, 15
Protein secretory pathways, 253-263
Pyruvate dehydrogenase, 246-247

Rat
    casein gene expression, 173-174
    lipid synthesis
        oxytocin, 265-266
        triiodothyronine , 269-270
    mammary apoptosis
        gene expression in, 50-54
        induction, 47
        localization, 47-48
    mammary involution, 103-104
    mammary microcirculation, 267-268
Regulated pathway of protein secretion, 254-256
Relaxin
    and mammary cell proliferation, 87-88
    and milk ejection, 212

Secretory cycle, 219
Snake venom gland, 193, 200, 229
Somatomedins, *see* insulin-like growth factors
Somatotropin, *see* growth hormone

Tammar wallaby, 153-170
    asynchronous concurrent lactation, 154,
        158-159
    feedback inhibition, 165-168
    feedback stimulation, 166-168
    lactation phases, 153-154
    late lactation protein, 157
    prolactin, 163-164
Teat-sealed rat model, 242
Tight junctions, 1, 197
Transcription factors
    mammary gland factor, 113, 149
    NF1, 126
Transforming growth factor-α
    and mammary proliferation, 14, 84

DATE DUE

APR 2 0 1999	